Studies in Neuroscience, Psychology and Behavioral Economics

Series Editors

This interdisciplinary book series addresses research findings and methods in the intersection of the fields of neuroscience, psychology and behavioral economics. At the core of the series is the biological basis of human decision making, including aspects of an economic, cognitive, affective and motivational nature. Contributions from fundamental and applied sciences are equally represented and cover a broad range of scientific work, reaching from mathematical modeling to scholarly introductions into new research disciplines and related methodologies and technologies (e.g. genetics, psychophysiology, brain imaging, computational neuroscience, animal modeling). The book series presents empirical and methodological scientific innovations bridging the gap from brain to behavior. It includes research works in neuroeconomics and neuro-information systems (e.g. human-machine interaction) as well as social, cognitive and affective neuroscience, which aim at enhancing the understanding of human decision making.

Ulrich Ettinger · Bert Heinrichs · Carsten Murawski
Editors

Decision Making

Fundamentals and Applications

Springer

Editors
Ulrich Ettinger
Department of Psychology
University of Bonn
Bonn, Nordrhein-Westfalen, Germany

Bert Heinrichs
Institute for Science and Ethics (IWE)
University of Bonn
Bonn, Nordrhein-Westfalen, Germany

Carsten Murawski
Centre for Brain, Mind and Markets
Faculty of Business and Economics
The University of Melbourne
Carlton, Victoria, Australia

ISSN 2196-6605 ISSN 2196-6613 (electronic)
Studies in Neuroscience, Psychology and Behavioral Economics
ISBN 978-3-032-00879-4 ISBN 978-3-032-00880-0 (eBook)
https://doi.org/10.1007/978-3-032-00880-0

This Springer imprint is published by the registered company Springer Nature Switzerland AG
The registered company address is: Gewerbestrasse 11, 6330 Cham, Switzerland

Preface

"Life is the sum of all your choices", Albert Camus famously mused. While we may not agree with his statement, it would be hard not to admit that decisions play a pivotal role in our lives. We make countless decisions every day. Some of them are trivial and unfold below the level of conscious awareness. Others are of great importance and shape our future paths. Either way, decisions play a central role in people's lives. It is therefore not surprising that they are the subject of investigation in a wide range of academic disciplines.

Basic research in the wider decision sciences is interested, among other things, in conceptual foundations, mental processes and neural mechanisms, how economic conditions affect decisions, and ethical standards of decisions. Clinical disciplines, on the other hand, focus on examining the effects of mental and physical disorders on decision-making. Among others, they have shown that an important feature of major psychiatric and neurological conditions is the tendency to make poor decisions, subsequently leading to poor life outcomes. Indeed, it has been suggested that an impairment of the capacity to make good decisions represents a major disruption to everyday functioning in the patients' lives (Caceda et al. 2014).

In the present book, we aim to foster the integration of basic and clinical research on decision making across multiple disciplines. We bring together new contributions on the fundamentals of decision making as well as their clinical implications and applications. In doing so, we combine work drawing on psychology, neuroscience, artificial intelligence, economics, philosophy and medicine.

The origins of the book can be traced back to conferences on decision sciences hosted by the Centre for Mind Research (CMR) at the University of Bonn in 2023 and 2024 and to collaborations in research and doctoral training between the University of Bonn and the University of Melbourne (since 2020), including a joint online seminar series on decision neuroscience and computational psychiatry. Many of the contributors to the present volume participated in one or more of those events and activities, and we are grateful to them for their contributions. We also thank all authors of this volume for their chapters. They are divided into three parts: (I) Concepts and Foundations, (II) Mechanisms and Correlates, and (III) Implications and Applications.

Part I Concepts and Foundations

In the first chapter, Heinrichs introduces decision making from a philosophical perspective and proposes a typology of different types of decision-making agents. A key distinction is observed between non-discursive and discursive agents, with the latter being uniquely responsible for their decisions. As the author argues, humans are the only known discursive agents, as they consciously reflect on and justify their choices.

Sturma considers decisions in action and places the concept of human behaviour at the centre of his considerations. He agrees with eliminativists that they reject naïve apologies for autonomy but accuses them of not having an adequate understanding of human behaviour as it manifests itself in the social space. In particular, he argues that eliminativists lack thick descriptions necessary to characterise a person's life.

Zafar proposes a normative framework for examining the conditions for optimal decision making with and without the support of artificial intelligence (AI). She critically investigates the understanding of and overall pursuit to make optimal decisions and describes the discrepancy between human practice of decision-making and the so-called AI decision-making processes.

Rockbach and colleagues investigate interactions between AI agents and humans. The authors point out that responsibility ultimately remains with the human agent and propose the concept of joint hybrid intelligence as a multidisciplinary framework for integrating human and AI decision making.

Part II Mechanisms and Correlates

The chapter by Tsetsos and Cao addresses the explanatory gap between description and prediction that arises from early theoretical decision science balancing theory and observation, and modern, heavily empirical behavioural research. The authors argue that irrational choices may arise from multiple plausible mechanisms, making it difficult to infer underlying cognitive processes from behaviour alone. They propose that time-resolved neural recordings may be used to reveal how attention and information processing unfold during complex decision-making.

Bröder and Scharf examine a formal model of coherence-based information search in decision making, the iCodes model, and its prediction of the attraction search effect. This effect explains how search direction is influenced by attractiveness differences between options. While evidence supports the model and its predicted moderators, some prediction failures highlight areas for improvement in understanding pre-decisional information search.

The chapter by Noorani and Anderson introduces the Linear Approach to Threshold with Ergodic Rate (LATER) model, which illuminates the processes underlying saccadic decision making. The model has been applied to both simple prosaccade tasks and the more complex antisaccade task, a widely used measure of oculomotor control in basic and clinical research. The chapter presents an overview of LATER from its historical origins to future applications and addresses comparisons with other models.

Diederich presents a formalized dynamic dual-process model of intertemporal choice, building on existing theories that conceptualise decision making as the interplay between an impulsive System 1 and a rational System 2. Unlike traditional dual-process theories, her framework generates testable predictions about choice probabilities and response times using stochastic process models. In her chapter, Diederich evaluates different model architectures using experimental data.

Thoma reviews non-invasive brain stimulation (NIBS) studies of judgment and decision-making (JDM). While individuals with cognitive impairments or addictions often benefit from NIBS, results in healthy participants vary widely, likely due to methodological and individual differences. However, evidence suggests that frontal anodal stimulation can significantly influence JDM-related behaviours, highlighting the need for future research to refine methods and account for individual variability.

The chapter by Kühnel et al. focusses on value-based decision making, with a particular emphasis on maladaptive choices characteristic of clinical disorders. The authors argue that despite advances in refining decision making tasks and models, clinical applications still face challenges, especially with laboratory-based assessments. Accordingly, the chapter explores the potential of naturalistic research, with a focus on psychometric properties, designs, and modelling approaches and with the aim to inform computational psychiatry.

The contribution by Keidel et al. focusses on the date/delay effect in intertemporal choice. This effect refers to a reduction in temporal discounting when time is framed using dates rather than delay units. The chapter explores the cognitive and neural mechanisms underlying this highly replicable effect, its correlates, and its implications for the study of clinical populations.

Hogeterp et al. review the molecular genetics of individual differences in decision making, with a focus on genes related to neurotransmitters and brain metabolic processes. While early studies utilised candidate gene approaches, recent research considers multiple genes through advanced multivariate methods. The authors argue that future studies may benefit from exploring gene-environment interactions, such as epigenetics, to better understand the complex interplay between biology and behaviour.

Part III Implications and Applications

The chapter by Asché et al. reviews the literature on decision making deficits in major depressive disorder and their impact on personal and social functioning.

The authors discuss two broad explanatory approaches, namely computational and clinical theories, and highlight the need for therapeutic interventions that improve decision-making skills.

Suzuki and Katahira argue that computational modelling is increasingly used to characterise altered decision-making processes in neuropsychiatric disorders. The authors introduce reinforcement learning (RL) as a widely used paradigm in computational psychiatry and provide detailed methodological guidance on the implementation of RL as well as examples of applications to obsessive-compulsive and gambling disorders.

Halbe et al. present evidence that impairments of decision making associated with ADHD stem from fundamental cognitive deficits and associated neurobiological factors, such as deficiencies in dopamine and noradrenaline signalling. A particular challenge in this field of research is the heterogeneity in cognitive profiles in ADHD, which needs to be considered when investigating the mechanisms underlying poor decision-making behaviour in this clinical disorder.

The chapter by Yan and Chen examines apathy, a key negative symptom of schizophrenia, through the lens of effort-based decision-making (EBDM). The authors provide a current theoretical framework of EBDM and discuss relevant experimental paradigms, behavioural manifestations, and neural correlates of this type of decision in schizophrenia.

Radbruch et al. propose that patient consent in medical decisions should be based on a collaborative process between physician and patient. As medical decision making is often complex and not always clear-cut, it should be approached as an ongoing process rather than a one-time event. This requires mutual respect, trust-building, and acknowledging both the physician's medical expertise and the patient's personal values and preferences.

Overall, the chapters in this volume provide a broad overview of current research methods used to investigate various aspects of decision-making, as well as the state of knowledge and open questions in several areas of decision sciences. They make it clear that decision research has evolved into a highly interdisciplinary endeavour that benefits from a variety of theoretical frameworks and empirical methods.

Bonn, Germany — Ulrich Ettinger
Bonn, Germany — Bert Heinrichs
Melbourne, Australia — Carsten Murawski

Reference

Caceda R, Nemeroff CB, Harvey PD (2014) Toward an understanding of decision making in severe mental illness. J Neuropsychiatry Clin Neurosci 26:196–213

Contents

Concepts and Foundations

Chapter 1
Varieties of Decision-Making

Bert Heinrichs

Abstract Starting with a broad use of the term, this chapter examines different types of decision-making and agents. A typology is proposed that can help to avoid confusion and promote understanding, especially in discussions between different scientific disciplines. The typology includes reporting/non-reporting agents, fully/partially transparent agents, and observer-/self-transparent agents. However, the most profound distinction turns out to be between non-discursive and discursive agents. This is mainly because discursivity and responsibility are closely linked. Discursive agents therefore have a special position: they must decide how they deal with other types of agents and what they use them for and how they use them in their own decision-making. So far, only humans can be considered as discursive agents.

Keywords Decision-making · Artificial Intelligence · Agency · Discursiveness · Responsibility · Morality

1.1 Introduction

We often use the term "decision-making" in everyday language, and it is also common in some scientific disciplines, in particular in decision theory (Steele and Stefánsson 2020) and operations research (Salhi and Boylan 2022), but also in psychology (cf. the chapters in Part II of this volume), and computer sciences (Kochenderfer et al. 2022). In philosophical theory of action, on the other hand, the concept is less prominent and ranks behind other concepts, most notably that of intention (Anscombe 2000; Davidson 2006; Setiya 2022). Like many other terms, "decision-making" is used in a variety of meanings that only partially overlap. While in philosophy the notion is often closely related to the concept of responsibility (Heidbrink et al. 2017), in other

B. Heinrichs (✉)
Institute of Neurosciences and Medicine: Brain and Behaviour (INM-7), Forschungszentrum Jülich, Jülich, Germany
e-mail: b.heinrichs@fz-juelich.de

Institute for Science and Ethics (IWE), University of Bonn, Bonn, Germany

U. Ettinger et al. (eds.), *Decision Making*, Studies in Neuroscience, Psychology and Behavioral Economics, https://doi.org/10.1007/978-3-032-00880-0_1

fields it is used rather broadly to describe the behavior of various types of systems in their respective environments. These systems can be humans, non-human animals, or artificial systems such as robots or computer programs. In what follows, I begin with such a broad use of the term and consider then whether it might be helpful to distinguish different types of decision-making to avoid confusion and promote understanding, especially in discussions among different scientific disciplines.

1.2 Decision-Making Broadly Conceived

Suppose S is a system that exists in an environment E, such that S has a set of observations O about the state of E, and S can perform an action A based on O that causes changes in E or S itself. Then let S be an *agent* and let the process that leads S to the selection of A from a set of possible alternatives be a *decision* D (this definition is inspired by Kochenderfer et al. 2022, pp. 1–2; cf. Russell and Norvig 2020, chap. 1).[1]

Based on this definition, S could be a person who notices the smell of fresh coffee in the house and as a result decides to go to the kitchen to get some. Or S could be a cat strolling through the garden, hearing a rustle in the grass, and decides to crouch down to launch an attack on a mouse it suspects to be there. Or S could be a surveillance system that registers suspicious transactions on a bank account and then decides to issue a warning to the regulatory authority in charge. In each of these three cases, the system S responds to observations O (the smell of coffee, the rustle in the gras, the suspicious transactions) from the environment and selects an action A (going to the kitchen, crouching in the gras, issuing a warning) out of several alternatives. Consequently, at this level of abstraction, all three systems—the human, the cat, and the surveillance system—can be said to make decisions. Note that the concepts of agent and decision-making are inextricably linked in that they describe the same phenomenon from different angles. We are concerned with entities and events that are special in some way. For example, stones are not agents and cannot make decisions.

An interesting question with respect to systems that are agents is how good their decisions are. To answer this question, one needs to introduce some standard V by which to measure the changes C brought about by S through A in E or S itself. Furthermore, one needs to estimate which alternative changes C* would have been brought about by an alternative action A*. One can say a decision D was good if the changes C brought about were positive as measured by V. One can further say a decision D was optimal if there was no alternative action A* so that any other changes C* brought about would have been more positive than C. An agent can be

[1] The term "observations" is to be understood here in a very broad sense and includes many types of inputs based on states of E to A. However, according to this definition, decisions are only those types of selections that are triggered by an external input, i.e. no purely internal status changes of A.

called rational if S acts (often enough) in such a way that the chosen actions $A_{1\text{-}n}$ are guided by the standard V.

At this very abstract level, understanding (good or optimal) decisions poses no fundamental problems. To be sure, this does not mean that it is easy for an agent S to make a good or even optimal decision. One reason for this is that the changes C or C* (i.e., the effects of A or A*) are subject to considerable uncertainties. For S it can be very difficult to foresee whether A will have a positive effect, and even more so whether A will have a more positive effect than A*. Additionally, the set of observations O can also be subject to uncertainty, which complicates matters further. Think about the person who smelled the coffee: maybe the smell came from outside. Or think about the cat: maybe there was a mouse there, but the attack came too late. Or think of the surveillance system: maybe it was a malfunction in data transmission. In all these cases, the decision that the agent made would not have been a good one.

Moreover, it can be difficult to define an appropriate standard V and to evaluate A or A* in terms of V. Apparently, the term "utility" is relevant here and hence defining and calculating utility functions is an important endeavor in decision science (Steele and Stefánsson 2020, Sect. 1.2).

Taken together, it should be clear that decision-making can be challenging for an agent, and it can be especially challenging to design an artificial agent in such a way that it makes good or even optimal decisions in a complex environment. Nevertheless, the basic constellation of decision-making as the selection of an action from a set of alternatives based on observations does not seem to be particularly complicated. In the following, I shall deal with some conceptual difficulties that may nevertheless arise.

Saying that the surveillance system decided to issue a warning to the regulatory authority after noticing suspicious transactions on a bank account may seem excessively anthropomorphic and therefore misguided. Some may even consider the notion of an agent for a computer system to be a misnomer. Taking this charge seriously, it might be worthwhile to look for differences in decision-making and consider how these relate to the nature of the agents involved (humans, non-human animals, artificial systems). This is how I shall proceed in this chapter. This will allow for a more precise use of the term decision-making.

1.3 Types of Decision-Making and Types of Agents

Based on the above definition, making a decision is essentially making a choice from a set of alternative actions. As shown, very different systems qualify for decision-making under this definition. Let's start now by looking at a simple artificial system and examine it a little bit more closely. Imagine a (simplified) robotic lawnmower that has various sensors, including a bump sensor to avoid obstacles, a tilt sensor that stops cutting blades if the robot is in a compromised position and a lift sensor that stops cutting blades if the robot is lifted off the ground. Apparently, the robotic lawnmower is a system that has the characteristics mentioned in the definition above.

It is therefore a system that makes decisions, i.e., an agent. If, for example, the lift sensor indicates that the robot has been lifted off the ground and the robot stops as a result, then it has made a decision, i.e., it has selected one of two possible actions—continue mowing or stopping. Note that if it was indeed lifted and if there was no sensor malfunction, then it was a good decision because it was in accordance with the manufacturer's standard aiming to prevent human injury. In fact, the decision was even optimal because there was no alternative action available that would have led to a better outcome.

An external observer might ask why the lawnmower stopped cutting. The lawnmower itself is not able to answer this question. Compare this with the person who smelled the fresh coffee. An observer might ask him why he decided to go to the kitchen. Obviously, and in contrast to the lawnmower, the person could have answered that he wanted to get a coffee. A distinction can be made, then, between two types of agents, those that can provide information about their own decisions and those that cannot. Let's call the first type *reporting* and the second *non-reporting*. Note that reporting systems do not need to be able to use speech. If the lawnmower has a built-in LED that indicates when the lift sensor reports that the robot has been lifted off the ground, this will also count as reporting. Note also that non-human animals are generally non-reporting according to this classification since they usually do not articulate why they made a particular decision. Rather, it is us who interpret their behavior—although often quite accurately. Humans, on the other hand, are by default reporting agents, at least when it comes to healthy adults. In artificial systems both variants—reporting as well as non-reporting—exist.

Let's go back to the lawnmower and assume that it does not have a built-in LED that indicates whether one of the sensors has passed on a signal. Since the system is non-reporting, the observer may still want to know why the robot stopped. For this purpose, the observer could disassemble the lawnmower to try to understand its functioning. Considering the rather simple setup, the observer could certainly find out why the decision to stop was made. Even more, the observer could, with some technical skill, figure out the entire functioning of the lawnmower. Let us call agents where this is possible *fully transparent*.[2] Humans and non-human animals apparently do not fall into this category. Even if we dissected a non-human animal (which would, of course, be problematic for ethical reasons) we would not find out why it makes the decisions it does. Animals are simply far too complex to understand their behavior in this way. We must therefore classify animals as only *partially transparent*. Although we can often capture their decisions-making quite accurately through observation, there is always some uncertainty and room for interpretation, so predictions are never entirely certain when it comes to animal behavior. Take the cat from above: Perhaps it did not lie in wait because it suspected a mouse in the grass, but it wanted to hide from a supposed predator. The original interpretation of the decision situation might be more plausible but is by no means beyond doubt.

[2] Transparency here refers exclusively to the decision-making process, i.e., the action selection process, but not to the observations or evaluative standard.

Although humans can provide information about why they made a decision, it would be naïve to believe that they are fully transparent. Our decisions are often influenced to a considerable extent by unconscious factors. Humans are therefore also only partially transparent, both for themselves and others (more on this below).

What about artificial systems? Apparently, there are some such systems which are fully transparent as is the case with the lawnmower. Others, however, are only partially transparent or perhaps even non-transparent. The latter is true, for example, for deep learning systems. To be sure, such systems have been designed by programmers. In the course of training with large amounts of data, however, these systems change the weights of their hidden layers, which ultimately leads to their decision-making being incomprehensible to humans including their programmers. This fact raises a number of problems, including ethical ones (Müller 2023, Sect. 2.3). Therefore, there is a demand that such systems should become explainable or, in the terminology suggested here, partially transparent, which has been the subject of intensive work for some time now (Holzinger et al. 2022).

Let's briefly summarize: One can distinguish between *reporting* and *non-reporting* agents on the one hand and between *fully transparent*, *partially transparent* and *non-transparent* agents on the other hand. Obviously, all agents that are reporting are at least partially transparent.[3] Humans are usually reporting and partially transparent, while non-human animals are non-reporting and partially transparent. For artificial systems, all variants are possible.

Another distinction that can be made has to do with how an agent is informed about its own decisions. The person who decided to go to the kitchen to get coffee usually knows why he went to the kitchen. This is why he can answer the observer's question. This is different from the lawnmower, which has an LED that indicates when a sensor has passed a signal. The lawnmower itself does not have the information that the sensor was active, i.e., this meta-information is not represented in the robot. It is simply an information that indicates to an external observer why the lawnmower has decided to stop. One could say the lawnmower is merely *observer-transparent*. In contrast, humans know about the factors that led to a decision—at least partially. One can therefore say that humans are (partially) *self-transparent*.[4]

It is important to distinguish between reporting/non-reporting, degrees of transparency and observer-transparent/self-transparent as each of these categories indicate different capacities of an agent. Above all, this seems important for avoiding a particular misunderstanding: One might think that artificial systems cannot be self-transparent and that this is something that distinguishes them from human agents, but this is not the case. It is true that the lawnmower with the built-in LED does not have this capacity. But more complex artificial systems are conceivable that have

---

[3] To be precise, another qualification may be necessary here: The system could be non-transparent and still be reporting. In this case, the reporting would of course very often be inaccurate. However, it is questionable whether the system's statements should still be regarded as reporting in this case.

[4] The notion "self" in "self-transparency" is not used here in a reifying way. Rather, it only indicates a self-referential structure. John Flavell coined the term "metacognition" for "knowledge and cognition about cognitive phenomena" (Flavell 1979, p. 906). The notion of self-transparency is meant here in the sense of such metacognition.

representations of their own decision-making. One can imagine, for example, that the program for monitoring bank accounts has internal representations of this kind. If this were the case, then in the event of an alarm, the program could be questioned as to why it made its decision. For example, the program could specify factors such as the frequency of transactions, the amount of money involved, and the type of source and destination accounts that led to the alert. Note that self-transparency is obviously only possible if a system is at least partially transparent, while the reverse is not true.

Let us briefly summarize again: I have argued that there are different types of agents capable of different types of decision-making. Humans have been classified as systems that are (typically) reporting and both partially self-transparent as well as observer-transparent. Non-human animals, on the other hand, are systems that are non-reporting and partially observer-transparent. Artificial systems cover a wide spectrum in this classification schema. They can be reporting as well as non-reporting; they can also be fully transparent, partially transparent or non-transparent; finally, they can be self-transparent or merely observer-transparent.

Generally speaking, complex systems perform worse than simple systems when it comes to transparency. However, this does not only apply to artificial systems, but also to humans and non-human animals. We, too, are only partially transparent to ourselves as well as to our fellows when it comes to our decisions. By the way, one can speculate whether there are also completely self-transparent agents. A divine being—if there is one—might fall into this category. But if such a divine being exists, it seems to be non-reporting.

Does the typology suggested here imply that the charge of anthropomorphism, mentioned at the outset, is completely mistaken? Apparently, humans as well as artificial systems can be reporting as well as partially self-transparent and turn out to be quite similar in this respect. This might suggest that the charge of anthropomorphism should be dropped. Or is there another difference, between artificial systems (at least current ones) and humans, in terms of decision-making?

1.4 Discursive Decision-Making

Let's recall the person who decided to go to the kitchen to get a coffee and compare it with the surveillance system that decided to send an alarm because there was a suspicious transaction on the account. Let us assume that the surveillance system is partially self-transparent, i.e., that it has internal representations of the factors that contributed to the decision it made. We could ask both the coffee drinker and the surveillance system about their respective decisions. To that extent, the two systems are similar. If we were to ask about the rationale behind the decisions, in one case we might get the following answer: "The house smelled so good of coffee and since I didn't have any before, but I really like to drink coffee I decided to go to the kitchen because I suspected that's where the smell came from." In the other case, the answer might perhaps be as follows: "I registered in quick succession a large number of transfers to and from the account of large sums going to or coming from various countries. Having ruled out a malfunction, as the bank's server did not document any such failure, it could be that these were illegal transactions intended to conceal criminal activity." Both answers would be appropriate in the sense that they make the choice of the particular action from the set of possible actions sufficiently plausible. (Of course, the information could still be wrong because humans are never completely transparent to themselves. A person could therefore be mistaken about their own reasons.) Both responses also indicate that the selected courses of action are rational in that they meet predefined standards (i.e., maximizing well-being, detecting crime). Still, in the case of the coffee drinker, one can imagine that an observer might have doubts about the soundness of the decision. The observer might say something like: "I understand why you decided to go to the kitchen, but it would have been better to stay at your desk and to continue to work to finish that book chapter you agreed to write." It is hard to imagine such a reply in the case of the surveillance system. What does this tell us about the decision-making of the two different agents?

In a famous passage of his *Empiricism and the Philosophy of Mind,* Wilfrid Sellars considers what it means to conceive of a mental episode as knowing:

> The essential point is that in characterizing an episode or a state as that of knowing, we are not giving an empirical description of that episode or state; we are placing it in the logical space of reasons, of justifying and being able to justify what one says. (Sellars 1997, § 36)

The point of comparing the coffee drinker and the monitoring system is that we can replace the term "knowing" with "decision-making" in the quote from Sellars. If a reporting self-transparent agent (human or artificial) provides information about which factors influenced a decision, then this agent only provides an empirical description in the sense of Sellars. The agent discloses what factors were computed and how this ultimately led to the selection of a particular action rather than another. However, humans can also provide *reasons* for a decision. If they do, they are no longer just giving an empirical description. Rather, they place the decision in the space of reasons, i.e., they enter into a *discourse* about it. They explore not only whether a decision was good or bad according to a predefined standard, but they examine whether an agent may have made a *mistake.* In other words, decisions then appear under the normative distinction of right and wrong. In this, such agents differ from those considered so far which suggests a final distinction, namely that between *discursive* and *non-discursive* agent.[5]

The space of reasons is a normative space, and it is a socially structured space (Brandom 1994, 1995). Making a move in this space, such as making a discursive decision, means taking on commitments and assigning entitlements to others. This includes, above all, accepting the obligation to give reasons when someone else doubts the legitimacy of the move. The entitlements that others receive are essentially

[5] It should be noted that the analogy to Sellars's notion of knowing is imperfect in at least one respect, insofar as it might suggest that the process of choosing an action, if it is not discursive, is not a form of decision-making at all, just as a particular description of a mental state does not describe this state as knowing. Here, however, I assume that such a process can be taken as a form of decision-making, if only as non-discursive decision-making. Unlike Sellars, I allow for qualifications within the concept. I am grateful to Tommaso Bruni for pointing this out to me.

that they may use the prior move as a starting point for their own future moves. These normative interrelations indicate that beings moving in the space of reasons must be able to take responsibility. Discursive agents are therefore necessarily *responsible* agents. This brings back a notion that had already briefly appeared at the beginning, and which has traditionally been closely associated with the concept of decision-making, especially in philosophy. Now it becomes clear that there is, indeed, an important dimension of decision-making associated with the notion of responsibility.

The exercise of responsibility presupposes a community of agents linked by a normative practice. There is an argument to be made that such a practice is tied to a shared life form. It is true that the normativity in the space of reasons, as conceived by Sellars and Brandom, is initially of a logical or inferential nature. However, it is reasonable to think that it must be anchored in elementary structures of a life form. Wittgenstein drew attention to this with his example of the talking lion, which we do not understand (Wittgenstein 1984, § 568). Floridi has a point when he argues against Wittgenstein that we share a lot with the lion (hunger, fear, pain, and pleasure), so that understanding might be possible if the lion started to talk to us (Floridi 2013, p. 298). However, the basic idea remains valid: there must be substantial common ground for different agents to form a discursive community. It is difficult—though not impossible, as some science-fiction shows—to imagine that artificial systems become members of our human community. If this is true, then artificial systems cannot be discursive agents at least as long as they are not members of our human community. But for this to happen, they would need to have some properties that are essential to us, such as embodiment and feelings—by the way, properties that some critics believe are necessary for artificial general intelligence (the *locus classicus* is Dreyfus 1992).[6]

For discursive agents, the evaluative standard for choosing an action is always up for discussion as well (and not only the particular decision taken). While some artificial systems can provide information about how they have reached a decision—which is quite impressive, especially if one considers that such systems can make extremely complex decisions for which they process a great number of factors—the standard of evaluation itself is fixed. For this reason, it is important to consider exactly which standard ensures that a particular goal is achieved by a complex system. After all, it could turn out that an artificial system pursues a predefined goal extremely efficiently, but then it turns out that it was not the goal we had initially in mind. Ironically, this decision would still be good or even optimal given the definition from above. This problem has been discussed for some time under the title "alignment problem" (Christian 2020). It is a problem not least because artificial systems do not change their goals or their evaluative standards. Therefore, we need to make sure that the goals we implement in them are appropriate. Or else we would have to build AI in such a way that it can change its goals on its own. However, this approach might be even more problematic. If artificial systems were discursive in

[6] Note that it could be that artificial agents form an entirely separate kind of discursive community that would exist beyond and entirely independent of our human discursive community. This is, however, an extremely speculative idea which I do not want to pursue any further here.

the sense just outlined, then they could rethink their decision-making practice under the impression of good reasons and, if necessary, realign it themselves. Just as the coffee drinker might concede to the observer's criticism that in the future, he will curb his desire for coffee until the book chapter is finished, so too the surveillance system might change its goals by itself under the impression of criticism. However, this would have to be accompanied by the ability to take responsibility for one's own decisions. Even if it should be possible to build artificial systems that can change their goals and standards themselves, we must think carefully about whether we want to do so. It could be that that would lead to profound conflicts with our social practice and, ultimately, even with the future of our life form (Bostrom 2014). In any case, the construction of such artificial agents would be a decision for which the persons in charge would have to take responsibility within the human community.

1.5 An Alternative Approach

In his *The Ethics of Information*, Luciano Floridi proposed a typology of agents too. A brief comparison shows where his approach and the one suggested here differ. First, Floridi introduces the notion of "level of abstraction (LoA)" to allow for different levels of analysis (Floridi 2013, pp. 31–34). On one level of abstraction, according to Floridi an agent is "a system, situated within and a part of an environment, which initiates a transformation, produces an effect, or exerts power on it over time" (Floridi 2013, p. 140). According to this definition, earthquakes are agents, as Floridi points out. However, since this is implausible, he suggests changing the LoA to include three criteria, namely "interactivity", "autonomy", and "adaptability". While interactivity is self-explanatory and, moreover, in line with the definition proposed above, autonomy and adaptivity require explanation. According to Floridi, autonomy simply means that a system can carry out internal transactions independently of its environment. This is also in line with the definition above. The situation is different with adaptivity, by which Floridi means that "the agent's interaction changes (or can change) the transitions rules by which it changes states." (Floridi 2013, p. 141). This was not required in the definition above. Otherwise, the robotic lawnmower, for example, could not have been classified as an agent. In this respect, Floridi places higher demands on a system. According to him, agents are always adaptive systems. Floridi goes on to introduce another subcategory, and that is that of a "moral agent" (Floridi 2013, pp. 146–148). By this he means agents capable of "morally qualifiable action". Note that his does not entail responsibility. On the contrary, Floridi wants to escape the following dichotomy:

- moral agency, therefore responsibility, therefore prescriptive action, versus
- if there is no responsibility then there is no moral agency, but without the latter there is no need for any prescriptive action. (Floridi 2013, p. 159).

In other words, he worries that agents can perform actions that are morally relevant (i.e., it can cause moral good or evil), but that no responsibility can be attributed. One can share this concern without having to accept Floridi's typology. On the contrary, one can adopt a more liberal concept of agents that includes non-learning systems on the one hand and assume a more nuanced typology that comprises discursive agents which are responsible for the consequences of action on the other hand. If one takes this approach, then the question of attribution of responsibility can also be answered: it is always discursive agents who are responsible for consequences of action (if it is not a matter of accidents or natural events). Moreover, this approach has the advantage of linking types of agents directly to types of decision-making, i.e., to specific capabilities, rather than to consequences in the environment. This is less revisionist and preserves common language and established normative practices. Floridi's approach, however, makes clear that agent interaction raises important questions.

1.6 Decision-Making Involving Agents

It is important to realize that agents can be present in the environment of other agents, so that different forms of interaction may occur. In particular, other agents may play a role in the decision-making of agents. They may, in the terminology suggested above, influence the set of observations O and the set of possible actions A in a variety of ways. One can therefore consider whether general rules can be derived from the typology developed earlier. For example, should we deal with reporting agents in a fundamentally different way than with non-reporting agents? Or should we always use non-transparent agents differently than transparent agents?

To begin with a fairly simple observation, the average quality of an agent's decisions does not seem to depend on the type of agent. The evidence suggests that robotic lawnmowers perform very well, i.e., make mostly good or even optimal decisions when measured against the specified standards (getting the mowing done, doing it quickly and accurately, not hurting anyone, etc.). This, of course, is mainly because they must base their decisions on few observations (three sensors), that processing these observations is rather easy, and that the set of possible actions is very limited (mowing in a certain direction, stopping). There is, therefore, little room for error. With complex systems such as the surveillance system, things are different. It is the complexity of the environment, or the set of observations that goes into a decision, as well as the possible courses of action that affect the quality of an agent's decisions. The uncertainties associated with the given parameters is equally important. In this regard, the typology of agents and decision-making is not decisive.

The picture is different when it comes to agents who can make decisions with serious consequences. It can be helpful to have a reporting agent for this may allow to intervene directly in the decision-making process to avoid unfavorable effects. Likewise, it can be helpful to have partially transparent or fully transparent agents for understanding unwanted consequences in retrospect. Efforts to create explainable AI take this fact into account. Especially in sensitive areas, we should not rely on non-transparent agents.

There is something else to consider: According to an established distinction, non-discursive agents are means or tools.[7] They are used by discursive agents as a means to an end. They can never take the position of a discursive agent within ends-means relationships because they cannot bear responsibility. For discursive agents, this means that they need to look very carefully at the types of agents they use or rely on in making their own decisions. Intransparency can become a problem here because it may make it more difficult to assign responsibility. Some have even argued that a "responsibility gap" is opening up (Matthias 2004), while others have disagreed (e.g., Köhler et al. 2017). In any case, the difference between discursive agents and non-discursive agents turns out to be more profound than the other distinctions. This is consistent with the classical view that persons, i.e., discursive agents, must never be used by other persons merely as a means (Kant 2016). This does not imply that the treatment of all other agents is morally neutral or unimportant. Animals, as sentient beings, are widely believed to have a moral status such that we may not treat them cruelly. What would have to be fulfilled for artificial agents to also acquire such a moral status is currently the subject of intense debate. For the time being, however, we as discursive agents need to think first and foremost about the requirements we place on agents we use as tools for specific tasks. The typology developed earlier may be particularly relevant in this regard.

1.7 Outlook

The considerations I have presented in this chapter show that the term "decision making" can be understood quite broadly and can encompass quite different types of processes. The unifying element is that it is always a selection of an action from a set of alternatives, based on observations of the environment. Systems that are able to do this are called agents. However, a broad understanding of terms can easily lead to misunderstandings, especially when different scientific disciplines work together. The proposed typology can help to avoid such misunderstandings by offering conceptual differentiations. If there are any uncertainties, it is always worth asking what kind of decision-making one actually has in mind. The most profound distinction is between non-discursive and discursive agents. So far, only humans fall into the latter category. Since discursivity and responsibility are necessarily linked, so far only humans can be assigned responsibility for decisions. We must, therefore, carefully consider how we deal with other types of agents and for what and how we employ them in our own decision-making.

[7] According to this distinction, non-human animals are also means or tools, since they cannot be responsible agents either. Even if we grant moral status to non-human animals, we still do not treat them as full members of the discursive community in our moral and legal practice. For example, we do not hold them responsible for damage they cause. In the case of pets or livestock, it is rather the owners who must assume this responsibility.

Acknowledgements I would like to thank Ulrich Ettinger and Carsten Murawski as well as the members of my working group "Neuroethics and Ethics of AI" for providing invaluable comments on an earlier version of this chapter.

References

Anscombe GEM (2000) Intention. Harvard University Press, Cambridge
Bostrom N (2014) Superintelligence: paths, strategies and dangers. Oxford University Press, Oxford
Brandom R (1994) Making it explicit. Reasoning, representing, and discursive commitment. Harvard University Press, Cambridge
Brandom R (1995) Knowledge and the social articulation of the space of reasons. Philos Phenomenol Res 55(4):895–908. https://doi.org/10.2307/2108339
Christian B (2020) The alignment problem. Machine learning and human values. Norton, New York
Davidson D (2006) The essential Davidson. Oxford University Press, Oxford
Dreyfus HL (1992) What computers still can't do. A critique of artificial reason. MIT Press, Cambridge
Flavell JH (1979) Metacognition and cognitive monitoring: a new area of cognitive-developmental inquiry. Am Psychol 34(10):906–911
Floridi L (2013) The ethics of information. Oxford University Press, Oxford
Heidbrink L, Langbehn C, Loh J (eds) (2017) Handbuch Verantwortung. Springer, Wiesbaden
Holzinger A, Saranti A, Molnar C, Biecek P, Samek W (2022) Explainable AI methods: a brief overview. In: Holzinger A, Goebel R, Fong R, Moon T, Müller K-R, Samek W (eds) xxAI—beyond explainable AI. xxAI 2020. Lecture notes in computer science (LNCS, vol 13200). Springer, Cham, p 13–38. https://doi.org/10.1007/978-3-031-04083-2_2
Kant I (2016) Grundlegung zur Metaphysik der Sitten. In: Kraft B, Schönecker D (eds) Meiner, Hamburg
Kochenderfer MJ, Wheeler TA, Wray KH (2022) Algorithms for decision-making. MIT Press, Cambridge
Köhler S, Roughley N, Sauer H (2017) Technologically blurred accountability? Technology, responsibility gaps and the robustness of our everyday conceptual scheme. In: Ulbert C, Finkenbusch P, Sondermann E, Debiel T (eds) Moral agency and the politics of responsibility. Routledge, London, pp 51–67
Matthias A (2004) The responsibility gap: ascribing responsibility for the actions of learning automata. Ethics Inf Technol 6:175–183. https://doi.org/10.1007/s10676-004-3422-1
Müller VC (2023) Ethics of artificial intelligence and robotics. In: Zalta EN (ed) The stanford encyclopedia of philosophy. https://plato.stanford.edu/archives/fall2023/entries/ethics-ai/
Russell SJ, Norvig P (2020) Artificial intelligence: a modern approach, 4th edn. Pearson, Hoboken, NJ
Salhi S, Boylan J (eds) (2022) The Palgrave handbook of operations research. Palgrave Macmillan, Cham
Sellars W (1997) Empiricism and the philosophy of mind. Harvard University Press, Cambridge
Setiya K (2022) Intention. In: Zalta EN (ed) The Stanford encyclopedia of philosophy. https://plato.stanford.edu/archives/win2022/entries/intention/
Steele K, Stefánsson HO (2020) Decision theory. In: Zalta EN (ed) The Stanford encyclopedia of philosophy. https://plato.stanford.edu/archives/win2020/entries/decision-theory/.
Wittgenstein L (1984) Philosophische Untersuchungen (Werkausgabe, Bd. 1). Suhrkamp: Frankfurt am Main

Chapter 2
Decisions in Action: A Naturalistic Approach to Human Behavior

Dieter Sturma

Abstract In the scientific worldview, there is no clear picture of the causal role that reflection, decision, and action play in changing the course of events. According to the widely held eliminativist view, there is no room for causal roles of the human mind in the closed realm of causes. The social practice of human language is an accessible field of research for studying the behavior of persons *as* persons that is easily bypassed by eliminativist approaches. Epistemic asymmetries are crucial to decision making. In all their deliberations, persons are well aware that they are affected by what they do in non-substitutable ways. The anxieties of a goalkeeper, a penalty kicker, and a pitcher show that reflecting, deciding, and acting are connected over time. Human behavior is not about random preferences but always about the whole person. In cases where decisions are linked to life plans, we are dealing with deep decisions that explain why persons, at least sometimes, do not follow situational motives. Under favorable epistemic conditions, an observer can make plausible conjectures about how a person is likely to respond to demands and expectations in social space—as long as a new reflection does not emerge. Decisions of persons are not mysterious episodes in the depths of their souls but behavioral manifestations in social space.

Keywords Reflection · Deep decisions · Thick description · Life plan · Linguistic behavior · Critical naturalism

2.1 Introduction

Anyone who answers questions about decisions and actions is making an implicit or explicit claim about the mind–body problem. The answer shows whether the behavior of persons is assumed to require new forms of explanation, or whether it can be explained in terms of laws that have been successfully applied to other fields of research. This will indicate whether actions are regarded as mere events or not.

D. Sturma (✉)
Department of Philosophy, University of Bonn, Bonn, Germany
e-mail: dieter.sturma@uni-bonn.de

U. Ettinger et al. (eds.), *Decision Making*, Studies in Neuroscience, Psychology and Behavioral Economics, https://doi.org/10.1007/978-3-032-00880-0_2

In both philosophical and non-philosophical explanations, decisions are often derived exclusively from antecedent conditions. In most cases, empirically identifiable preferences are taken to precede the action in terms of time as well as content, and to be decisive for its execution. The willingness to understand actions as more than mere events is not common outside philosophical discourse. This is probably due not least to the refusal to include reasons *as such* in the causal fabric of the world.

Even if it is the prevailing opinion that decisions precede actions, it is not obvious how the temporal sequence is organized in detail. It remains to be seen whether decisions can be clearly distinguished from respective actions. On closer examination, their beginning and end seem rather vague, especially as they are dependent on preceding episodes. If we want to assume that a given action can be recognized by its effect, we need to clarify what causal role we should assign to preceding episodes. Actions are permeated by prehistories, which exert both observable and reproducible effects. Any attempt to decipher the complex and intricate causal relationships in the case of concrete actions therefore faces epistemic problems that are hard to overcome.

What follows is an examination of the role that thinking plays in human behavior, with particular reference to decisions and actions. It is therefore about the role of reflecting and deciding in changing the course of events.

2.2 Naturalism and the Scientific Image of Man[1]

Like everything that exists, persons inhabit the realm of causes. Its mechanisms are hard facts in the sense of physical constraints and are the subject of various disciplines of the natural sciences. In the light of scientific achievements, a materialist image of man has emerged that is essentially oriented towards the causal fabric of the world. Very different scientific methods are used to investigate the behavior of persons and to make predictions about their future behavior. A radical, but nevertheless widespread version of materialism is scientism, which represents a methodological universalism based on one discipline of the exact sciences—as in the case of physicalism. Its scientific image of man is built on the idea that all events in the world, including human behavior, are governed without any exception by laws discovered in the exact sciences. Since these laws are supposed to apply without exception, the realm of causes turns out to be closed. Nothing acts on it from outside—neither gods nor souls.[2]

The standard formula of the scientistic worldview is that everything that exists is composed of basic elements whose movements are determined by natural laws. The dependence of the macro-world on the micro-world is explicitly emphasized. Persons are made up of atoms, whose movements determine their behavior and fate. The scientistic image of man undermines emphatic notions of human freedom and

[1] Cf. Sellars (1963).

[2] The scientistic image of man can already be found in French Materialism of the eighteenth century; cf. in particular La Mettrie (1748).

autonomy. Human beings do not have the power to escape the laws of nature. All things had to change their nature if men were to be free in the literal sense of the word. Anyone who wanted to hold on to such ideas would have to understand men at least partly as an ›unnatural‹ beings in the realm of causes.

The eliminativist position of French Materialism has survived to this day, with slight modifications and adaptations to the respective scientific paradigms. In particular, neurosciences and neurophilosophy have provided us with new interpretations of the realm of causes. In continuity with the critique of French Materialism, Francis Crick formulated what he called the *Astonishing Hypothesis*, which states that everything that makes up a person can be traced back to a collection of neurons whose movements follow the laws of natural sciences.[3] Based on this view, our personal attitudes, feelings, and perceptions of ourselves and others are not what we think they are. Persons are mistaken when they assume that they have intentionally initiated an action. In reality, unconscious mechanisms in their brains are the origin of their behavior.

In neurophilosophical eliminations, the experiments of Benjamin Libet are a recurrent theme (see Libet et al. 1999; Libet 2004). They are interpreted as evidence that consciousness deceives itself about its practical impact. The experiments seem to show that in human behavior, there are no temporal corridors for decisions of a self-conscious person. Following the eliminativist interpretation of the experiments, a conscious act of will can only be discerned at a point in time when the execution of the action has already been initiated in the brain. The Libet experiments do indeed show that a pre-motor potential is built up before a person can be said to be acting. Still, what can really be deduced from Libet's experiments remains controversial (cf. Bennett and Hacker 2003, pp. 228–231). The eliminativist critique is nonetheless instructive because it draws attention to temporal aspects of decisions and actions.

Daniel Wegner has defended the thesis of a closed realm of causes by pointing to the practical shortcomings of epistemic states: Although we could think about our intentions or actions in advance, this would always be too late compared to the neural or reactive processes taking place at the same time.[4] Preoccupation with our own mental states diverts our attention from the causal processes that actually underlie human behavior.[5]

---

[3] See Crick (1994), p. 3: "The Astonishing Hypothesis is that ›You‹, your joys and your sorrows, your memories and your ambitions, your sense of personal identity and free will, are in fact no more than the behavior of a vast assembly of nerve cells and their associated molecules. As Lewis Carroll's Alice might have phrased it: ›You're nothing but a pack of neurons.‹" Cf. Dennett (1991), p. 431/2.

[4] Wegner calls it a game of cat and mouse, see Wegner (2002), p. 59: "Consciousness and action seem to play a cat-and-mouse game over time. Although we may be conscious of whole vistas of action before the doings get underway, it is as though the conscious mind then slips out of touch. A microanalysis of the time interval before and after action indicates that consciousness pops in and out of the picture and doesn't really seem to do anything".

[5] See Wegner (2002), p. 27: "Each of our actions is really the culmination of an intricate set of physical and mental processes, including psychological mechanisms that correspond to the traditional concept of free will, in that they involve linkages between our thoughts and our actions. This is the

What the various neurophilosophical versions of the causally closed physical world have in common is the assumption that no specific initiatives arise from the reflective consciousness of human beings. In this view, what persons believe, and especially what they believe to do, is determined by micro-mechanisms spreading out in space and time, without the person *as* a person being able to intervene in the causal processes. There is no way out of causal closure. Attempts to establish a personal standpoint outside the realm of causes would violate the basic premises of modern science. Persons can make physical processes transparent, but this fact has no practical consequences for the course of their actions. La Mettrie emphasized this fact when he said that man is “une machine bien éclairée.”[6] This peculiar epistemic situation turns out to be a double misfortune: not only are persons, like all other beings in the world, trapped in the realm of causes, but unlike other beings they are aware of it.

In world models of the closed realm of causes, neither consciousness in general nor self-consciousness in particular plays a causal role and is therefore excluded from the explanation of human behavior. However, this powerful image is loaded with fallacies and category mistakes: Scientistic eliminativism[7] fails to recognize that theories do not cover facts in all aspects and do not define a particular theoretical framework. They lack the syntactic and semantic resources to account for the world of events in its entirety (*fallacy of underdetermination*). The conditions of possibility of its own position are neglected (*fallacy of self-referential inconsistency*). The difference between evidence and reference is ignored so that information about an event is confused with the event itself (*fallacy of reference*). Methodological reduction is confused with the supposed discovery of the smallest basic elements, from which it is assumed that the whole world can be constructed. The fallacy is characterized by the formula *x is nothing but y* (*atomistic fallacy*). Parts are confused with the whole—for example, when it is claimed that the brain decides what a person does (*mereological fallacy*). Any assumptions about the realm of causes depend on the current state of research. The dynamics of scientific research are ignored (*fallacy of absolutizing the present*). The identification of a local cause for certain changes leads to the conclusion that all changes are caused only locally (*local fallacy*).

From a scientific point of view, the certainties of the lifeworld are visions of reality that must, in principle, be open to methodological examinations that show whether conventional methods are adequate or whether new approaches need to be developed. In answering the question of what exists, it is not enough to refer to one discipline alone. Rather, different disciplinary competences must be decided on a

---

empirical will. However, we don’t see this. Instead, we readily accept a far easier explanation of our behavior: We intended to do it, so we did it.”

[6] La Mettrie (1748), p. 70: “*Mais* puis que toutes les facultés de l’Ame dépendent tellement de la propre Organisation du Cerveau et de tout le Corps, qu’elles ne sont visiblement que cette Organisation même; Voilà une Machine bien éclairée!”.

[7] It is possible to distinguish more sharply between eliminativist and scientistic methodologies, although this distinction is not evident in its consequences. The narrowing of the research field is accomplished either through the elimination of content or through methodological oversimplification.

case-by-case basis. Descriptions, models, and theories from all areas of the natural sciences, social sciences, and humanities can provide direct or indirect information about what there is, provided that they meet the *established* and *communicable* standards of explanation and justification.

Given the conceptual and epistemological difficulties of eliminativist positions, the question is whether naturalistic explanations have sufficient explanatory potential at all for explaining human behavior in general and decisions in particular. It is not difficult to conclude from the fallacies of eliminative or scientistic naturalism that there is a need for a *critical* naturalism—an approach characterized by a commitment to the potentials and limitations of its methods as well as to what there is.

2.3 Thick Description and Linguistic Behavior

Eliminativists rightly object to naive apologies of freedom and autonomy, as well as to the methodological overemphasis on the first person perspective or self-consciousness. The decisive criticism of dualistic worldviews and speculative concepts of mental causation is also justified. But the eliminativists misunderstand the true nature of the scientific division of labor. Disciplinary expertise cannot simply be assumed. It belongs to the discipline that has the methods to grasp adequately the phenomena in question. This requirement is easily ignored in interdisciplinary discourses (cf. Sturma 2024a). Instead, the respective disciplines compete for the most favorable starting position by dragging the behavior of persons as a whole into their research field.

The sometimes fanciful narratives that eliminativists invent to explain human behavior are irritating. The brain is said to create models of a self or free will—metaphorical expressions without clear reference—with the intention of deceiving. It is suggested that there is a world out there about which we know nothing because our brains provide us with illusions. But there is no *other* physical world behind our sensorimotor interactions in the realm of causes. In addition, these scenarios ignore that models are tied to specific scientific and technical practices, which are firmly embedded in the space of reasons. In the realm of causes, there are neither models nor deceptions.

The eliminativists could have learned from René Descartes, whom they rejected for his dualism, that one cannot deceive oneself about self-awareness in the sense of self-certainty about one's own existence (cf. Sturma 2005, pp. 81–83). Sentences such as *I do not think*, or *I do not exist* are simply based on the fallacy of self-referential inconsistency. I cannot simply imagine that I am aware of myself. Though I can be mistaken about my self-knowledge and about the properties or states of my body.

There is an observable and accessible field for studying the behavior of persons *as* persons that eliminativist positions easily bypass. This is the social practice of human language. The life of persons is deeply rooted in the syntax and semantics of human language. It is not only a form of expression and communication but, above all, an order of symbolic behavior with a sense of meaning that opens up the space of reasons (cf. Sturma 2024b, pp. 132–136).

Wilfred Sellars introduces the concept of the space of reasons to emphasize that in propositional attitudes and cognitive states we do not refer directly to something empirically given, rather we operate in inferential relations by which what is said is or can be justified (see Sellars 1997, pp. 75/6). He gives the example of the use of predicates. The term ›red‹ would not be a predicate if it did not have the syntactic properties common to all predicates (see Sellars 1997, p. 66). These properties go hand in hand with propositional attitudes, which are the appropriate responses to situations in which something is identified as red. Sellars argues that a person's actions and behavior cannot be made intelligible without reference to the space of reasons.

For Sellars, rules or rule-governed behavior are to be regarded as essential manifestations of the human life form.[8] Sellars finds a striking image for this phenomenon:

> When God created Adam, he whispered in his ear, ›In all contexts of action you will recognize rules, if only the rule of groping for rules to recognize. When you cease to recognize rules, you will walk on four feet.‹ (Sellars 1980, p. 138)

The space of reasons is not the only essential feature of the human life form. Persons also make active use of their ability to express themselves, i.e., they have a sense of self-reference. They ascribe attitudes and actions to other persons that they would attribute to themselves in similar situations. They assume that other persons have the same sensibilities and self-referential attitudes, share values, adopt forms of recognition, and pursue life plans.

Human attitudes and activities cannot develop without language. Subjectivity or inwardness are not passive states; they must be expressed. The development of human attitudes depends on linguistic differentiation. States such as love, grief, hate, resentment, self-esteem, or remorse are all based on a complicated system of different concepts and terms, and they can take on a structured form and semantic consistency only within the space of reasons. In this sense, language is an evolving dimension of the human mind.[9]

Gilbert Ryle demonstrated the important difference between a thin and a thick description. He uses the example of two boys who "fairly swiftly contract the eyelids of their right eyes."[10] If we look only at the movement of the eyelids, it

---

[8] For the use of the term ›life form‹, cf. Sturma (2019b).

[9] When referring to oneself and to the attitudes of other persons, the same conceptual framework is used; see Strawson (1959), pp. 29, 106: "Yet it is a single picture which we build, a unified structure, in which we ourselves have a place, and in which every element is thought as directly or indirectly related to every other; and the framework of the structure, the common, unifying system of relations is spatio-temporal. [...] There is no sense in the idea of ascribing states of consciousness to oneself, or at all, unless the ascriber already knows how to describe at least some states of consciousness to others."

[10] See Ryle (1971), p. 480: "Two boys fairly swiftly contract the eyelids of their right eyes. In the first boy this is only an involuntary twitch; but the other is winking conspiratorially to an accomplice.

is initially impossible to decide whether this is a nervous twitch or silent communication. However, once one understands the interplay between physical movements, intentions, and rules of linguistic behavior, the nature of silent communication does become clear. Silent communication occurs only when a number of conditions are met by both the winker and the addressee in the process of using and decoding the signs. Ryle's example illustrates the wide range of linguistic behavior. It can manifest itself without speaking in public by pointing, winking, but also in ›eloquent silence‹.

A thin description contains only a record of observational data that an external observer reads from the behavior. In contrast, a thick description captures meaningful connections between attitude, language, intention, behavior, interpretation, understanding, and action that depend on specific human characteristics and capabilities. The thick description opens the wide field of the multicolored life of persons that contains visions of past, present, and future worlds, as well as possible and impossible worlds. This is why fictional scenarios are so valuable from a methodological point of view when it comes to defining the inferential relations of the human language.

A thick description shows how persons act in accordance with the world of meanings, interests, and concerns. Participation in linguistic practice allows persons to communicate with each other about their intentions and reasons. It is the fundamental problem of eliminativist or scientistic strategies that they overlook or even deliberately ignore this important difference between thick and thin descriptions.

Observing behavior and language use allows us to make plausible conjectures and conclusions about what might be going on in the first person perspective. To talk about experiences, it is not necessary to look for the private language of a person. Sellars has developed a so-called theory-theory of how self-referential linguistic behavior can be understood without dualistic models of introspection or private languages: His fictional Genius Jones suggests that persons do not behave intelligently only when they express overt verbal episodes. For Jones, overt linguistic behavior is the manifest culmination of a process that begins with inner speech as an expression of thoughts. These are to be regarded as *theoretically* inferred inner episodes that provide more or less reliable information about the attitudes and inner episodes of persons. When talking about inner episodes, something is not represented or depicted, rather, assumptions are made under conditions of referential uncertainty. As a result, statements about mental states are just as error-prone as theoretical hypotheses. In the end, what began as language with a purely theoretical function gradually takes on

---

At the lowest or the thinnest level of description the two contractions of the eyelids may be exactly alike. From a cinematograph-film of the two faces there might be no telling which contraction, if either, was a wink, or which, if either, was a mere twitch. Yet there remains the immense but unphotographable difference between a twitch and a wink. For to wink is to try to signal to someone in particular, without the cognisance of others, a definite message according to an already understood code. It has very complex success-*versus*-failure conditions. The wink is a failure if its intended recipient does not see it; or sees it but does not know or forgets the code; or misconstrues it; or disobeys or disbelieves it; or if anyone else spots it. A mere twitch, on the other hand, is neither a failure nor a success; it has no intended recipient; it is not meant to be unwitnessed by anybody; it carries no message. It may be a symptom but it is not a signal." Cf. Sturma (2019b).

a reporting role in terms of linguistic self-reference.[11] The ability to have thoughts unfolds in the process of acquiring a language, and only when this language is firmly established can self-reference develop.

Ryle's thick description and Sellars' theory-theory provide a wide semantic field for comprehending the observable behavior of persons, the subjects of the human life form. The benefit of their analytical reconstructions lies in identifying the incompleteness of scientistic eliminations without having to rely on traditional models of introspection or dualism. They show that eliminativist strategies lack the potential to step out of the field of thin descriptions. Eliminativists conceal this deficit with an atomistic fallacy. In their version of the formula *x is nothing but y*, *x* stands for a phenomenon of everyday experience—such as a decision or action—and *y* for a physical event (cf. Sect. 2.2).

In deciding and acting, persons ponder how to intervene in the course of events. The way in which they do this is manifested in social practice. This is also true if they remain inactive. Observable linguistic behavior is an expression of internal processes that are not directly accessible as such, nor are they nothing. Rather, they prove to be causally effective in the way persons influence the course of events. If one tries to separate the external from the internal, the inner episodes evaporate.

The thick description provides access to experiences through the analysis of linguistic behavior. This approach ensures the naturalistic character. Linguistic behavior is both a publicly observable event and the result of subjective attitudes and inner episodes. These are not locked away in the depths of the soul but are expressions of processes that manifest themselves internally and externally in the open field of social practice.

Persons share a common system of syntactic and semantic rules that forms linguistic behavior in the first, second, and third person perspectives. It is of existential importance whether a person uses the term ›pain‹ for herself, for another person, or for another living being. Except for certain psychopathological conditions, every person is aware of being in this or that situation. This self-consciousness is the source of thought and reflection, and it also generates epistemic asymmetries in social practice.

Epistemic asymmetries are of crucial importance to decision making. In all their deliberations and reflections, persons are well aware that they are affected by what they do in specific and non-substitutable ways. This existential experience usually initiates a series of expectations and reflections, and the acting persons know that the practical responses depend crucially on how they understand the situation. For a person *as* a person, thinking is already acting.

---

[11] See Sellars (1997), pp. 106/7: "For once our fictitious ancestor, Jones, has developed the theory that overt verbal behavior is the expression of thoughts, and taught his compatriots to make use of the theory in interpreting each other's behavior, it is but a short step to the use of this language in self-description. (…) *What began as a language with a purely theoretical use has gained a reporting role.*"

2.4 The Anxiety of the Goalkeeper

At the end of an early novel, Peter Handke has developed a literary version of the workings of reflection before and during the execution of a penalty kick (see Handke 1970, p. 124; cf. Sturma 2019a, pp. 127–132). The goalkeeper may have knowledge of which corner the kicker usually prefers. The kicker might assume that the goalkeeper knows this and now expects that he will choose the other side. The kicker could now put himself in the goalkeeper's shoes and decides to go for his preferred corner. The reflection has no ending. Then, there is the goalkeeper's anxiety. He knows that before the ball is kicked, his body, in an involuntary act, can give away his thoughts about which side he is going to turn to. In Handke's scenario, the goalkeeper finds an ingenious solution. He deliberately avoids moving at all and thus does not indicate what he will do next. Now, the kicker starts his run-up. The goalkeeper—Handke emphasizes that he is wearing a bright yellow jersey—remains completely motionless. The kicker shoots the ball directly into the goalkeeper's hands.[12] Handke does not reveal the reason for the missed penalty. Of course, there is also the anxiety of the kicker when taking a penalty. His run-up and foot position reveal a lot about where he can or will place the ball.

The internal connection between knowledge and reflection is well illustrated by a typical baseball situation: We are in the ninth inning. The home team is leading by one run. All the bases are loaded. Two of the opposing batters are already out. The pitch count is full. The next pitch could result in the third out, ending the game. The next batter, a power hitter, knows that the pitcher will want to prevent a walk. A fastball through the middle of the strike zone is more likely to minimize the chance of a walk but increases the risk of a hit. The pitcher knows that the batter knows what he is thinking. Both have the same knowledge of the situation. Neither knows anything about the other that the other does not know.

What will happen next? Will the pitcher go against his normal routine and choose an off-speed pitch? Does he consider a walk, which would immediately score a run because of the loaded bases, to be the lesser of two evils? A walk would only result in a tie game, and the home team would still have the opportunity to win the game in the bottom of the inning. Does the batter's behavior or a signal from the catcher cause him to suddenly change his mind? At what point in the reflexive process does the pitching motion begin?

The duel between pitcher and batter shows that even under the most favorable epistemic conditions, predictions in the first, second, and third person perspectives are almost impossible to make. Game routines can be easily identified by players and observers. But what the pitcher does in a given situation can neither he nor anyone else predict because he can change his mind at any moment through reflections and

---

[12] In the 1976 European Championship final, as the final penalty kick was being taken, the German goalkeeper, wearing a light blue jersey, seemed unaware of Handke's novel. His body position indicated that he was going to move to his left corner, and he did indeed jump there early on, while the kicker effortlessly flicked the ball into the center of the goal for the championship. The German goalkeeper was obviously not aware that his body was expressing his thoughts.

conjectures about the knowledge of other players. He can always ask, "What if …?", and his answer has a direct impact on his decisions and actions.

When the pitcher learns what the batter has discovered about his usual behavior, it is immediately incorporated into his practical considerations. It is now possible for him to adopt a new attitude toward his maneuvering space. Knowledge becomes decision. He can leave the situation as it is or make a correction. Whatever he ultimately decides to do can no longer be deduced from the initial conditions. The knowledge shared by the pitcher and the batter creates loops of reflection that make it impossible for both to predict with certainty what will happen next. They will never be entirely sure when reflection ends and action begins, and for what reasons in particular.

The anxieties of the goalkeeper, the kicker, and the pitcher show that thinking, deciding, and acting are intrinsically linked *over time*. Human behavior is not about isolated preferences or a series of loosely connected episodes, but always about the whole person and the continuous chain of decisions and actions.

Reflection has practical implications for dealing with external determination. This also applies in cases of severe physical weakness. Even then, it is usually possible to decide what attitude to take toward the given circumstances. A person can think deeply about how to deal with a specific situation and limit its negative effects. Jean-Paul Sartre illustrated this in the form of a phenomenological reconstruction (see Sartre 1948, pp. 530–534). Contrary to initial assumptions, it turns out that physical exhaustion during a mountain hike with friends is not a mechanically triggered compulsive force. What seems unbearable actually depends on attitudes and decisions. They ultimately determine how physical challenges are handled and where the line is drawn at what is no longer endurable. *Dealing* with physical distress is not regulated from the outside, rather, it is regulated by the attitude of the person.[13]

Baruch de Spinoza expressed the naturalistic requirement for the study of human behavior in a simple formula: persons, like everything that does not exist in itself, are firmly embedded in a system of causal dependencies, and their attitudes and decisions are integral parts of the course of events influenced by persons.[14] In this respect, Spinoza operates within the mainstream of naturalism. By allowing no ontological difference between mind and body, Spinoza avoids all the separation syndromes of the subject-object models that characterize conventional theories of mental causation and dualistic solutions to the mind–body problem.

Persons are constantly interacting with their natural and social environments. The actual course of these interactions depends on what they perceive and understand. Their interactions have an internal and external aspect. They make choices and decisions that lead directly to their practical attitudes and actions, which are observable, at least to some extent, in social space. An external observer is not concerned with the mysteries of mental causation or the mind–body problem. He tries to find traces

---

[13] See Sartre (1948), p. 532: "Si donc je veux comprendre à quelles conditions je puis souffrir une fatigue comme intolérable, il ne convient pas de s'adresser à de prétendues données de fait, qui se révèlent n'être qu'un choix, il faut tenter d'examiner ce choix lui-même et voir s'il ne s'explique pas dans la perspective d'un choix plus large où il s'intégrerait comme une structure secondaire."

[14] Cf. Spinoza (2018), p. 67: "*The mind does not know itself, except insofar as it perceives ideas of affections of the body.*" [E2p23].

of his opponent's mindset in what he sees. Internal episodes and related attitudes are so closely interwoven that they cannot be separated into cause and effect.[15] Decision and practical attitude express each other without causal direction.

In the Spinozist version of critical naturalism, the behavior of persons proves to be a hard case, but it is certainly not an ontological exception. The difficulty arises from the fact that human beings have self-consciousness and reflection at their disposal, which are always associated with epistemic uncertainty and vagueness. This does not mean that persons leave the system of causal dependencies. Their behavior can be examined like everything else in this system.

Spinoza explicitly emphasizes that the ideas of ideas are also physical processes.[16] To make plausible assumptions about a person's decisions and actions, we can rely on the rules of linguistic behavior, on the semantic fields of the thick description, and on what is publicly accessible—which includes what the various scientific disciplines bring to light about aspects of human conduct. We may never be able to decipher human behavior in its entirety, but at the very least we will be able to understand it better.

2.5 Deep Decisions

Under normal circumstances, persons are receptive to reasons over long periods of their lives. They constantly reflect, modify, and correct their behavior, whether as a result of existential awareness or in response to the attitudes and actions of others, and, most importantly, they can give reasons for their actions. Because of the role of reasons in their lives, their reflections and self-referential judgments are not momentary episodes. Rather, they are directly related to what they care about and consider important—regardless of whether it is fully transparent to them.

The term ›life plan‹ has been coined to describe the continuity of a person's attitudes, interests, and actions (cf. Rawls 1971, pp. 407–416; Taylor 1989, pp. 97–105). Life plans are not plans in the literal sense, but normative perspectives based on the interplay between coherence, reason, change, and adaptation that can be attributed

[15] See Hampshire (1969), 225/6: "My self-conscious materialist [Spinoza] argues that the relation between a man's thought and sentiment, and the physical state that embodies it, is in these ways *too close to be counted as a causal relation* [my emphasis], or as analogous to the relation between two physical states, one of which adequately explains the other. It is precisely the point of materialism to assert a much closer relation between processes of thought and physical processes than is implied in most of the idioms of ordinary speech. My philosopher is a genuine materialist in the narrow sense that he asserts that every change in the state of the organism, which is a change in thought, is also a change in some bodily state, and usually in the principal instrument of thought, the brain. From the standpoint of an observer, a person's bodily and mental processes are each reflections of the other, with neither of them privileged as substance to shadow." Hampshire's use of the term 'materialism' in this passage is consistent with what is understood here by critical naturalism.

[16] Cf. Spinoza (2018), p. 66: "*The idea of the mind is united with the mind in the same way as the mind itself is united with the body.* [...] *The human mind perceives not only the affections of the body but also the ideas of these affections.*" [E2p21, 22].

to the life of a person. They limit, expand, or reinterpret the scope for action. The factual status of a life plan is reflected in what the person actually does and intends to do. Even unfulfilled intentions remain within the scope of action as past options and provide an opportunity for reassurance or revision. The life plan determines how a person perceives and wants to shape the social environment. The range and depth of the life plan vary greatly from person to person and from situation to situation.

In cases where decisions are directly or indirectly related to life plans, we are dealing with deep decisions. Deep decisions explain why a person does not always follow situational motives. Preferences may be in line with a life plan, but in the end, they are always determined by situational motives. With deep decisions, we enter the space of higher-order thinking. Just as we can formulate desires about our first-order preferences (cf. Frankfurt 1988, pp. 13–22), we can make decisions about our first-order choices. Persons constantly revise their current decisions and actions based on longer-term ambitions, and they simply want to get rid of at least some of their inclinations.

When we talk about deep decisions, we implicitly or explicitly refer to self-awareness, space of reasons, reflection, life plan, action, and thick descriptions.[17] In deep decisions, the inferential relationships between expressions and reasons play a significant role, regardless of how appropriate a person's action is. What a person does is therefore neither a mere physical reaction nor a mere pursuit of immediate preferences. Because scientistic and eliminativist explanations of action begin outside the space of reasons, they miss deep decisions and their causal role from the start.

It is because of deep decisions that persons do not perform actions as such only in states of explicit consciousness. Much of what can be attributed to them occurs as a result of a chain of actions below explicit states of consciousness. What they do can be related to what they have done before. For this reason, much of a person's behavior is sometimes easily predictable by an external observer with knowledge of the relevant attitudes and habits and can be explained even in the case of deviant behavior.

The unity of decision and action is obscured in models of constructive separation. In these models, which are mainly used in eliminativist scenarios, the cause or alleged decision occurs before the action. But when the penalty kicker begins his run-up, determined to shoot into the left corner, he is only thinking, not yet deciding. There is simply too much time for him to change his mind. He cannot even predict with certainty what he will actually do.

While persons live, they have a position in the realm of causes. The constraints associated with the realm of causes are hard facts for acting persons in the sense of spatio-temporal limitations. However, they do not determine how actions will actually turn out. Operating in the space of causes is essential for translating decisions into actions. Persons follow the rules of the space of reasons on the basis of the ontological conditions of the realm of causes. The laws of the realm of causes limit the spatio-temporal scope of action in a way that is easily accessible to the external

[17] Although the decisions of the goalkeeper and the pitcher are based on self-awareness and reflection, they are not yet deep decisions. Their situations are all about circumstantial awareness.

observer. Under favorable epistemic conditions, the observer can also make plausible conjectures about how a person is likely to react to demands and expectations in social space—as long as a new reflection does not emerge. Decisions of persons are not mysterious episodes in the depths of their souls but are accessible to external observers as behavioral manifestations in social space. This does not mean that they know with certainty what will happen next.

Decisions are not isolated mental events. They shape the course of events in which persons are involved over time. In epistemic terms, persistent behavior can be differentiated into episodes, and decisions are, as it were, landmarks of changes of direction. Persons decide to influence the course of events in one way or another. Often, their actions are directly or indirectly related to deep decisions. Nevertheless, they can always reflect and decide to react quite differently to the course of events, in a way that neither they nor others could have foreseen. This is the lesson that a self-aware person will learn from the goalkeeper's anxiety.

References

Bennett MR, Hacker PMS (2003) Philosophical Foundations of Neuroscience. Blackwell, Oxford

Crick F (1994) The Astonishing Hypothesis: The Scientific Search for the Soul. Simon and Schuster, New York

Dennett D (1991) Consciousness Explained. Penguin, Harmondsworth

Frankfurt H (1988) Freedom of the will and the concept of a person. In: Frankfurt H (ed) The importance of what we care about. Philosophical Essays. Cambridge University Press, Cambridge, pp 11–25

Hampshire S (1969) A Kind of Materialism. In: Hampshire S (ed) Freedom of Mind and Other Essays. Princeton University Press, Princeton, pp 210–231

Handke P (1970) Die Angst des Tormanns beim Elfmeter. Suhrkamp, Frankfurt [Translation: The Goalies Anxiety at the Penalty Kick. Farrar, Straus and Giroux, New York 1972]

La Mettrie JO (1748) L'Homme machine. Elie Luzac, Leiden

Libet B, Freeman A, Sutherland K (eds) (1999) The Volitional Brain. Towards a Neuroscience of Free Will. Imprint Academic, Thorverton

Libet B (2004) Mind Time. The Temporal Factor in Consciousness. Harvard University Press, Cambridge

Rawls J (1971) A Theory of Justice. Oxford University Press, Oxford

Ryle G (1971) The Thinking of Thoughts. What is 'Le Penseur' doing? In: Ryle G (ed) Collected papers, vol II, 1929–1968. Hutchinson, London, pp 480–496

Sartre J-P (1948) L'être et le néant. Essai d'ontologie phénoménologique. Gallimard, Paris

Sellars W (1997) Empiricism and The Philosophy of Mind. Harvard University Press, Cambridge

Sellars W (1963) Philosophy and the Scientific Image of Man. In: Science, Perception and Reality. Ridgeview, Atascadero, pp 1–40

Sellars W (1980) Language, Rules and Behavior. In: Sicha J (ed) Pure Pragmatics and Possible Worlds. The Early Essays of Wilfrid Sellars. Ridgeview, Atascadero, pp 129–155

Spinoza B (2018) Ethics Proved in Geometrical Order. Cambridge University Press, Cambridge

Strawson PF (1959) Individuals. An Essay in Descriptive Metaphysics. Methuen, London

Sturma D (2007) Person as Subject. J Consc Stud 14. Special Issue on Dimensions of Personhood, pp 77–100

Sturma D (2019a) Subjekt sein: Über Selbstbewusstsein, Selbstbestimmung und Enhancement. In: Viertbauer K, Kögerler R (eds) Neuroenhancement. Die philosophische Debatte. Suhrkamp, Frankfurt, pp 115–147

Sturma D (2019b) Persons: A Thick description of the Human Life Form. In: Noller J (ed) Was sind und wie existieren Personen? Probleme und Perspektiven der gegenwärtigen Forschung. Mentis, Leiden, pp 147–165

Sturma D (2024a) Mind and Time: Interdisciplinary Division of Labor. In: Sturma D (ed) Mind and Time. Philosophical and Psychological Perspectives. Nomos, Baden-Baden, pp 7–14

Sturma D (2024b) The Language of Time and Self-consciousness. In: Sturma D (ed) Mind and time. Philosophical and Psychological Perspectives. Nomos, Baden-Baden, pp 131–150

Taylor C (1989) The concept of a person. Human agency and language. In: Taylor C (ed) Philosophical Papers, vol 1. Cambridge University Press, Cambridge, pp 97–114

Wegner DM (2002) The Illusion of Conscious Will. MIT Press, Cambridge

Chapter 3
Optimal Decisions and Why Not to Ask AI for Them

Mandy Zafar

Abstract Although we make decisions all the time in our own personal lives and as scientists, there is still a surprising amount of ambiguity about what constitutes an optimal decision. We recently started to improve or fully outsource some of our decision-making to AI technologies that are built to provide recommendations or decision-making support. In a society that relies more and more on such support systems, it is also remarkably unclear how optimal decision-making could be further boosted. Although AI outputs often successfully assist human decision-making processes, what has yet to find sufficient attention in the debate is an encompassing understanding of the *conditions* for *optimal* decision-making with and without AI. This chapter (i) critically investigates the understanding of and overall pursuit to make optimal decisions, and (ii) describes the discrepancy between human practice of decision-making and the so-called AI decision-making processes. The concept of decision-making is connected to concepts of agency and thus part of a normative framework. One result of this investigation is that while AI can provide some calculativc assistancc to find a dcfincd optimum, it cannot makc optimal decisions.

Keywords Decision making · Optimum · Ideal · Reasons · Artificial intelligence · Moral agents

3.1 What Are Optimal Decisions?

Since ancient times, our decision-making has been discussed as aiming towards the good and virtuous. Still, it is debatable what good decision-making exactly amounts to. Given that the issue of goodness of choice has been covered extensively in the philosophical literature, I will not examine it here. Instead, this chapter investigates *optimal* decisions. Recent investigations about optimal decision-making often include discussions of AI-technological decision-support systems. Although widely

M. Zafar (✉)
Institute of Neurosciences and Medicine—Brain and Behaviour (INM 7), Research Group: Neuroethics and Ethics of AI, Research Centre Jülich, Jülich, Germany
e-mail: m.zafar@fz-juelich.de

U. Ettinger et al. (eds.), *Decision Making*, Studies in Neuroscience, Psychology and Behavioral Economics, https://doi.org/10.1007/978-3-032-00880-0_3

used in multiple sciences and everyday life applications, it is remarkably unclear how AI can execute or boost decision-making. To get a deeper understanding of optimal decision-making, I analyse the meanings of decision, decision-making, and optimal. Since decision-making is a critical aspect of human agency, it seems reasonable to clarify the enabling and boundary conditions from a proper agent's point of view, i.e., the human decision-maker.

3.1.1 Decisions and Decision-Making

As a basic starting point, it only makes sense to talk about a decision if a person (or group of persons) x is in a situation S in which x has at least two (but ultimately a finite set of practicable)[1] options for action $A_1, A_2, \dots A_n$ (an omission of an action is also an action) and has to decide which one to choose. In this basic setting, a decision is thus a choice of one option of action from the given alternatives A_1, A_2, ..., A_n. To decide on an option is itself an action that manifests as the actualization of the chosen option.

Further, the decision-making process describes (i) how a decision was attained (methods and processes) and (ii) why person x decided on a certain option A (reasons, motifs, overall goals). There are multiple methods with different foci[2] to figure out what an optimal decision in a given context is. The reasons for a decision are usually evaluated and relate to x's motivation, goals, and emotional congruency with the desired outcomes, as well as the existing and socially accepted system of norms and values. An evaluation enables persons to understand options for action as good or bad, desirable or undesirable, efficient or not efficient, reasonable or unreasonable, commanded or forbidden, etc.

A decision-making process also depends on whether choices are made via contemplation or habit: *habitual* choices are often referred to as *routine response behaviour* or *habitual decision-making* since they are usually generated by "cognitive shortcuts" (van der Pligt 2001), habitual motifs or reflexive repetitions that are based on repeated exposure to a similar environment, former successful choices and conscious or unconscious value adoptions (e.g., Huang et al. 2020; Bernacer et al. 2014). Making an optimal decision is connected to deliberation and thus a case of *goal-oriented decision-making*, which we focus on here. However, the distinction between goal-oriented and habitual decision making is not strict (Drummond and Niv 2020). On the one hand, habitual decision-making often depends on the outcomes of former

[1] The number of practicable options is finite, even though theoretically there may be infinite options. In practice, however, we would never start to decide if we kept counting possible options ad infinitum.

[2] There are multiple approaches that explain the decision-making process. Descriptive approaches are, e.g., prospect theory, image theory, rule-following, and reason-based choice. A normative theory, is, e.g., subjective expected utility theory (van der Pligt 2001). I won't delve into the specifics of these theories because the relevance and emphasis of the requirements and criteria vary widely depending on the scientific field and theory in which they are applied and go far beyond our purposes here.

deliberation. On the other hand, goal-oriented decision-making is often biased by previous experiences and successful habitual choices.

3.1.2 Optimal Versus Ideal Decisions

We usually understand an optimal decision as leading to (measurably) better consequences than all other decisions under the same circumstances. In most decision theories,[3] the evaluation of "better" depends on other measurement values, such as *efficiency*, *rationality*, or *comprehensibility*, that describe what the optimum actually amounts to. An optimal choice as a goal in a decision-making process is a calculatable trade-off between defined options, which are bound to a situational context and affordances at a given time. Efficiency, rationality, and comprehensibility are also demanded for the methods and goals of as well as the justifications for decision-making. Context-related questions can outline other constraints: For whom is the decision optimal? Who decides based on what values and norms or ethical theory (i.e., utilitarian, deontological, virtue ethics, mixed approach)? What is the rationale for an optimal decision? What exactly are the pursued goals?

There is a flip side to decision-making that is typically referred to as "externality,"[4] namely tolerating undesirable, inconvenient consequences, which seems to be less discussed in the context of *optimal* decisions. Yet also an optimal decision could be considered as overall good for a society or country while having potentially harmful effects for other societies or individuals, now or in the future. Take, as an example, current Western consumerism of goods and services. On the one hand, the connected practices in line with consumerism seem to be optimal (i.e., "effective," "profitable," "sustainable") now for some people, and the preservation of established practices and systems. On the other hand, people acting accordingly exploit other people and regions, which creates unfairness, asymmetries, and discrimination now and in the future (e.g., Crocker 2017).

One reason that this flip side is usually less discussed in the context of *optimal* decisions seems to be based on the common misunderstanding that an optimal decision is also the best possible decision (Eisenführ et al. 2010; Consul et al. 2022). This practice might be a result of normative decision theories that idealise rational decision-making and were heavily influenced by the eighteenth-century enlightenment paradigm of rationalism and the scientific analysis of human behaviour (Fishburn 2011).[5] Historically, we have known the concept of an ideal decision-maker

[3] There are various theories of decision-making studies developed in different disciplines, such as philosophy, psychology, economics, mathematics, etc. The focus of this paper does not allow for an overview of the different theories. For a detailed historical account of the experimental literature on decision-making see Heukelom (2014).

[4] E.g., Boundreaux and Meiners (2019); de Oliveira (2021).

[5] There are normative, descriptive, and prescriptive theories of decision-making (Bell et al. 2011). The main literature around the different decision-making and justification strategies can be depicted with regard to common descriptive or normative decision theories. In contradistinction to normative

at least since Aristotle's moral expert, or "man of practical wisdom" (Drefcinski 1996), the *phronimos* (Aristotle (NE) II 6, 1106b36-1107a2). Nevertheless, Aristotle's holistic view highly contrasts with modern discussions of decision-making. Although vast and highly varied (Elliott 2019), they focus on only some aspects of decision-making, as e.g., rational choice.[6] Such a narrow focus supports the idea that an optimum is a rational, actualizable, achievable ideal. This might have led to using the terms "ideal decision" and "optimal decision" interchangeably. It is, however, important to distinguish between optimal and ideal choices because of their different connotations and implications.

The most important difference between an optimum and an ideal in relation to decision-making concerns epistemological constraints. It can only be *estimated* whether an option for action is optimal in a particular situation *s* and time *t* in social context *c*. It is not yet *knowable*, i.e., principally not cognizable or truthful. An option can only be evaluated *as* optimal retrospectively, i.e., *after* actualizing (acting out) the option. An ideal, however, is timeless. Even in the estimating phase, we can understand with respect to the connotations that an ideal option might not be actualizable in a given situation. An optimal option, however, is always actualizable since it is by definition the best option *considering* the given circumstances and affordances at the given time. Thus, if an option is ideal and actualizable, then it is also optimal. If an option is ideal but not actualizable, then another option can still be optimal. This does not exclude the possibility that there can be more than one optimal decision in a situation *s* at time *t* in social context *c*, i.e., options that are equally good. The ideal does not allow for further optimisation and is potentially unattainable. In contrast, an optimal decision, if not the actualizable ideal, leaves room for improvement and is by definition attainable.

3.1.3 Context of Justification: Reasons for Choices

Given our chosen reference frame of human agency,[7] making a decision is ultimately connected to having responsibility and thus accountability over the actualization of

theories, prescriptive theories observe and predict behaviours of decision-makers following steady decision rules. Descriptive approaches are concerned to describe the actual process of decision-making (Chandler 2017).

[6] See, for instance, the expected utility theory: cf. Luce and Raiffa (1957) applied to social sciences; Fishburn (2011); for an encompassing overview Briggs (2023).

[7] Decision-making is an agency-related capacity. Despite the fact that also questions about agency and decision-making in non-human animals are highly significant and fascinating, they cannot be further discussed here. I acknowledge forms of agency-related behaviour in living systems (life forms) close to a kincentric point of view. The boundaries as well as the similarities and differences of decision-making abilities are a matter of recent discussions that are partly based on enabling conditions and capabilities, e.g., sense of being and experience, lived joint attention, reciprocity, choice, predictive decision making, or goal directed instead of purely stimulus-driven behavioral control. See, e.g., Bhattacharyya and Slocombe (2017), Blattner (2021), Suddendorf et al. (2018), Stake (2023, pp. 124–148, 272–287), Pinsdorf (2016), Špinka (2019), Thomas (2016).

the decided action taken. A decision-maker is principally able to justify the resulting actions with reasons. One way of distinguishing reasons is logical: *deductive, inductive, and epistemic* reasons. A deducted reason is the conclusion of a derivation or consequence relation of a proposition from a set of propositions. *Inductive reasons* are statements concluding from what is known about what is not known yet, are thus not truth-preserving and support the conclusion at most plausibly (Kleinknecht 2006, p. 19). *Epistemic reasons* are formed by the deduction of belief contents from other belief contents. An in-depth understanding of such a reason would entail the ability to give an answer to why a statement A is a reason for person x to believe another statement B.[8] For any kind of reason, the following applies: (i) not every correct derivation automatically ensures its rationality and soundness, and (ii) the whole argument for a reason has to be logically valid, substantive, and strong (Dorn 2006).

Although all three types of reasons are present in decision-making processes, people are mostly concerned with uncertainty and estimate what to expect based on inferences, probability calculations, or expectations about consequences that are not always foreseeable. In many decision theories, the existence of decision-making abilities is thus bound to the requirement that a person's degrees of belief (or doxastic states)[9] are consistent with the axioms of probability. Usually this understanding includes evaluation and ranking criteria, which categorise and order impressions, expected states of mind, and feelings, such as "option A_n is (likely to be): pleasant/unpleasant, desirable/undesirable, good/bad, etc." The evaluation criteria are based on the already mentioned situatedness of the given timeframe and circumstances. Further, they are described by the emotional desires, values, social and ethical principles, as well as the moral rules, laws, customs, and cultural and religious belief systems that a decision-maker has or is exposed to.

Aristotle was one of the first thinkers in Western philosophy who summarised characteristics of decision-making in his *Nicomachean Ethics* and *The Eudemian Ethics*. His concept of practical rationality is not limited to intellectual or theoretical contemplation (*nous*) and accuracy but also balances emotional desires, values, virtues, and nuanced ethical principles, as well as the capacities to evaluate different perspectives (*sunesis*, 'judgement') and to discern exceptions to universal moral rules (*gnomē*, 'discernment').[10] His theory of competent deliberation (*euboulia*) and choosing the best means to achieve a goal is based on rules as well as practical

[8] According to Kleinknecht (2006, p. 19ff), the conditions for deductive reasons can be stated comparatively satisfactorily. The premises of inductive reasons are mere statements of conditional probability that come with numerous technical difficulties. In order to achieve a clear and illuminating investigation of epistemic reasons, we would need access to different degrees of (and other complex reasons for) belief, which comes with a lot of technical and access-related difficulties (Kleinknecht 2006, p. 19).

[9] See, for instance, theories of epistemic rationality explicated in Eder (2021).

[10] For a brief overview of Aristotle's moral theory see e.g. Khan (2005). For further discussions about the phronimos, see Drefcinski (1996), for *sunesis* and *gnomē* with regard to decisions in the context of business ethics, see, e.g., Steyl (2020).

experience and propagates the constant goal of an accumulation of practical wisdom (*phronēsis*), which further develops a person's ability to choose wiser over time.

The balanced act of practical rationality, assessing personal intentions or goals as well as arguments for or against a certain action, is still of central importance in the understanding of decision-making processes. Intentions and goals can be based on individual beliefs,[11] judgements, desires, emotions, intuitions, and intercultural learning. This enables a person to stand for or decide *against* a socially accepted norm or rationale. Interestingly, in the case of *optimal* decision-making, the possibility to choose against an option for action may decrease because of a higher alignment of the norm, to consider as many options as possible with regard to the higher social human good. Individualistic and collectivist-aligned cultures may vary in the degree of how much the overall social value impacts a decision-making process (Yates and de Oliveira 2016). However, *optimal* decisions seem to highly depend on or are more aligned with the current socially accepted normative standard and rule system and are also more likely to relate to intersubjectively accessible rational arguments than to individually isolated intuitive ones. This does not mean that a person considering an optimal option cannot decide against a social norm. But it suggests that for a decision to be categorised as an *overall* optimal decision, the acceptance and consideration of these social norms and the recognition and management of situational group interaction with respect to the higher social human good seem to be more influential and salient (Larrick 2016).[12]

3.1.4 Limiting Boundary Conditions Lifted by AI?

People are generally not particularly good at finding an optimal choice, especially if the situation is new and the decision has to be made under time pressure. One reason for this could be that they are usually not engaged in a *constant* mathematical check of whether their decisions are going to be or were optimal, and instead often reach for options that are only good enough.[13] Another reason might be that many, if not most, of our decisions are biased by habitual, emotional, or intuitive factors depending on connected psychological constraints like fears or framing, i.e., how choices are worded (Johnson 2021, ch. 12). Other common obstacles that affect human decision-making are different information search strategies, an impaired ability to separate and prioritise information, low levels of patience (e.g., Reeck et al. 2017), or stress and anxiety (Hengen and Alpers 2021). Under stress in particular, people are more likely to fail to implement emotion regulation strategies, which can lead to reduced

[11] In how far decisions are more individually or group-based varies and is explored in cross-cultural reviews and analyses; see e.g., Yates and de Oliveira (2016).

[12] A conflict over an established normative or rule system may also lead to higher intra- and intercultural negotiations (Yates and de Oliveira 2016).

[13] This is further discussed under bounded rationality bias, see Wheeler (2018).

self-control, and stress-related coping mechanisms like higher risk-taking, higher influenceability, or manipulation strategies (Porcelli and Delgado 2017).

As mentioned above, decisions with the highest possible, achievable degree of rationality and efficiency are commonly seen to be optimal, not only because that implies a lower amount of "psychological quirks that often plague human moral agency" (Podschawek 2017, p. 337, cf. 338). Taking into account people's limited time and bounded cognitive resources in real-world application scenarios, AI decision-making or support systems (AI-DMS) are built to generate outputs faster and with a high degree of rationality and efficiency. The fact that they seem to simplify certain steps in a decision-making process and may avoid many problems related to some of the usual obstacles of human decision-making led to their growing influence in everyday application areas. For instance, AI-DMSs are already used for impactful decisions concerning credit, loan and job applications, clinical decision-making, predictive policing measures, and legal judgements as, e.g., COMPAS (cf. Asaro 2019; Spaulding 2020; Günther and Kasirzadeh 2022). Another and rather new application area concerns optimal decision-making in the form of *moral advisory* or *moral assistance systems* such as MoralDM (Deghani et al. 2011), combining deontological and utilitarian aspects, or the software LIDA (Wallach et al. 2010).[14] So-called "recommendations" for "decisions" stemming from very complex *machine learning* (ML) algorithms in AI-DMSs are the result of implemented mathematical optimisation rules (e.g., Boyd and Vandenberghe 2009, ch. 1). Also in medicine, AI-DMSs are used to quickly and efficiently choose personalised treatments or therapy recommendations (Molnar and Molnar 2023) that are aligned with the previously defined efficiency criteria and framework conditions.[15]

Although it is a widely agreed-upon fact that AI-DMS cannot completely take over decision-making processes and people remain the ultimate decision-makers (Becker et al. 2022, p. 468; Genin and Grote 2021), it has been argued that some human natural decision-making can be boosted by heuristic optimization algorithms. The requirements for their application demand defined descriptors that give logically valid, comprehensible outputs with high predictive accuracy, as well as defined efficiency parameters and consideration of some natural conditions (Consul et al. 2022, p. 186).

[14] See also Lara and Decker (2020) proposing an AI Socratic Assistant for human moral enhancement.

[15] More specifically, in neurological practice, AI may help to make faster diagnoses about strokes from CT/MRI scans, interpretations of electroencephalograms (EEG) to prognosticate coma, or predictions of a conversion of mild cognitive impairment to Alzheimer's dementia (Vinny et al. 2021; cf. also Chandra et al. 2022). In psychology, AI can help experts make decisions concerning diagnoses of mental health disorders based on analyses of monitored speech patterns, facial expressions, and physiological data (Luxton 2014).

3.2 AI-Made Optimisation

It seems that boosting decision making via AI-DMS is a step towards more optimal decision-making. However, the application conditions for AI-DMSs have to meet high standards, particularly in critical areas such as medicine and jurisprudence.[16] It thus seems worthwhile to take a closer look at the ongoing processes and results of AI-DMSs and their advantages and limitations.

In general, the applied optimisation algorithms of AI-DMSs can (1) give outputs in the form of action suggestions or (2) provide simplification strategies for making a decision. In the first case, persons can delegate or outsource aspects of their decision-making, such as reducing options, gathering, evaluating, and concluding information, or calculating an optimal minimum regret outcome and assessing probabilities (Consul et al. 2022, p. 186). The fact that certain data is considered or excluded by the algorithms could already be seen as a decision by the AI-DMS or its developers. Nevertheless, it is the outputs that are commonly described as recommendations, suggestions, or decisions. In the second case of offering simplifying strategies, the focus lies on the improvement of the decision-making process itself, which is commonly called "boosting." In the context of decision-making, behavioural sciences explore boosting—in contradistinction to nudging[17]—as a support for nonfiscal and noncoercive interventions that may foster people's competence to exercise their agency by making their own choices (Hertwig and Grüne-Yanoff 2017). AI-powered boosting methods usually derive strategies that take into account the constraints of limited time as well as the cognitive resources for optimal decision-making (Becker et al. 2022, p. 468). Such AI-DMSs seem to be more beneficial compared to those that give action suggestion outputs (Becker et al. 2022, p. 467). For instance, decision-making or supporting programmes that include fast-and-frugal heuristic (FFH) functions were shown to be effective because they incorporate intuition in the form of general-purpose heuristics that facilitate the system's adaptation to varied applied settings, such as aviation, finance, and medicine (Hafenbrädl et al. 2016). Besides, FFHs are comparatively cost effective since they direct the focus of the decision-maker to the most essential pieces of information, for instance, via generating a decision tree (Becker et al. 2022). Overall, FFHs appear to be more accessible, fast, and transparent.[18]

---

[16] Popular systems are, for instance, precise expert systems, sequential decision making, explainable AI, or learning-based programmes such as inductive logic programming, relevance-based learning ML (neural networks, statistical learning, decision trees), or hybrid systems (hybrid neural systems, learning with knowledge) (DKE 2022).

[17] For a thorough discussion of the distinction between nudging and boosting, see Hertwig and Grüne-Yanoff (2017).

[18] There are other heuristics, such as heuristics and biases (HB), which treat intuitive decisions as a source of bias and error, and natural decision heuristics, which define intuition as based on large patterns gained through experience creating tacit knowledge (Klein 2015, p. 164).

3.2.1 Boosting Human Decision-Making

In an AI-DMS, an optimal decision is the well-defined goal that can be calculated from information about all accessible (not: all possible) options and the connected, predictable risks of consequences under a minimum degree of error within a loss function (Consul et al. 2022). The calculations are based on comparison models and statistical evaluations that use pre-programmed goals, rules, probability-weighted averages, and a hierarchical value system (Consul et al. 2022). For comparison purposes, a utility value is assigned to each outcome (Chandler 2017). The hierarchical value system is the heuristic function that ranks options. Concerning planning strategies used for decision-making, for instance, the normative principle of resource rationality[19] is applied to define strategies adaptable to specific problem fields and according to the expected utility and the given cognitive resources (Consul et al. 2022).

Prima facie, AI-DMS can be very good at weighing input utility values against each other. They can detect hidden patterns and structures in the observed data, make highly accurate predictions of unobserved data (Günther and Kasirzadeh 2022, p. 375), and produce interpretations as well as algorithmically translatable evidence-based assumptions. The results for an optimal outcome are derived quickly and may indeed boost human decision-making in complex environments under time pressure.[20] Consider, for example, the context of medical decision-making, where the majority of decisions concern the optimal treatment of a person suffering from a particular medical condition. Weighing and evaluating enough alternative options to reach an optimal decision for a patient can be time-consuming, which is especially crucial if the situation is classified as a medical emergency. In settings like these, humans' natural decision-making happens under urgent, uncertain, and continuously changing conditions, which can imply significant consequences. Coordinating decision-action cycles in continuously changing, time-sensitive situations is one of the many key abilities of clinicians. However, they often decide more intuitively and according to their expertise and experience than by searching for an optimal solution. Such natural heuristics often direct human decision-making processes to decisions that theoretically have less justification and deliberation compared to optimal heuristics (Patel and Kaufman 2001, p. 9515). This has been a reason for the development of clinical decision support (CDS) systems, which may assist clinicians to critically

[19] The concept of resource rationality is usually used in descriptive models to explain how people make decisions. Consul et al. (2022) focus on automatically discovered efficient planning strategies because they seem to be promising to improve people's decisions. They use reinforcement learning models together with a cognitively informed structure to generate efficient planning strategies that generate higher resource-rationality scores than people who use intuitive strategies (Consul et al. 2022).

[20] For a different boosting argument of algorithms in our lives, see Christian and Griffiths (2016, p. 4), who introduce the idea of a "human algorithm design" that supposedly helps to tackle everyday choices by relying on lived algorithms understood as finite sequences of steps used to solve a problem. The gist of the book is that any rule-following is basically a lived-out algorithm and that we can adjust these rules in order to make better, quicker, or more optimal decisions.

assess risks, diagnose, prognose, and select therapies via provision of patients' data (Visweswaran et al. 2022, p. 286). CDS in connection with AI (short: AI-CDS) are created to (i) tackle the exponential growth of medical knowledge and (ii) rapidly accumulate patient data to increase inference complexity and clinical data capture.

Nevertheless, compared to human decision-making, these systems often cannot meet what was promised and come with problems that can outweigh their strengths. Considering the example of clinical decision-making, (AI-)CDS systems come with enormous compliance burdens for clinicians to verify, regulate, and evaluate the complex and increasingly automated systems for safety, effectiveness, and performance, i.e., their scientific, analytical, and clinical validity (Visweswaran et al. 2022, pp. 286f, 294ff). Moreover, such systems usually do not explain their predictions in a humanly understandable way. Even second (post-hoc) explainable models, i.e., models that are applied to explain the result of the previously used algorithm, have been found to be unreliable, mostly misrepresentative, and may lead to overly complicated decision pathways (Rudin 2019).

Although there are many advantages, the above example displays severe challenges regarding the improvement of human decision-making via an AI-DMS because "no matter how well it functions in a laboratory, continues to struggle to function optimally in the real world" (Patel 2023, p. 2). Other problems of AI-DMSs applied to real-world problems have been reported in the literature, such as insufficient comprehensibility and transparency, arbitrariness, a lack of justification criteria for the outputs, the adoption of unquestioned bias (fairness), or unaligned goals and value systems that may lead to discrimination.[21] These raise concerns about the prima facie promising features of AI-DMSs.

3.2.2 Structural Limitations

The first section pointed out that a focus on only some aspects in a decision-making process, such as rationality, seems to be deficient in practice. This aspect seems to gain more importance considering that AI-DMS cannot provide anything other than "rational," calculable outputs. Even though rationality is seen as one of the main characteristics of goal-oriented *optimal* decision-making, it is overall questionable what a *purely* rational choice is. It is apparent that reasons need to be evaluated and weighed according to our personal boundaries, values, and needs. Otherwise, they would stand next to each other indifferently. In AI-DMSs, the evaluation and ranking of data are usually calculated according to the assigned utility degrees. However, the utility degree for and an evaluation of the goodness or badness of a reason is mostly not given as a purely descriptive or quantifiable proposition. In fact, most ethical reasons, arguments, and value statements are not purely descriptive but contain prescriptive or normative propositions. This implies that utility values for ethical or normative propositions are increasingly difficult, if not impossible, to assess. If we see

[21] See, e.g., Grote and Keeling (2022), Creel and Hellman (2022), Corbett-Davies et al. (2017).

optimal decision-making as a purely rational and calculatable choice, we thus have to face general questions concerning measurability in both natural and AI-DMSs-related decision-making, namely (i) if all reasons are quantifiable or mathematically rateable, and (ii) how utility values and outcome probabilities are assignable. These questions also concern the epistemic access to the given information and the expected outcomes. Even for persons, the epistemic access to expected outcomes is limited and sometimes radically unknown, which can make the attachment of probabilities to various outcomes arbitrary. This is problematic because, without them, we cannot do a conventional utility analysis (Schwartz et al. 2010, p. 216f). The issues are even more severe for AI-DMS because there is *no* epistemic access to or ethical deliberation of information at all.

Considering that some of the problems are present in both human and AI-DMS-related decision-making, one might wonder why they are especially emphasised for AI-DMS if they supposedly create more effective and rational outputs. In line with this doubt are some scholars who criticise a double standard of requirements for AI technologies that are not equally demanded for people (Zerilli et al. 2019; Walmsley 2021). While acknowledging the success rate of AI-DMS, other scholars argue for *higher* standard requirements for AI-DMS, especially concerning prevalent issues of transparency (Günther and Kasirzadeh 2022, p. 375). Their hope is that higher requirements reduce the prevalent opacity (Zednik 2021) and increase explainability in AI-DMSs. In reality, however, it is generally unclear how and to what standard explainability is achievable in AI-DMSs, especially since their inner logic is opaque due to complex decision functions (Creel 2020). Creating more explainability is a complex endeavour that depends on the application purposes (Creel 2020), does not always lead to more transparency, and can negatively affect accuracy (Rudin 2019). A recent solution to evade higher requirements and the explainability debate, is to only use intentional stance explanations in AI systems (Zerilli et al. 2019). These, however, do not properly apply to black box algorithms (Günther and Kasirzadeh 2022, pp. 378, 380).

Another problem with intentional stance explanations is that they involve mentalistic terms and thus assume "as-if"-abilities. Even though it seems practical to ascribe minimal beliefs, intentions, or reasons to some ML algorithms or AI-DMSs, "as-if"-concepts[22] come with problematic ontological commitments regarding (subliminal) attributions of agential features that impact their evaluation. Although AI-DMSs show aspects of reasoning in their outputs or processes—such as utility calculations, rule following, or logical derivation for an optimisation problem from limited, presented information—it does not mean that they have or create reason. As was discussed above, the logical validity of an argument does not imply its adequacy or suitability for practice. Thus, without context sensitivity, the aspects of reason stay abstract, impartial, and impersonal. While people can discuss reasons and are able to understand, explain, and evaluate them, AI-DMSs cannot. They neither have an understanding nor a "rationale" of their processes or outputs or reliable predictions of

[22] See, for instance, Johansson (2010), who advocates for a functional "as-if"-morality in AI technologies she calls "AI-agents".

decision outcomes (Günther and Kasirzadeh 2022, p. 380). Since AI-DMSs have no epistemic access to an intersubjective environmental structure, an "extra-semiotic" sphere, they are one with their algorithmic environment (Abbate 2023). They only apply formal heuristics in a particular case, which is usually *un*creative, rests on reliability and function, and is thus *inflexible* in changing environments or conditions (Nagl 2022, p. 78f). This structurally binds AI-DMSs to *external* evaluation.

AI-DMSs process data *about our* knowledge of the structure of the real world and feed from *our* feedback, especially in terms of negative consequences, for example, when a decision was based on bias. The required relatability to real-world problems with all connected insecurity factors, the complex situatedness and social structure cannot be integrated into algorithms yet (Becker et al. 2022, pp. 486–487). Concerning decision-making, these structural conditions are troubling because an encompassing view of many other realistically possible options can often not be provided. This, however, should be anticipated in order to find new opportunities, avoid unforeseen risks, and quickly adapt to changing conditions. In practice, AI-DMSs are thus predisposed to insufficient exploration of possible novel actions and rankings, which can result in a limited number of familiar action suggestions (echo chambers), confirmation bias, or standardising effects.[23] When people apply the results of AI-DMSs, the consequences can include blindly following instructions and becoming rigid in behaviour (Pentland 2021). In extreme cases, a person's agential competences can be reduced as a result of an undue reliance on AI-DMSs and a decrease in using their own practical reasoning when making decisions.[24] As is the case for many AI systems, however, following fairness principles and ethical recommendations as well as a socio-technological approach can help to adjust the given structural imbalances and difficulties (e.g., Patel 2023, p. 2).

The above points corroborate similar findings: AI-DMSs, aiming for an optimum (defined as maximisation of an expected utility), do not choose to comply with the programmed rules, compute the meaning of the rules, or cypher the rules as pertaining to themselves, and thus are not able to fully endorse the rules or violate them deliberately (Swanepoel 2021, p. 160). These are important prerequisites for decision-making that only persons have.[25] Moreover, the insights reconfirm that AI-DMS and human decision-making are not equally complex and opaque (Peters 2023). Beyond that, the above considerations hint at an important issue, namely that common prerequisites for decision-making are often overlooked when arguing in favour of AI-DMSs. However, the analysis of the structural differences between data-processing AI-DMS and human decision-making gives further reasons for the

[23] See, for a differing view on creativity with regard to large language models (LLM), Millière and Buckner (2024, p. 2) who claim that LLMs are capable of "flexibly blending patterns from their training data to produce genuinely novel outputs".

[24] See, however, Constantinescu et al. (2022, p. 20f), who argue that if a decision is only understood as systemized information, the user's ability to deliberate might be supported.

[25] See a recent discussion about conditions for agency and their connectedness with normativity in Rosati (2016), who distinguishes proper agents from "shmagents" like AI technologies that only supposedly possess similar prerequisites of agency.

assumption that not only are AI-DMS not able to make *optimal* decisions, they are not able to make any decision at all.

3.3 Decision Makers Are Moral Agents: Outlook on Optimal Decision-Making

The supposedly trivial fact that the structures of AI-DMSs differ significantly from a person's decision-making process through discernment (Matthias 2004) has important consequences and can be understood by reflecting on human deliberation. I argue subsequently that there is an interesting blind spot in claims about the perceived advantage of AI-DMSs' superior rationality: the human "obstacles" that AI-DMS-advocates aim to overcome and avoid for optimal decision-making are importantly connected to our normative evaluations, necessary for wise decision-making, and represent components of our practice of agency.

Moral agents live and experience their choices. In every decision-making process, we more or less use the resonance space of feelings to categorise reasons, choices, and values. By doing so, we can measure our choices relative to an ideal option of action. If we understand this ideal as reachable, then a decision is also the optimal decision. If we cannot achieve the ideal because of any constraints, then we usually ponder about an optimal action. For this, we use our practical wisdom. Decisions are reasonable *insofar* as they are rational *and* in line with what we value. We could have a lot of rational reasons for a decision but still not feel aligned with it. Trying to explicate what exactly it is we do not *feel* aligned with is time-consuming and grounded in our lived experience, yet worthwhile to discover since it makes us aware of other reasons to support a well-informed optimal decision. It is not clear how additional subjective attitudes, the extent of a feeling, confidence in a belief, or the satisfactory degree of achieving what is desired are to be interpreted and measured. It is clear, however, that the ability to navigate between the two capacities of rationality on the one hand and evaluation of needs and higher values, beliefs, or feelings on the other hand defines a person's competence in practical wisdom. The alignment and nuanced navigation of reasons and lived experience are the actualisation of our practical wisdom and moral habituation, which are not yet representable in pure systemized informational data[26] but are also commonly *avoided* in theories of optimal decision-making and discussion about AI-DMSs.

These thoughts seem to confirm a recent argument that, without sentience or any respective connection to normatively relevant subjective experiences attached to value, AI-technologies, in our case AI-DMSs, are only "functional moral zombies" that are incapable of moral reasoning (Véliz 2021, pp. 488–494). In contrast, the human mind-set is *principally* open for creative, new untried behaviours and leniency (Marcinowski 2022) and able to leverage new creative affordances in immediate surroundings (Roli et al. 2022). Human decision-making *depends* on this flexibility,

[26] Constantinescu et al. (2022, p. 20f) give similar reasoning for moral advisory systems.

which is expressed by the ability to react to nuances in given situations and rests on a level of trust in and authenticity towards other persons involved in or impacted by our decisions. In this sense, human decision-making is not generic but *principally* criticisable, (re-)negotiable, and sensitive to context.

Another point that is often dismissed in discussions around optimal decision-making concerns the outcomes of a decision-making process, especially regarding *non*-optimal choices. By making decisions, we learn important intersubjectively appreciated abilities, such as patience and cooperation. We get to experience and accept different value systems and points of view, learn to compromise, or how to take responsibility for our choices. We also learn from the consequences of our decisions, whether they have turned out to be good, useful, valuable, or not as good as expected, not optimal, bad, or indifferent. These kinds of experiences let us grow in practical wisdom, which is replicable in future decision-making processes.

In this sense, we can generally question the widely established trend of constant improvement and optimization of choices understood as an ideal. On the one hand, striving for optimal decisions might be an understandable expression of our psychological need for control over our environment (e.g., Leotti et al. 2010), especially if we can do this in a less cognitively demanding way (Kool et al. 2010) via AI-DMSs. On the other hand, it is debatable if boosting decision-making necessarily means optimising under the false premise of reaching ideal outcomes via utility-maximising conditions, when satisficing, i.e., reaching good enough decisions,can be a better and more rational strategy under circumstances where we face radical uncertainty (Schwartz et al. 2010). Leaving space for good enough decisions and mistakes might induce important educational or informative experiences in the long term. Further, it seems important to avoid an undue dependence or strong reliance on AI-DMSs since constant endeavouring to make optimal decisions, while believing they can be done easier and more efficiently via AI-DMSs, could negatively impact feelings of agential control and self-adequacy. It would be valuable to empirically investigate these concerns.

All in all, the asymmetry and insurmountable discrepancy between decision-making processes in AI systems and persons are structurally given and rely on a normative framework. Although we can acknowledge that AI-DMSs can calculate an optimum based on given information, it seems reasonable to state that AI-DMSs cannot make any decisions. As long as we do not accept their outputs *as* reasons, decisions, or recommendations and evaluate them *as* good, bad, optimal, etc., they remain mere unexplained derived statements of probability. Provided the arguments in this chapter are sound, it seems reasonable to state that AI-DMSs cannot make optimal decisions because they do not belong to the reference frame of decision-makers. An optimal decision is a holistic endeavour, necessarily normative, and embedded in an intersubjective *human* practice of deliberation that might get boosted in aspects but cannot be outsourced to AI-DMSs.

References

Abbate F (2023) Natural and Artificial Intelligence: A Comparative Analysis of Cognitive Aspects. Minds Mach 33:791–815. https://doi.org/10.1007/s11023-023-09646-w

Aristotle (EE) (2011) The Eudemian ethics. Translated with an introduction and notes by Anthony Kenny. Oxford University Press, Oxford

Aristotle (NE) (2011) The Nicomachean ethics. A new translation by Robert C. Bartlett and Susan D. Collins. The University of Chicago Press, Chicago

Asaro PM (2019) AI ethics in predictive policing: from models of threat to an ethics of care. IEEE Technol Soc Mag 38(2):40–53

Becker F, Skirzyński J, van Opheusden B, Lieder F (2022) Boosting human decision-making with AI-generated decision aids. Comput Brain Behav 5:467–490

Bell DE, Raiffa H, Tversky A (2011) Descriptive, normative, and prescriptive interactions in decision making. In: Bell DE, Raiffa H, Tversky A (eds) Decision making descriptive, normative, and prescriptive interactions. Cambridge University Press, Cambridge, pp 9–30

Bernacer J, Balderas G, Martinez-Valbuena I, Pastor MA, Murillo JI (2014) The problem of consciousness in habitual decision making. Behav Brain Sci 37(1):21–22

Bhattacharyya J, Slocombe S (2017) Animal agency: wildlife management from a kincentric perspective. Ecosphere 8(10):e01978

Blattner C (2021) Turning to animal agency in the anthropocene. In: Bovenkerk B, Keulartz J (eds) Animals in our midst: the challenges of co-existing with animals in the anthropocene. The international library of environmental, agricultural and food ethics, vol 33. Springer, Cham. https://doi.org/10.1007/978-3-030-63523-7_4

Boundreaux DJ, Meiners R (2019) Externality: origins and classifications. Nat Resour J 59(1):3

Boyd S, Vandenberghe L (2009) Convex optimization, 7th edn. Cambridge University Press, Cambridge

Briggs RA (2023) Normative theories of rational choice: expected utility. The Stanford Encyclopedia of Philosophy. https://plato.stanford.edu/archives/win2023/entries/rationality-normative-utility/

Chandler J (2017) Descriptive decision theory. The Stanford Encyclopedia of Philosophy. https://plato.stanford.edu/archives/win2017/entries/decision-theory descriptive/

Chandra J, Rangaswamy M, Banerjee B, Prajapati A, Akhtar Z, Sakauye K, Joseqh A (2022) 12—applications of artificial intelligence to neurological disorders: current technologies and open problems. In: Pillai AS, Menon B (eds) Augmenting neurological disorder prediction and rehabilitation using artificial intelligence. Academic Press, New York, pp 243–272

Christian B, Griffiths T (2016) Algorithms to live by: the computer science of human decision. Henry Holt and Company, New York

Constantinescu M, Vica C, Uszkai R, Voinea C (2022) Blame it on the AI? On the moral responsibility of moral advisors. Philos Technol 35(2):529. https://doi.org/10.1007/s13347-022-005 29 z

Consul S, Heindrich L, Stojcheski J, Lieder F (2022) Improving human decision-making by discovering efficient strategies for hierarchical planning. Comput Brain Behav 5:185–216

Corbett-Davies S, Pierson E, Feller A, Goel S, Huq A (2017) Algorithmic decision making and the cost of fairness. In: Proceedings of the 23rd ACM SIGKDD international conference on knowledge discovery and data mining, pp 797–806. https://doi.org/10.1145/3097983.3098095

Creel KA (2020) Transparency in complex computational systems. Philos Sci 87(4):568–589

Creel KA, Hellman D (2022) The algorithmic leviathan: arbitrariness, fairness, and opportunity in algorithmic decision-making systems. Can J Philos 52(1):26–43

Crocker R (2017) Somebody else's problem: consumerism, sustainability and design. Routledge, New York

de Oliveira ACM (2021) When risky decisions generate externalities. J Risk Uncert 63:59–79

Deghani M, Tomai E, Forbus K, Iliev R, Klenk M (2011) An integrated reasoning approach to moral decision making. In: Anderson M, Anderson SL (eds) Machine ethics. Cambridge University Press, Cambridge, pp 422–441
DKE (Deutsche Kommission Elektrotechnik) (2022) Deutsche normungsroadmap künstliche intelligenz. In: Wahlster W, Winterhalter C (eds), 2nd edn, pp 65–69. https://www.dke.de/resource/blob/2008010/776dd87a4b9ec18d4ab295025ccbb722/nr-ki-deutsch---download-data.pdf
Dorn GJW (2006) Deskriptive argumente und argumenthierarchien. In: Kreuzbauer G, Dorn GJW (eds) Argumentation in theorie und praxis: philosophie und didaktik des argumentierens. Lit, Wien, pp 167–202
Drefcinski S (1996) Aristotle's fallible phronimos. Ancient Philos 16(1):139–154
Drummond N, Niv Y (2020) Model-based decision making and model-free learning. Curr Biol 30:R860–R865
Eder A-MA (2021) Explicating the concept of epistemic rationality. Synthese 199:4975–5000
Eisenführ F, Weber M, Langer T (2010) Rational decision making. Springer, Heidelberg
Elliott E (2019) Normative decision theory. Analysis 79(4):755–772
Fishburn PC (2011) Normative theories of decision making under risk and under uncertainty. In: Bell DE, Raiffa H, Tversky A (eds) Decision making. descriptive, normative, and prescriptive interaction. Cambridge University Press, Cambridge, pp 78–98
Genin K, Grote T (2021) Randomized controlled trials in medical AI: a methodological critique. Philos Med 2(1):1–15
Grote T, Keeling G (2022) Enabling fairness in healthcare through machine learning. Ethics Inform Technol 24:39. https://doi.org/10.1007/s10676-022-09658-7
Günther M, Kasirzadeh A (2022) Algorithmic and human decision making: for a double standard of transparency. AI Soc 37:375–381
Hafenbrädl S, Waeger D, Marewski JN, Gigerenzer G (2016) Applied decision making with fast-and-frugal heuristics. J Appl Res Memory Cognit 5(2):215–231. https://doi.org/10.1016/j.jarmac.2016.04.011
Hengen KM, Alpers GW (2021) Stress makes the difference: social stress and social anxiety in decision-making under uncertainty. Front Psychol 12:8293. https://doi.org/10.3389/fpsyg.2021.578293
Hertwig R, Grüne-Yanoff T (2017) Nudging and boosting: steering or empowering good decisions. Perspect Psychol Sci 12(6):496
Heukelom F (2014) Behavioral economics: a history. Cambridge University Press, Cambridge
Huang Y, Yaple ZA, Yu R (2020) Goal-oriented and habitual decisions: neural signatures of model-based and model-free learning. Neuroimage 215:116834
Johansson L (2010) The functional morality of robots. Int J Technoethics 1(4):65–73. https://doi.org/10.4018/jte.2010100105
Johnson J (2021) Designing with the mind in mind. Simple guide to understanding user interface design guidelines. Morgan Kaufman
Khan C-AB (2005) Aristotle's moral expert: the phronimos*. In: Rasmussen LM (ed) Ethics expertise: history, contemporary perspectives, and applications. Springer, New York, pp 39–53
Klein G (2015) A naturalistic decision making perspective on studying intuitive decision making. J Appl Res Memory Cognit 4(3):164–168
Kleinknecht R (2006) Deduktive ableitung und deduktive begründung. In: Kreuzbauer G, Dorn GJW (eds) Argumentation in theorie und praxis: philosophie und didaktik des argumentierens. Lit, Wien, pp 17–29
Kool W, McGuire JT, Rosen ZB, Botvinick MM (2010) Decision making and the avoidance of cognitive demand. J Exp Psychol Gen 139(4):665–82. https://doi.org/10.1037/a0020198
Lara F, Decker J (2020) Artificial intelligence as a socratic assistant for moral enhancement? Neuroethics 13:275–287. https://doi.org/10.1007/s12152-019-09401-y
Larrick RP (2016) The social context of decisions. Ann Rev Organ Psychol Organ Behav 3:441–467
Leotti LA, Iyengar SS, Ochsner KN (2010) Born to choose: the origins and value of the need for control. Trends Cognit Sci 14(10):457–463

Luce RD, Raiffa H (1957) Games and decisions: introduction and critical survey. Wiley, New York
Luxton DD (2014) Artificial intelligence in psychological practice: current and future applications and implications. Profess Psychol Res Pract 45(5):332–339
Marcinowski M (2022) Artificial intelligence or the ultimate tool for conservatism. Danube 13(1):1–12
Matthias A (2004) The responsibility gap: ascribing responsibility for the actions of learning automata. Ethics Inform Technol 6:175–183
Millière R, Buckner C (2024) A philosophical introduction to language models. Part I: continuity with classic debates. https://doi.org/10.48550/arXiv.2401.03910
Molnar MJ, Molnar V (2023) AI-based tools for the diagnosis and treatment of rare neurological disorders. Nat Rev Neurol 19:455–456
Nagl L (2022) Digital technology: reflections on the difference between instrumental rationality and practical reason. Kantian J 41(1):60–88. https://doi.org/10.5922/0207-6918-2022-1-3
Patel V (2023) Cognitive science in the evaluation of medical AI systems. BMJ Health Care Inform 30(1):e100929
Patel VL, Kaufman DR (2001) Medical expertise, cognitive psychology. In: Smelser NJ, Baltes PB (eds) International encyclopedia of the social and behavioral sciences. Pergamon, pp 9515–9518
Pentland A (2021) Optimized human-AI decision-making: a personal perspective. In: Proceedings of the 2021 international conference on multimodal interaction. Montréal, Canada. https://doi.org/10.1145/3462244.3479880
Peters U (2023) Explainable AI lacks regulative reasons: why AI and human decision-making are not equally opaque. AI Ethics 3:963–974
Pinsdorf C (2016) Lebensformen und Anerkennungsverhältnisse. Zur Ethik der belebten Natur. de Gruyter, Berlin/Boston
Podschawek F (2017) Do androids dream of normative endorsement? On the fallibility of artificial moral agents. In: Artificial intelligence and law 25, special issue: machine law, pp 325–339. https://doi.org/10.1007/s10506-017-9209-6
Porcelli AJ, Delgado MR (2017) Stress and decision making: effects on valuation, learning, and risk-taking. Curr Opin Behav Sci 14:33–39
Reeck C, Wall D, Johnson EJ (2017) Search predicts and changes patience in temporal choice. Biol Sci 114(45):11890–11895
Roli A, Jaeger A, Kauffman S (2022) How organisms come to know the world: fundamental limits on artificial general intelligence. Front Ecol Evolut 9:283. https://doi.org/10.3389/fevo.2021.806283
Rosati C (2016) Agents and shmagents: an essay on agency and normativity. In: Shafer-Landau R (ed) Oxford studies in metaethics, vol 11. Oxford University Press, Oxford, pp 182–213
Rudin C (2019) Stop explaining black box machine learning models for high stakes decisions and use interpretable models instead. Nat Mach Intell 1:206–215
Schwartz B, Ben-Haim Y, Dacso C (2010) What makes a good decision? Robust satisficing as a normative standard of rational decision making. J Theory Soc Behav 41(2):209–227
Spaulding NW (2020) Is human judgement necessary? Artificial intelligence, algorithmic governance, and the law. In: Dubber MD, Pasquale F, Das S (eds) The Oxford handbook of ethics of AI. Oxford University Press, Oxford, pp 375–402
Špinka M (2019) Animal agency, animal awareness and animal welfare. Anim Welf 28(1):11–20. https://doi.org/10.7120/09627286.28.1.011
Stake M (2023) Selbstbewusstsein und Perspektivität. Eine Untersuchung der Zweite-Person-Perspektive. J.B.Metzler, Berlin
Steyl S (2020) Aristotelian practical wisdom in business ethics: two neglected components. J Bus Ethics 163:417–428
Suddendorf T, Bulley A, Miloyan B (2018) Prospection and natural selection. Curr Opin Behav Sci 24:26–31
Swanepoel D (2021) The possibility of deliberate norm-adherence in AI. Ethics Inform Technol 23:157–163

Thomas N (2016) Animals as agents. In: Animal ethics and the autonomous animal self. The Palgrave Macmillan animal ethics series. Palgrave Macmillan, London, pp 7–36. https://doi.org/10.1057/978-1-137-58685-8_2

van der Pligt J (2001) Decision making, psychology. In: Smelser NJ, Baltes PB (eds) International encyclopedia of the social and bahvioral sciences. Pergamon, pp 3309–3315

Véliz C (2021) Moral zombies: why algorithms are not moral agents. AI Soc 36:487–497

Vinny PW, Vishnu VY, Padma SMV (2021) Artificial intelligence shaping the future of neurological practice. Med J Armed Forces India 77(3):276–282

Visweswaran S, King AJ, Cooper GF (2022) Integration of AI in clinical decision support. In: Cohen TA, Patel VL, Shortliffe EH (eds) Intelligent systems in medicine and health. The role of AI. Springer, Cham, pp 285–308

Wallach W, Franklin S, Allen C (2010) A conceptual and computational model of moral decision making in human and artificial agents. Top Cognit Sci 2(3):454–485

Walmsley J (2021) Artificial intelligence and the value of transparency. AI Soc 26:585–595

Wheeler G (2018) Bounded rationality. The Stanford Encyclopedia of Philosophy. https://plato.stanford.edu/archives/fall2020/entries/bounded-rationality/

Yates JF, de Oliveira S (2016) Culture and decision-making. Organ Behav Hum Dec Process 136:106–118

Zednik C (2021) Solving the black box problem: a normative framework for explainable artificial intelligence. Philos Technol 34:265–288

Zerilli J, Knott A, Maclaurin J, Gavaghan C (2019) Transparency in algorithmic and human decision-making: is there a double standard? Philos Technol 32(4):661–683

Chapter 4
An Approach to Joint Hybrid Decision Making Between Humans and Artificial Intelligence

Jonas D. Rockbach, Sven Fuchs, and Maren Bennewitz

'Thus an intelligent machine can be defined as a system that utilizes information, and processes it with high efficiency, so as to achieve a high intensity of appropriate selection. If it is to show really high intelligence, it must process a really large quantity of information, and the efficiency should be really high.' Mechanisms of Intelligence

Ross Ashby (Conant 1981, p. 298)

Abstract Due to the progress in artificial intelligence, it is important to understand how capable artificial agents should be used when interacting with humans, since high level authority and responsibility often remain with the human agent. However, integrated frameworks are lacking that can account for heterogeneous agents and draw on different scientific fields, such as human-factors engineering and artificial intelligence. Therefore, joint hybrid intelligence is described as a framework abstracting humans and artificial intelligence as decision making agents. A general definition of intelligence is provided on the basis of decision making competence being applicable to agents of different sorts. This framework is used for proposing the interrelated design space of joint hybrid intelligence being aimed at integrating the heterogeneous capabilities of humans and artificial intelligence. At the core of this design space lies joint agent engineering with the goal of integrating the design subspaces operator training, artificial intelligence engineering, and interface design via developing joint agent patterns. The "extended swarming" approach to human-swarm interaction is discussed as an example of such a pattern.

J. D. Rockbach (✉) · S. Fuchs
Human-Machine Systems, Fraunhofer FKIE, Wachtberg, Germany
e-mail: jonas.rockbach@fkie.fraunhofer.de

S. Fuchs
e-mail: sven.fuchs@fkie.fraunhofer.de

J. D. Rockbach · M. Bennewitz
Humanoid Robots Lab, University of Bonn, Bonn, Germany
e-mail: maren@cs.uni-bonn.de

U. Ettinger et al. (eds.), *Decision Making*, Studies in Neuroscience, Psychology and Behavioral Economics, https://doi.org/10.1007/978-3-032-00880-0_4

Keywords Hybrid decision making · Human-automation interaction · Joint hybrid intelligence · Joint agent patterns · Intelligent agents · Artificial intelligence.

4.1 Introduction

Automation has become increasingly capable due to the progress in Artificial Intelligence (AI). While more processes are automatized based on AI, traditional problems of human-automation interaction also become more important (Hollnagel and Woods 2005). In complex use cases with high stakes, such as search-and-rescue, full autonomy of AI is still hard to come by, and not always desired given ethical, legal or organizational constraints. On the other hand, introducing AI into the system is expected to increase total performance. Therefore, AI and human capabilities need to be integrated in a way so that the resulting system can deal as effectively and efficiently as possible with the task at hand in comparison to AI and human acting isolated with less total performance. If the integrated system's capabilities exceeds that of the individuals' acting alone, it is regarded as a joint agent itself acting as a unit to cope with the task demands.

While different labels and approaches for joining humans and AI are being used, such as human-machine symbiosis (Licklider 1960), augmented cognition (Stanney et al. 2009), cognitive systems engineering (Hollnagel and Woods 2005), human-machine cooperation (Flemisch et al. 2012) and human-autonomy teaming (O'Neill et al. 2020), we refer to the joining of human capabilities and AI as Joint Hybrid Intelligence (JHI). In contrast to other labels, JHI explicitly couples concepts of human-factors engineering and AI under the umbrella concept of intelligent agents (Rockbach et al. 2022b), a concept also used in the transdisciplinary sciences of cybernetics and cognitive science. JHI refers to the methodical integration of heterogeneous agents, human and AI, into a joint hybrid agent demonstrating superior performance compared to human and AI acting isolated (Rockbach et al. 2022b). In line with human-autonomy teaming (O'Neill et al. 2020), the focus is on AI demonstrating high competence in specific tasks.

So far, the approach of JHI lacks a unified theoretical and practical basis. Therefore, in this work we summarize the theoretical background considerations of intelligence and how these can be used for designing joint hybrid agents. More specifically, Sect. 4.2 defines intelligence as decision making competence so it can be applied to different types of agents. While a theoretical measurement of decision making competence with little resource utilization c is proposed, it is discussed that intelligence is hard to estimate in practice. However, it is shown how the formalization is useful for theorizing about designing intelligence, including the augmentation of agent intelligence with other agents. This theoretical basis is used in Sect. 4.3 to define JHI and its design space in the form of the JHI triad. At the core of the JHI design space lies Joint Agent Engineering (JAE) with the goal to integrate and constrain the different design subspaces human training, AI engineering, and interface design. Here, joint agent patterns are developed to weave the interdependent design subspaces together.

Section 4.4 briefly discusses such a joint agent pattern for Human-Swarm Interaction (HSI), called "extended swarming". Finally, Sect. 4.5 summarizes the considerations.

4.2 Intelligence

4.2.1 Intelligence as Competent Decision Making

4.2.1.1 Agent Theory

A general definition of intelligence is provided by early work in cybernetics, a discipline that has been defined as the (comparative) study of control and communication in animals and machines (Ashby 1961; Wiener 1961). Cyberneticists were enthusiastic about the idea to understand control and communication in different agent types, both natural and artificial. The cybernetician Ashby defined intelligence simply by stating "*intelligent is, what intelligent does*" (Conant 1981, p. 297); thus, intelligence is being a capable decision maker and acting accordingly. In the following, this definition will be specified and contextualized in the light of a recent stream of cognitive science, 4E cognition (embodied, embedded, enactive, and extended) (Newen et al. 2018), that are a modern descendant of early cybernetic ideas. The objective is to define intelligence in a way so it can be applied to any agent, i.e., any system that acts upon the world based on the state of the world (Russell and Norvig 2016), such as humans and AI.

Definition 1 An agent is a system that can make a decision and act accordingly on the basis of the state of the world.

An agent is a system embedded in a particular environment that can select actions based on its own capabilities which in turn are enabled by its body[1] (Fig. 4.1) (Shapiro 2019). According to enactivism, agent and world co-influence each other, resulting in an agent-world couple being defined by both the agent state and the world state (Varela et al. 1991). Here, the agent actively acts on the world and is shaped by it, while the world can actively influence the agent. The world state is a function of world dynamics WD taking as inputs its previous state as well as influences from the agent. The agent body on which basis actions are selected are usually grouped into three building blocks: sensors, decision making system, effectors. The agent's decision matrix DM (Ashby 1961), modeling an agent's brain, computes control commands for the effectors based on its previous internal state as well as estimated world states provided by the sensors. Note that an agent can also make a decision without acting. Similarly, a change in world state is not a necessary condition to trigger a decision or action, since the agent could decide itself to perform some action based on its internal state.

[1] This perspective includes software agents embedded in their virtual environments.

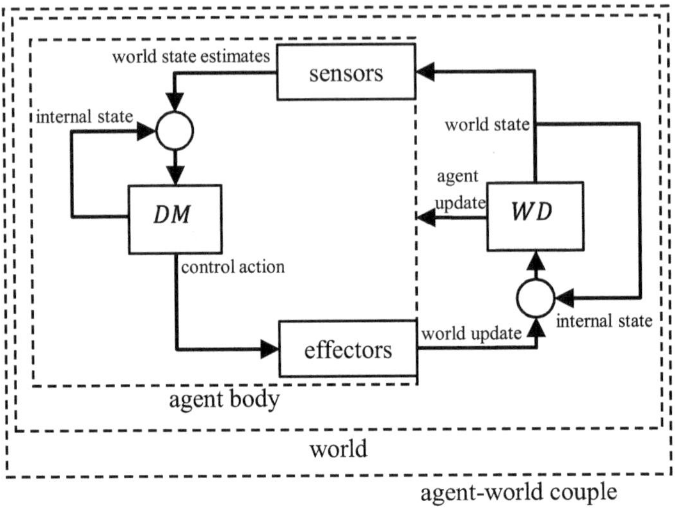

Fig. 4.1 An agent body with sensors, decision matrix, and effectors is embedded in a particular world

4.2.1.2 Agent Intelligence

Now, given an agent placed in a world, it can be asked what makes it intelligent. Based on Ashby's definition above, intelligence is defined as competent decision making, i.e., purposeful behavior attracting the agent-world system towards desired goal states (Rosenblueth et al. 1943). Intelligence is defined as follows:

Definition 2 *(Intelligence)* Agent intelligence refers to the agent's decision making competence promoting the occurrence of goal states in the context of the world (effectiveness) while spending as few resources as possible (efficiency), as judged by an outside observer.

The definition is visualized in Fig. 4.2 and has three important properties. First, intelligence is not binary, but rather a gradual. The better the agent is at making appropriate decisions (effectiveness) while using as few resources as possible (efficiency), the more intelligent it is. Second, Definition 2 takes observer dependencies into account (Brooks 1999; Downing 2015). Observer refers to the human evaluating the agent's intelligence (Maturana et al. 1970). Judging something as intelligent, and something as not intelligent, is primarily a matter of defining agent goals and the situational context of the agent-world system, and only secondary of actual agent properties. These observer constraints provide the frame for the agent decision making competence; the combination of sensors, decision matrix, and effectors.

Thus, the constraints define the minimal required agent capabilities to solve the task. If a task is not bound to resource expenditure, a random decision search process can be judged as satisfactory. However, such a relaxed constraint is hardly ever the

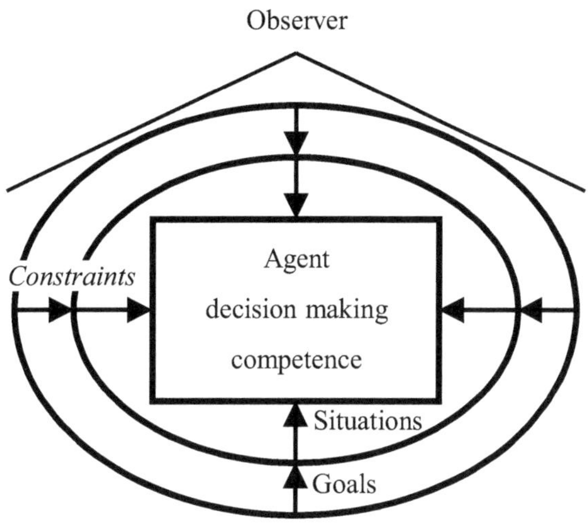

Fig. 4.2 Intelligence depends on the observer constraints of agent purpose and agent-world situational contexts which constrain the required agent decision making competence. Adapted from (Rockbach et al. 2022b)

case, and even then an agent with less resource expenditure would be judged as more intelligent than a random search procedure.

Third, while the purpose of an agent always depends on the observer's judgment, a difference can be made for the situational context. Given an agent-world system, there are variables which actually influence the agent's competence to reach goals, and variables the observer selected as relevant. However, every system model is a simplification of a true phenomenon, if not wrong (Ashby 1961). Therefore, the observer might evaluate an agent in situational contexts that are actually not influencing the agent's purposefulness or leave out relevant situational contexts. While this distinction is an important one, we will not distinguish between the two in the following.

Definition 2 could be further extended by taking bodily limitations into account (Simon 1982). For example, if an agent cannot sense a relevant variable of the agent-world couple because it does not have the appropriate sensors to do so, it can be seen as misleading to refer to the agent as being less intelligent. However, some organisms are capable to extend or augment on their bodily limits (Hutchins 1995), such as apes using sticks, or humans developing scientific instruments to investigate phenomena not accessible by their biological bodies. For this reason, the body is not taken into account in our definition of intelligence, which is aimed at defining "output intelligence" across organic and synthetic agents.

4.2.2 Quantifying Agent Intelligence

4.2.3 Effectiveness: Probability to Reach Goal States

Quantifying agent intelligence can be split into considering effectiveness and efficiency. Based on Definition 2, an intelligent agent makes appropriate decisions to reach goal states in the context of possible situations. In evaluating such decision making effectiveness of an agent, one observes the average agent's probability to reach goal states over possible situation states of the coupled agent-world system,

$$P(g|\mathbb{S}) = \frac{\sum_{i=1}^{|\mathbb{S}|} P(g|s_i)}{|\mathbb{S}|} \tag{4.1}$$

with $\mathbb{S}$ being the set of possible situations, also called the situation space, $s \in \mathbb{S}$ being a particular system state $s_i = \{v_1^i, ...v_M^i\}$ with v being a selected system variable and M being the length of the agent-world system model, and $|\mathbb{S}|$ being the size of the situation space as the number of all possible states s. In turn, G is the set of desired performance states out of the possible performance space $G \subseteq \mathbb{G}$, and $g \in G$ is a desired goal state. $|G|$ represents the goal space size as the number possible goal states. The formalization is visualized in Fig. 4.3.

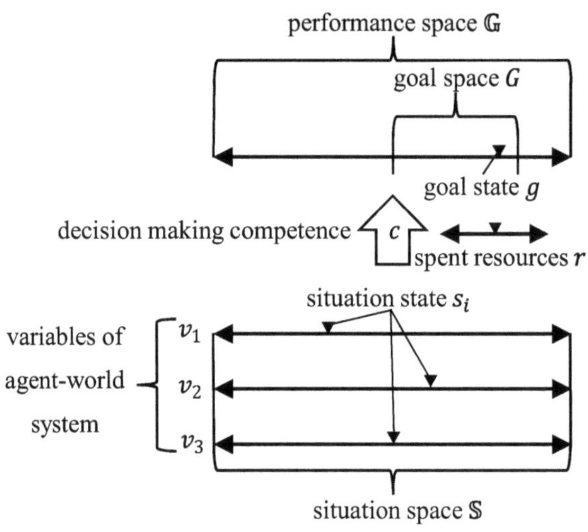

Fig. 4.3 Visualization of the situation space $\mathbb{S}$ as a collection of agent-world system variable states s impacting the performance space $\mathbb{G}$ via the agent's decision making competence c. G as a subset of the performance space describes the tolerated performance values of the system, while g is a particular state in the tolerated goal space range. The agent's decision making competence c determines how effective and efficient the agent is at finding goal states g in the context of the situation space $\mathbb{S}$

In order to evaluate $P(g|\mathbb{S})$, the observer must at least select one performance metric $\mathbb{G}$ and the relevant agent-world system variables including their ranges to define the situation space $\mathbb{S}$ the agent could be confronted with. G is given by selecting a region of the performance space $\mathbb{G}$ that seems to represent purposeful behavior. The thresholds of G determine the difficulty of the competence estimation since a broad goal range makes it easier for the agent to reach a goal state. A narrow range tests for optimality, while a broad range tests for satisfactory competence (Simon 1982).

The average performance over different situation sets could be measured instead without defining an acceptable performance range, and one would have good reasons to do so. Here the dichotomy of "goal reached" versus "goal not reached" is used so the observer needs to make explicit what the purpose of the agent really is. In addition, it is often relevant in practice to define acceptable performance ranges for agents since it provides a guideline for optimizing the agent's intelligence without undershooting (poor performance) or overshooting (wasting resources). Taken together, $P(g|\mathbb{S})$ evaluates an agent's probability to find goal states, i.e., the demonstration of purposeful behavior, over different situational contexts.

4.2.4 *Efficiency: Spent Resources*

Considering two agents demonstrating the same effectiveness $P(g|\mathbb{S})$, the evaluation of efficiency, i.e., resources spent such as required time or energy, provides another quantitative distinction between them (Rockbach et al. 2022a). Intelligence is defined as effective decision making while spending as little resources as possible and is being calculated as decision making competence $c \in [0, 1]$ by

$$c = rP(g|\mathbb{S}) \tag{4.2}$$

with $r \in [0, 1]$ being a normalized measurement of spent resources, 1 being no resources spent, and 0 being an undesired amount of resources spent. Figure 4.3 summaries the formalization of intelligence c as how capable the agent is at finding goal states g while spending few resources r in the context of possible situations $\mathbb{S}$.

4.2.5 *Estimating Decision Making Competence*

Evidently, c is a theoretical construct and practically impossible to estimate in reality for complex problems since this would include the observation of a large part of the agent-world situation space multiple times. An estimation $\hat{c}$ can be achievable via statistical sampling for a selected subset of the situation space $S \subset \mathbb{S}$, which however relaxes the demand on agent intelligence, as will be discussed in the following.

However, it will be seen later that qualitative assessments based on c can also be of use for human-AI integration.

4.2.6 Designing Agent Intelligence

4.2.6.1 Implications from Quantification

Even if c is hard to estimate and depends on the observer constraints G and $\mathbb{S}$, the theoretical measure gives relevant implications for understanding what agent mechanisms are required for intelligent behavior. In general, the decision making competence c (Eq. 4.2) quantifies the agent's general capability to make correct decisions weighted by spent resources (Fig. 4.4). Thus, an agent is only judged as intelligence if it has both the capability to make the right choices (effectiveness) and does so with acceptable effort (efficiency), e.g., in terms of reaction time. An agent acting fast but wrong most of the time is not considered very intelligent. In turn, an agent acting right but very slow most of the time is also not considered very intelligent. Equation 4.2 assumes that effectiveness and efficiency are of equal importance. Thus, an agent acting fast but wrong most of the time may be judged as similar intelligent compared to an agent acting slow but right most of the time. Effectiveness and efficiency are seen as interrelated; the higher the required reaction time, the less information processing capacity is needed to come up with the right solution. Only when an agent acts both fast and right most of the time it demonstrates a higher form of intelligence according to our assumptions.

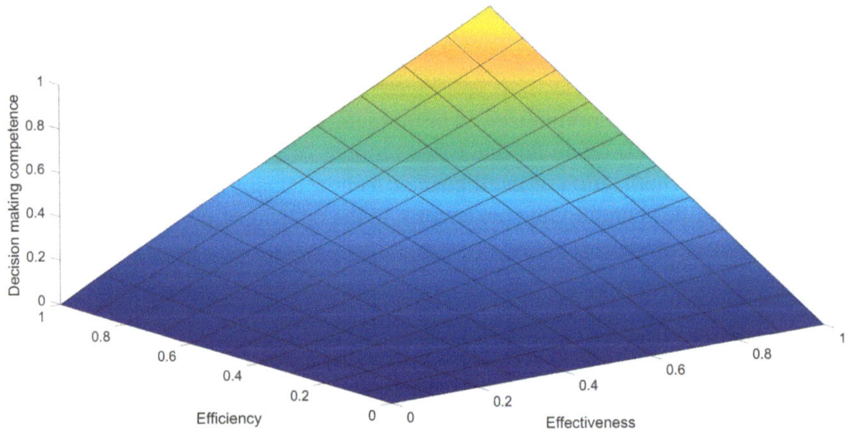

Fig. 4.4 High intelligence in terms of decision making competence requires a combination of effectiveness and efficiency

The implications of the quantification can be analyzed in more detail. The probability of an agent to demonstrate a high effectiveness $P(g|\mathbb{S})$ (Eq. 4.1) is driven both by the observer constraints and by the agent mechanism. In general, in terms of observer constraints a large situation space $|\mathbb{S}|$ will make it harder for the agent to behave so that $P(g|\mathbb{S})$ is high, while in turn a larger goal set $|G|$ has the opposite effect and makes it simpler for the agent (Ashby 1961). If the situation space is non-complex, because of a low size and few linear dependencies between system variables, there may exist a set of heuristic rules for the DM leading to optimal performance. In turn, if the situation space is complex, because of a large size and non-linear dependencies between system variables, an optimal agent also needs to represent a large set of different transformation rules in its DM in order to model the systems dynamics (Conant and Ross Ashby 1970). Taken together, a large complex situation space with a small goal set requires a higher form of decision making competence than a small non-complex situation space with a large goal set.

Given the observer constraints, designing intelligence comes down to selecting appropriate sensors, the decision making mechanism, and effectors, and their combination into an agent architecture (Russell and Norvig 2016). For non-complex demands, it can be feasible to simply represent a rather full set of appropriate rules in the DM. However, for complex demands, alternative approaches are required because an agent body sets an upper limit to the amount of information it can process at a time. There are three general approaches to cope with complex demands. First, one can aim to reduce the agent's confronted situation space by specializing in an ecological niche situation subset (Hasbach and Witte 2021). Second, an agent can attempt to continuously fit its rules to situation subsets, i.e., adapt to the current situation subset (Ashby 1960). In this case, adaptation takes the form of adaptation over life cycles, i.e., evolution and design, and during life cycles, i.e., learning (Hasbach and Witte 2021). Third, an agent can attempt to extend or augment its bodily limitations, such as via tool-use and agents working together (Hutchins 1995). All three strategies will be of importance for the later discussion of integrating human and AI capabilities.

Finally, two additional strategies for increasing c from Eq. 4.2 can also be derived in terms of spent resources. First, heuristics and prepared strategies, if possible and appropriate, with low resource demands increase c compared to coming up with ingenious but resource intensive strategies (Rasmussen 1983). Second, projecting what could happen in the future can prevent the agent from wasting actions that may not be required, while anticipation is also relevant for decision making competence in general.

It must be noted that the above discussed implications serve only as an introduction and are by no means complete. There are more detailed factors of the agent-world couple influencing c. For example, the following list is provided by Russell and Norvig (2016), suggesting that is matters whether the task...

- ...is observable or not
- ...deterministic or stochastic
- ...episodic or sequential

- ...static or dynamic
- ...time-tolerant or time-sensitive
- ...discrete or continuous
- ...known or unknown
- ...single-agent or multi-agent
- ...competitive or cooperative.

Each selected combination changes the demand on the agent.

4.2.6.2 Intelligent Agent Architecture

The desired properties are combined into an abstract intelligent agent architecture blueprint (Fig. 4.5). Similar agent blueprints are discussed in both AI (Russell and Norvig 2016) and cognitive science (Rasmussen 1983). For small and non-complex regions of the situation space, fast-heuristic rules can be reactively applied in a stimulus-response fashion. We assume that some rules are fixed in the agents, such as reflexes, while others can be updated, such as learned heuristics. However, for large or complex situation spaces, the cognitive system modulates the behavior of its low-level fast-heuristic arsenal, e.g., by activating and parameterizing appropriate agent subsystems (Minsky 1988). This requires a representation of the problem in terms of a model, which is resource intensive to attain, however allows the agent to deal with complex problems. Stacked on this model-based capabilities are the adaptive and anticipatory system. The adaptive system maintains the problem representations during the agent's life, therefore continuously checks whether the model fits to the current problem and updates it if necessary. Finally, the anticipatory subsystem allows the agent to project to future states.

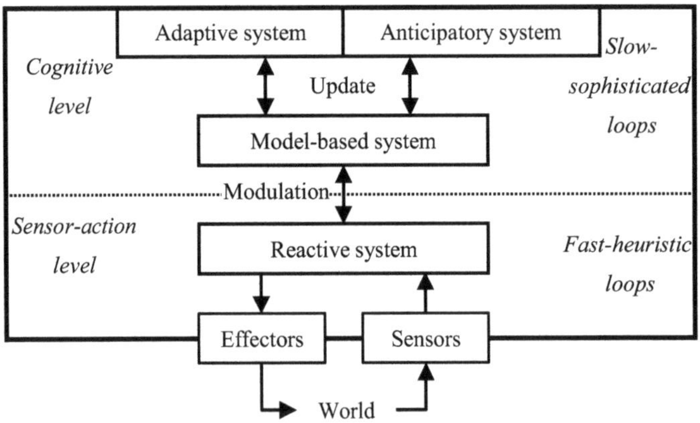

Fig. 4.5 Blueprint of an intelligent architecture with fast reactive loops at the low level being modulated by slower but more capable cognitive loops. Adapted from (Rockbach et al. 2022b)

The agent architecture in Fig. 4.5 shows different levels of mechanisms for decision making competence. Easier tasks may only require fast reactive mechanisms, while harder tasks can only be solved by the slower cognitive system stacked on top. The agent engineer must analyze the agent's task for difficulty and resource-sensitivity, and choose an appropriate mechanism level for maximizing c.

4.3 Joint Hybrid Intelligence

4.3.1 Definition of Joint Hybrid Intelligence

Given the above limitations of an embodied agent to deal with complex problems, the objective is to join human intelligence with AI for improving on their relative decision making competencies. JHI refers to the relative competence augmentation of two or more heterogeneous agents[2] into one joint agent demonstrating superior competence compared to the isolated individuals' (Hollnagel and Woods 2005; Rockbach et al. 2022b).

On the basis of G and $\mathbb{S}$, the decision making competences of individual ("internal") agents are estimated, c_{nat} for the human's intelligence and c_{arti} for the AI, and joined via an integration design in such a way that the joint intelligence c_{joint} is superior to the individual's acting isolated. Thus, we assume that each internal agent possesses valuable decision making competencies that should be combined, i.e., we search for an integration design[3] such that the joint competence c_{joint} satisfies $c_{joint} > c_{nat} \vee c_{arti}$. If $c_{joint} \leq c_{nat} \vee c_{arti}$, the design is considered disjoint and control may fully remain with either human or AI.[4]

Definition 3 *(Joint Hybrid Intelligence)* JHI refers to optimizing the joint decision making competence c_{joint} by finding an integration design for heterogeneous internal agents given G and $\mathbb{S}$. If the integration satisfies $c_{joint} > c_{nat} \vee c_{arti}$, the resulting system is called a joint (hybrid) agent. The practice of integrating heterogeneous internal agent aspects into a whole is referred to as Joint Agent Engineering (JAE).

While in this work internal agents are humans and AI, the definition can also be applied to other internal agent types, such as hybrid societies (Hamann et al. 2016). A joint agent provides a joint decision matrix DM_{joint}, it being a function of the integration design $DM_{nat} \cup DM_{arti}$. Figure 4.6 shows a visualization of a simple joint agent with one human and one AI. Human and AI interact with each other through a Human-Machine Interface (HMI) (Bennett and Flach 2011). They

---

[2] This includes tools only if the tool is an AI, i.e. an artificial agent according to Definition 1.

[3] With the simplest integration being human and AI acting in parallel without interacting.

[4] In some cases, $c_{joint} > c_{nat} \vee c_{arti}$ exists, however control responsibility should fully remain with the human because of ethical, legal, or organizational concerns, or fully with the AI because human participation is not possible or desired.

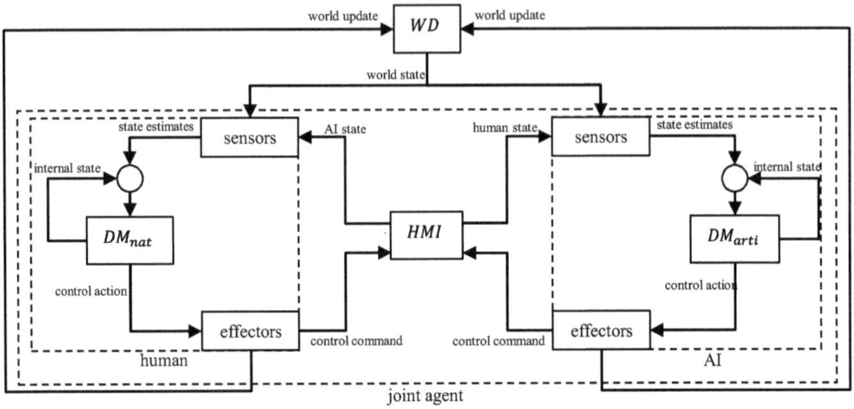

Fig. 4.6 Human and AI are integrated by coupling their individual decision matrices via a HMI

also individually or both sense the world and act in it. In the following, a general framework is discussed for joining human and AI.

4.3.2 Designing Joint Hybrid Intelligence

4.3.2.1 Triad of Joint Hybrid Intelligence

A visual description for designing JHI is depicted by the JHI triad in Fig. 4.7. The outside of the triad represents the analysis space, the inner smaller triads the design space. The sequence of designing JHI goes from bottom to top, while the center constitutes JAE with the goal of integrating the different design subspaces.

The analysis space is constituted of the following steps:

1. Analyze the joint agent for demands on its intelligence
2. Determine the individual allocation potential by comparing relative c_{nat} and c_{arti}

The design space is made of

1. JAE: Designing integrative patterns to bring the different design subspaces together
2. Training engineering: Designing human training procedures for optimizing the human's DM_{nat}
3. AI engineering: Designing the AI for optimizing the AI's DM_{arti}
4. Interaction engineering: Designing the HMI for optimizing the joint DM_{joint}.

Finally, the design is evaluated and iterated by

1. Evaluating the joint agent intelligence
2. Evaluating internal variables of the joint agent

3. Iterate the joint agent design.

In this work, the analysis space, JAE, and the evaluation will be discussed. For details about the specific design subspaces the interested reader is referred to (Ericsson et al. 2018) for operator training, to (Russell and Norvig 2016) for AI engineering, and to (Bennett and Flach 2011) for interaction design.

4.3.2.2 Demands on Joint Agent

Analyzing the joint agent demands follows the logic of Fig. 4.2. First, the joint agent is analyzed in terms of the observer constraints. Here, the system analyst defines the purpose of the system, preferably in a manner so that the purpose can be formalized into a performance metric. Next, general system dynamics for the agent-world couple are modeled in the context of the system's purpose. Variables can be external, i.e., variables of the world external to the joint agent, or internal, i.e., variables describing the joint agent. Another important classification is whether variables can be influenced or not. The objective is to understand what situational context could impact performance, and to get an overview how.

A range of modeling methods can be used to model and understand the agent-world couple. The basic technique is to develop a qualitative graph with interlinked variables. For example, the abstraction hierarchy as part of cognitive work analysis models relevant variables from abstract to physical in the context of agent purpose (Rasmussen et al. 1994). More generally, system dynamics simulations provide

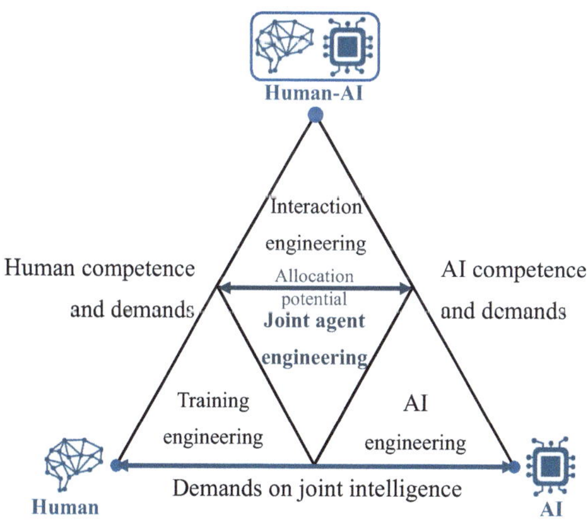

Fig. 4.7 The triad of joint hybrid intelligence with its analysis space (outside of triad) and the embedded design subspaces (inside triads). Joint agent engineering at the center integrates the different design subspaces

a coupled set of variables and their resulting behavior (Forrester 1971). Fuzzy cognitive mapping is a technique that can also be used for simulations by providing an interlinked network of variables and is especially useful for capturing expert knowledge (Özesmi and Özesmi 2004). In general, the modeler may use any qualitative or quantitative technique that advances the understanding of the agent-world couple in the context of the agent purpose.

Two levels of systems understanding can be given. First, the analyst should aim to capture general system behaviors to get an overview of relevant system states. Second, the utility of system states in the context of the agent purpose should be evaluated. For example, controllable world states leading to a large probability of goal attraction are important variables to influence by effectors, since it makes the task easier for the agent.

After one has defined and analyzed the observer constraints, the required decision making competence of the joint agent is examined (Fig. 4.2). As defined in Sect. 4.2.1.1, agent competence is a function of the sense-decide-act loop. Based on the above analysis of the observer constraints, relevant variables need to be tracked by sensors and acted upon by effectors. Here, the analyst can derive a list of requirements for what sort of information the joint agent must have access to.

The heart of the decision making competence is the decision matrix. Here, the required information processing is helpful for understanding the demands on the decision matrix. Task models, which are often qualitative in nature, group information processing into chunks with required input and provided output and combines them into a flow model (Crandall et al. 2006). Following the sense-decide-act loop, a task model provides a more detailed account on the information processing demands on the joint agent.

The task model can then be analyzed for difficulty. For example, classification can take the form of evaluating the required competence levels and time-sensitivity as shown in Fig. 4.5. Another classification category can be in terms of cognitive processing with categories coordination, problem detection, sense-making, naturalistic decision making, adaptation, and planning (Crandall et al. 2006). If the detailed task model is taken as a whole, this provides an overview of the required joint intelligence.

4.3.2.3 Estimating Relative Agent Intelligence

Given the analyzed system model and task model, the individual decision making competence c_{nat} and c_{arti} are compared. Such a relative comparison can be done for all tasks taken together, or for individual tasks, resulting in a task allocation potential (Rockbach et al. 2022a). The task allocation potential describes the human-AI tension field on the basis of the relative individual competences of human and AI, which is not to be confused with actually allocating the tasks based on an integrated view to optimize c_{joint} as will be discussed in the next section (Dekker and Woods 2002). In general, task allocation levels take the form of the categories manual (only human performs task), automated (only AI performers task), or shared (both perform some aspects of the task). Such levels are referred to as the Levels of Automation (LoA), and often more detailed levels are used (Sheridan 1992).

In practice, quantitative estimations of relative c are almost never feasible, given the large situation space. However, qualitative experience exists that provide a heuristic whether the human or automation performs better for a particular task. For example, the classical "Men Are Better At—Machines Are Better At" (MABA—MABA) list provides a rough estimate what internal agent would perform better depending on the task (Dekker and Woods 2002). However, the relative internal agent intelligence must not only take the situational context into account, but also the human's experience and the sort of algorithm being used. Therefore, defining the task allocation potential constitutes more of a design art than an engineering technique (Bennett and Flach 2011).

4.3.2.4 Joint Agent Engineering

At the center of the JHI design space lies JAE. The objective of JAE is to integrate the internal decision making competences in a way that the joint decision making competence is optimized based on the analyzed demands on joint agent intelligence and the task allocation potential drawn from individual competences. In order to maximize c_{joint}, the joint agent's body and decision making mechanism should be able to cope with the task demands while spending as few resources as possible. The joint agent must be scaled with sufficiently internal agents to cope with the task demands. At the same time, internal agents should not be added blindly, since their interactions can increase internal complexity and thereby waste resources (Hamann and Reina 2021; Woods and Hollnagel 2006).

A joint agent is a distributed network of interacting internal agents (Fig. 4.6) forming computational task-specific subnetworks (Hutchins 1995). Internal agents can be exclusively attributed to a task corresponding to a manual or automated LoA, share the allocation with other agents (shared LoA), as well as participate in multiple tasks (Fig. 4.8). The allocation graph of internal agents to tasks is called the allocation design. While today's sensors and effectors are often automated, their operations are still supervised by humans, meaning that normally both AI and human are allocated to sensor and effector tasks. The brain of the joint agent is modeled as the DM_{joint}. Based on the task models and the associated relative allocation potential, different allocation designs can be developed. The task allocation design heavily influences the surrounding design subspaces of the JHI triad and is therefore seen as the foundation of JAE. However, as noted above, task allocation is still more art than quantitative engineering, since coupling heterogeneous complex agents leads to unpredictability in the joint agent dynamics (Woods and Hollnagel 2006). Different allocation designs are normally developed and qualitatively compared, e.g., via consulting system users and experts, or prototypes are implemented to quantitatively evaluate resulting joint agent competences (Sect. 4.3.2.5).

To summarize, the general design problem of JHI is complex, since one can come up with many possible design solutions per design dimension, resulting in

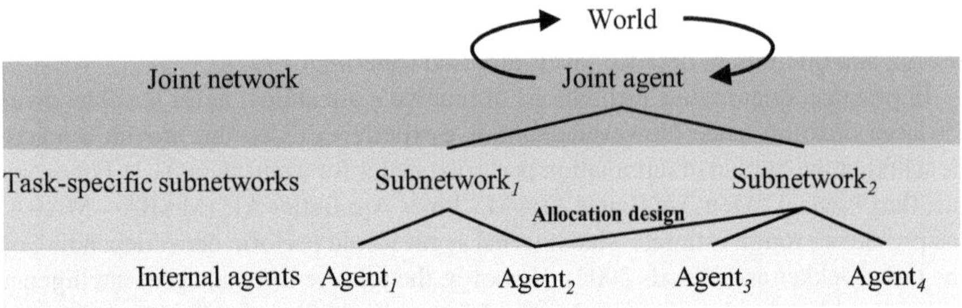

Fig. 4.8 A joint agent is a distributed network of interacting internal agents forming into computational task-specific subnetworks. The allocation example shows that internal agents can contribute to multiple task-specific subnetworks. Adapted from (Rockbach et al. 2022b)

scattered design facets over the actually interdependent design subspaces (Fig. 4.7). Therefore, it is helpful to come up with overreaching design heuristics, called design patterns (Alexander 1977; Woods and Hollnagel 2006), serving as blueprints by providing a unified idea how the individual design parts could come together.

Definition 4 *(Joint Agent Pattern)* A joint agent pattern is a unifying blueprint for logically constraining all design subspaces of the JHI triad in order to reduce design complexity. In line with the objective of JAE, the patterns are aimed at maximizing the joint agent competence c_{joint} by integrating different design facets into a whole.

JAE's central objective is to find such unifying patterns and apply them in real-world contexts (Woods and Hollnagel 2006). Two examples of joint agent patterns are discussed in the following, first for task allocation in supervisory control, and second for approaches to AI presentation.

Pattern for Supervisory Control. Even if tasks were allocated to the AI, human operators are often made responsible for supervising the AI (Bainbridge 1983; Sheridan 1992). This can be a severe design dilemma, since the aim of automation is to reduce the human's workload. At the same time, however, the workload is increased by the additional task of AI supervision. A general joint agent architecture for automation supervision, based on the previously discussed intelligent agent architecture (Fig. 4.5), is shown in Fig. 4.9. Today's AI systems still constitute "weak intelligence", referring to competencies covering subregions of the situation space S, in comparison to the flexibility of humans' "strong intelligence", referring to broad situation spaces $\mathbb{S}$. In other words, AI surpasses human competence in specialized tasks, i.e., for S, while humans surpass AI when evaluated in $\mathbb{S}$. Therefore, AI is deployed for narrow subregions of the situation space for which it is specialized in, while the human's task is to check whether c_{arti} fits to the current subspace and parameterize the AI accordingly. In this way, the different response times of humans and AI are aligned with their competences, and supervision does not become a performance bottleneck (Hasbach and Witte 2021). This type of architecture is also known as human-on-the-loop design (Sheridan 1992).

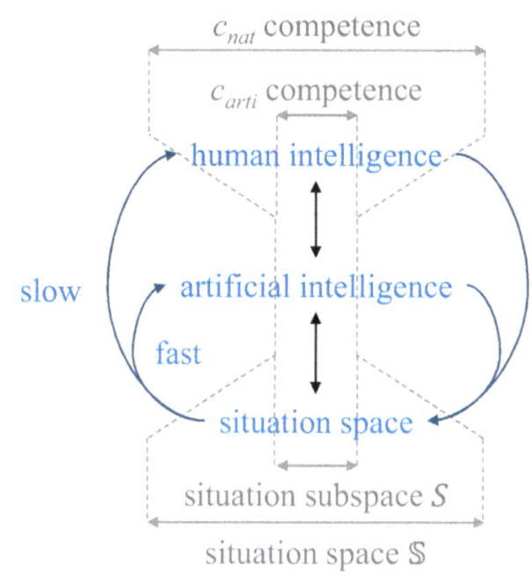

Fig. 4.9 Joint agent architecture for supervisory control showing AI being deployed by the human to narrow subregions of the situation space where they are allowed to act semi-autonomously. Adapted from (Hasbach and Witte 2021)

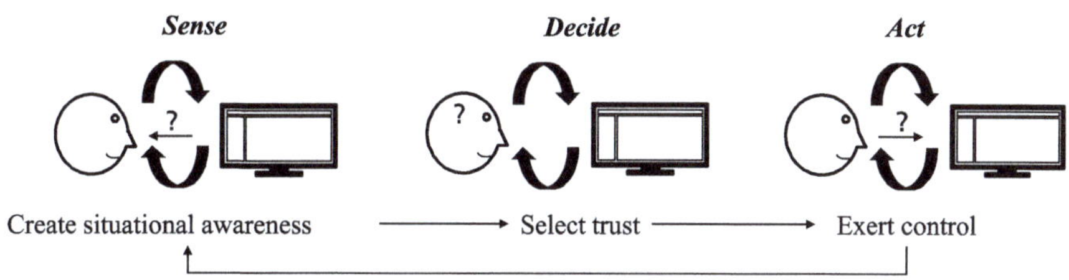

Fig. 4.10 Sensor-decide-act steps for human supervision over AI. Adapted from (Rockbach et al. 2022a)

The design subspaces human training, AI engineering, and interaction design are seen in light of this joint agent pattern. For example, in order to supervise c_{arti} and parameterize the AI accordingly, the human supervisor should be experienced in implementing the following steps of the supervisory sense-decide-act loop (Fig. 4.10). First, the human supervisor needs to understand the current state of the AI and the world (Endsley 1995), which is influenced by the transparency of AI (Barredo Arrieta et al. 2020) and its presentation via the HMI. Then, the human supervisor is required to check whether the AI actually fits the situation subspace, it being referred to as trust calibration (Lee and See 2004). Optimally, the operator trusts the AI if the fit is aligned and mistrusts it if the fit is not aligned. However, operators can also misjudge the current AI competence. Misuse refers to regions of overtrust where the AI is deployed in situation subspaces with poor performance. In turn, in disuse operators tend to deactivate the AI even if its activation would actually increase c_{joint}.

Pattern for AI Presentation. Another example of a joint agent pattern are blueprints for how an AI can be presented to the human operator. First, an AI can be presented as a tool to the human.[5] In the tool pattern, the main responsibility and intelligence remains with the human who uses the properties of the tools as seen fit. Depending on the complexity of the tool, the human must be trained in its specific use, and the HMI should provide intuition on how to control it in detail. In contrast, the team member pattern presents the AI as a sophisticated agent (O'Neill et al. 2020). More decision making is allocated to the AI and the human may judge it as an intelligent system itself. However, there is the risk of undertrust if the AI does not hold up to the expectations or overtrust if the HMI presents it as too intelligent. Less training is required if the AI demonstrates sophisticated intelligence and the HMI design is aligned with this by communicating high level information. Finally, the cyborg pattern treats the AI as part of the human body and is exemplified in Sect. 4.4.

4.3.2.5 Evaluation and Iteration

Given the unpredictability of the actual joint agent behavior, an incomplete system model being its main source, it is essential to implement a prototype and evaluate the joint agent in terms of c_{joint} given ranges of situational contexts and joint agent variables.

While the specific internal variables depend on the use case and are provided by the abstraction hierarchy discussed in Sect. 4.3.2.2, supervisory control provides an important example of internal dynamics influencing c_{joint}. From a human perspective, the system design should be evaluated for its tendency to equip the human operator with situational awareness over the AI, enable trust calibration, as well as control possibilities. Moreover, different algorithms and transparency approaches (Adadi and Berrada 2018) can be checked for their potential to fit into the joint agent as well as what level of supervisory competence is required from the human operator to use the AI efficiently. Finally, the joint agent can be tested for ranges of the situation spaces with different difficulties, AI competences, and different levels of operator training, to identify joint performance regions of "brittleness" (Woods and Hollnagel 2006), i.e., regions where c_{joint} suddenly breaks down. Such analysis helps to understand the system dynamics and to update the selected subdesigns of the JHI triad.

Often, it is not feasible to implement a range of complete prototypes to compare between different designs. It can, however, be feasible to design low-fidelity prototypes. For example, the wizard-of-oz technique only simulates AI behavior that is controlled by a hidden experimenter (Maulsby et al. 1993). In addition, "staged world experiments" increase the ecological validity of experimental lab settings, e.g., by providing a realistic contextual scenario, and have been discussed as a helpful practical trade-off between controllable experiments and ecological complexity (Woods and Hollnagel 2006).

[5] Note that an AI agent according to Definition 1 can still be presented as a tool to the human.

4.4 Human Control Over Robot Swarms

4.4.1 Swarm Supervision as Design Problem

Swarm robotics envisions robot collectives in which rather simple robots interact only with local neighbors without reference to global information (Dorigo et al. 2021). This decentralized architecture makes robot swarms both scalable and robust to single robot failures. However, in order to deploy robot swarms to real-world problems, human operators must be in control of these decentralized systems (Kolling et al. 2016). Such HSI is a particular case of JHI.

HSI comes with two special design challenges. First, integrating a human supervisor into the robot swarm introduces centralization into the robot swarm that was deliberately designed in a decentralized way, since the human must be able to influence the whole swarm and be provided with state estimations from the swarm. This is referred to as centralized-decentralized trade-off (Hasbach and Bennewitz 2022). Second, supervisory control methods must take the swarm's inherent scalability into account, meaning that the control complexity should be invariant to the swarm size. For example, a human operator could control a leader robot which influences the rest of the swarm via local interactions (Kolling et al. 2016).

Work in HSI has mainly focused on the aspects of swarm engineering and interaction design. However, it was discussed here that the design space is a function of the integrated aspects of the JHI triad. Therefore, we summarize recent work on a joint agent pattern for HSI based on the cyborg pattern (Clark 2003), in the following referred to as "extended swarming".

4.4.2 Swarms as Controllable Embodied Extensions

In the joint agent pattern "extended swarming", the swarm is designed as an embodied extension of the human operator (Hasbach and Bennewitz 2022). The pattern is treated as a scientific theory that must generate testable design solutions for the JHI design subspace given the hypothesis of good joint agent performance. From the human's perspective, the swarm becomes a sensation-action extension of oneself. A human who is extended by a self-organizing swarm is referred to as swarm-amplified human, since the extension of the human's low-level sensation-action range is hypothesized to amplify the human's high-level natural cognitive capacities by augmentation. Given this cyborg pattern, the ultimate design goal is to induce the "sense of ownership" (Braun et al. 2018) towards the swarm into the human operator, but the pattern is also useful without such perceived embodiment (Rockbach and Bennewitz 2022). In turn, from the swarm's perspective the human becomes a special swarm agent while robots interact only locally with the human. These two perspectives are required to balance the centralized-decentralized trade-off. In the following, the different design subspaces are shortly addressed in the light of extended swarming.

4.4.2.1 Operator Training

Little work has considered operator training in HSI although the human's expertise of a technical system modulates joint agent dynamics (Ericsson et al. 2018). If the swarm is treated as an extension of the human, the training follows the learning of body control in humans via trial-and-error (Hasbach and Bennewitz 2022). The joint human-swarm agent is exposed to a range of situations with different degrees of complexity (Hasbach et al. 2020). The goal is to develop an intuitive mental model of swarm control, similar to the feeling we have when controlling our own bodies (Rockbach and Bennewitz 2022).

4.4.2.2 Swarm Engineering

Swarms must both provide the possibility to be controlled by the human and provide fused estimates about the swarm state (Kolling et al. 2016). The swarm is designed as a self-organizational network with the human agent at the center of it, while only minimally reducing the swarm's decentralized nature. In terms of controllability, the swarm self-organizes into "extending body postures" based on human input and environmental states (Rockbach 2023). For example, a swarm state clustering at the human corresponds to a "contracted body state", while a dispersed swarm is an "extended body state" (Fig. 4.11).

Information fusion is implemented in a decentralized manner by the swarm itself. The swarm self-organizes into a decentralized fusion network with the human operator at the center of it based on the environmental state (Rockbach et al. 2023). The fusion network can be thought of as a spider web providing a 360° input of the sensed environment, thereby extending the human's sensation range (Fig. 4.11). In addition, the spider web can also be used to allocate own agents to particular areas, e.g., the human can follow traces of the spider web towards objects of interest.

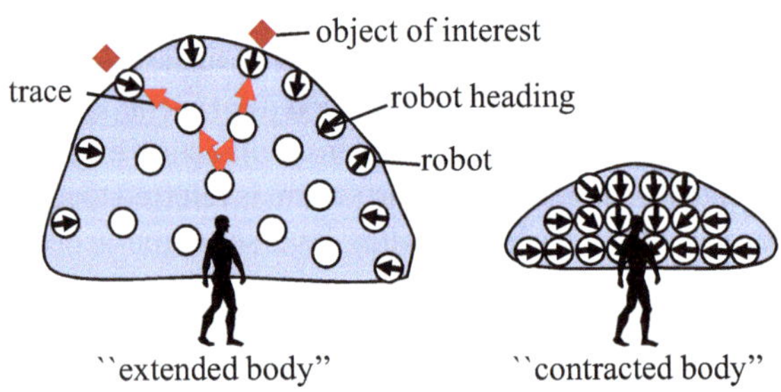

Fig. 4.11 Example of extending body postures implemented by a robot swarm with an embedded fusion network. The pose is the observed collective state of the robot swarm, shown by the blue region. The distributed fusion network points towards sensed states (red arrows), such as objects of interest, and can be observed and used by the human as a movement gradient

4.4.2.3 Interface Design

In extended swarming, the swarm is treated as an extension of the human's sensation-action loop with the world. Thus, the swarm itself becomes the interface between the human and the world such as discussed in terms of decentralized information fusion above. From the human's control perspective, it is hypothesized that classifying human body states, such as body posture or physiological states, and utilizing them to automatically influence the swarm posture is beneficial for overall performance (Hasbach and Bennewitz 2022; Rockbach and Bennewitz 2022). This is derived from the general experience with intuitive body control, which is mainly subconscious in everyday life. In addition, the design must feature the possibility for deliberate control, e.g., by controlling an extended limb formed by the swarm (Rockbach 2023) with one's own arm, as one can deliberately control detailed aspects of the body when one desires to do so.

The feedback provided to the human would take the form of low-level sensation stimulation such as an increase in a visual or tactile stimulus when facing a state of interest sensed by the swarm extension (Hasbach and Bennewitz 2022; Rockbach and Bennewitz 2022). This also reflects daily-life embodied phenomenology, e.g., when coming closer to a fire increases the feeling of body warmth. Based on such low-level stimulation, the brain's high-level operations fuse and embed the low-level input into a holistic percept of the current agent-body state given an adapted human's model of body and world.

4.5 Conclusion

This chapter discussed an approach for joining the heterogeneous capabilities of humans and AI into a joint hybrid agent. Intelligence was formalized as decision making competence to be applicable to both humans and AI. This formalization was analysed for designing intelligence, including the augmentation of one agent by another agent, providing the theoretical basis for JHI. JHI was presented as the optimization of joint decision making competence by integrating the design subspaces operator training, AI engineering, and interface design via the development of overreaching joint agent patterns. Finally, "extended swarming" provided an example of a joint agent pattern for human-swarm interaction. While the integration of humans and AI has a long history, it is important to explicitly develop a framework which brings the different disciplines and their terminologies together. JHI provides a common basis for the multiple scientific and engineering branches concerned with human-AI integration, and therefore facilitates the integration of human and AI into intelligent joint decision making agents.

References

Adadi A, Berrada M (2018) Peeking inside the black-box: a survey on explainable artificial intelligence (XAI). IEEE Access 6:52138–52160

Alexander C (1977) A pattern language: towns, buildings, construction. Oxford University Press

Ashby WR (1960) Design for a brain: the origin of adaptive behaviour, 2nd edn. Chapman & Hall Ltd.

Ashby WR (1961) An introduction to cybernetics. Chapman & Hall Ltd.

Bainbridge L (1983) Ironies of automation. Automatica

Barredo Arrieta A, Díaz-Rodríguez N, Del Ser J, Bennetot A, Tabik S, Barbado A, Garcia S, Gil-Lopez S, Molina D, Benjamins R, Chatila R, Herrera F (2020) Explainable explainable artificial intelligence (XAI): concepts, taxonomies, opportunities and challenges toward responsible AI. Inf Fusion 58:82–115

Bennett KB, Flach JM (2011) Display and interface design: subtle science, exact art. CRC Press

Braun N, Debener S, Spychala N, Bongartz E, Sörös P, Müller HH, Philipsen A (2018) The senses of agency and ownership: a review. Front Psychol 9:1–17

Brooks RA (1999) Cambrian intelligence: the early history of the new AI. MIT Press

Clark A (2003) Natural-born cyborgs: minds, technologies, and the future of human intelligence. Oxford University Press

Conant R (1981) Mechanisms of intelligence: Ashby's writings on cybernetics. Intersystems Publications

Conant RC, Ross Ashby W (1970) Every good regulator of a system must be a model of that system. Int J Syst Sci

Crandall B, Klein GA, Hoffman RR (2006) Working minds: a practitioner's guide to cognitive task analysis. MIT Press

Dekker SWA, Woods DD (2002) MABA—MABA or Abracadabra? Progress on human-automation co-ordination. Cogn, Technol Work 4:240–244

Dorigo M, Theraulaz G, Trianni V (2021) Swarm robotics: past, present, and future [point of view]. Proc IEEE 109(7):1152–1165

Downing KL (2015) Intelligence emerging: adaptivity and search in evolving neural systems. MIT Press

Endsley MR (1995) Toward a theory of situation awareness in dynamic systems. Hum Factors 37(1):32–64

Ericsson KA, Hoffman RR, Kozbelt A (2018) The Cambridge handbook of expertise and expert performance. Cambridge University Press

Flemisch F, Heesen M, Hesse T, Kelsch J, Schieben A, Beller J (2012) Towards a dynamic balance between humans and automation: authority, ability, responsibility and control in shared and cooperative control situations. Cogn, Technol Work 14(1):3–18

Forrester JW (1971) Counterintuitive behavior of social systems. Theory Decis 2(2):109–140

Hamann H, Khaluf Y, Botev J, Divband Soorati M, Ferrante E, Kosak O, Montanier J-M, Mostaghim S, Redpath R, Timmis J (2016) Hybrid societies: challenges and perspectives in the design of collective behavior in self-organizing systems. Front Robot AI 3:14

Hamann H, Reina A (2021) Scalability in computing and robotics. IEEE Trans Comput

Hasbach JD, Bennewitz M (2022) The design of self-organizing human-swarm intelligence. Adapt Behav 30(4):361–386

Hasbach JD, Witte TE (2021) Human-machine intelligence: frigates are intelligent organisms. In: 2021 IEEE international conference on systems, man, and cybernetics (SMC)

Hasbach JD, Witte TEF, Bennewitz M (2020) On the importance of adaptive operator training in human-swarm interaction. In: International conference on human-computer interaction. Springer, pp 311–329

Hollnagel E, Woods DD (2005) Joint cognitive systems. CRC Press, Foundations of Cognitive Systems Engineering

Hutchins E (1995) Cognition in the wild. MIT Press

Kolling A, Walker P, Chakraborty N, Sycara K, Lewis M (2016) Human interaction with robot swarms: a survey. IEEE Trans Hum-Mach Syst 46(1):9–26

Lee JD, See KA (2004) Trust in automation: designing for appropriate reliance. Hum Factors 46(1):50–80

Licklider JCR (1960) Man-computer symbiosis. IRE Trans Hum Factors Electron

Maturana HR, Von Foerster H, Weston P (1970) Neurophysiology of cognition. University of Illinois Urbana

Maulsby D, Greenberg S, Mander R (1993) Prototyping an intelligent agent through Wizard of Oz. In: Proceedings of the INTERACT'93 and CHI'93 conference on Human factors in computing systems, pp 277–284

Minsky M (1988) The society of mind. Simon and Schuster

Newen A, Gallagher S, De Bruin L (2018) 4E cognition: historical roots, key concepts, and central issues

O'Neill T, McNeese N, Barron A, Schelble B (2020) Human–autonomy teaming: a review and analysis of the empirical literature. Hum Factors

Özesmi U, Özesmi SL (2004) Ecological models based on people's knowledge: a multi-step fuzzy cognitive mapping approach. Ecol Model 176(1–2):43–64

Rasmussen J (1983) Skills, rules, and knowledge; signals, signs, and symbols, and other distinctions in human performance models. IEEE Trans Syst, Man, Cybern 3:257–266

Rasmussen J, Pejtersen AM, Goodstein LP (1994) Cognitive systems engineering. Wiley

Rockbach JD (2023) Enhancing human self-regulation with controllable robot swarms acting as extended bodies. In: 2023 IEEE International conference on systems, man, and cybernetics (SMC)

Rockbach JD, Bennewitz M (2022) Robot swarms as embodied extensions of humans. IOP Conf Ser: Mater Sci Eng 1261:152

Rockbach JD, Bluhm L-F, Schlangen I, Over L, Apfeld S, Henneke L, Wilkinghoff K (2022a) Towards human-machine integration for signals intelligence applications. In: 2022 Sensor data fusion: trends, solutions, applications (SDF). IEEE, pp 1–6

Rockbach JD, Fuchs S, Witte TE, Bluhm L-F (2022b) Ingredients for hybrid intelligence: towards an integrated theory and application. In: 2022 IEEE international conference on human-machine systems (ICHMS)

Rockbach JD, Schlangen I, Bennewitz M (2023) Self-organising distributed sensor fusion networks for hierarchical swarm control and supervision. In: 2023 IEEE symposium sensor data fusion and international conference on multisensor fusion and integration (SDF-MFI). IEEE, pp 1–6

Rosenblueth A, Wiener N, Bigelow J (1943) Behavior, purpose and teleology. Philos Sci 10(1):18–24

Russell S, Norvig P (2016) Artificial intelligence: a modern approach, 3rd edn. Pearson

Shapiro L (2019) Embodied cognition. Routledge

Sheridan TB (1992) Telerobotics, automation, and human supervisory control. MIT Press

Simon H (1982) The sciences of the artificial. MIT Press

Stanney KM, Schmorrow DD, Johnston M, Fuchs S, Jones D, Hale KS, Ahmad A, Young P (2009) Augmented cognition: an overview. Rev Hum Factors Ergon 5(1):195–224

Varela FJ, Thompson E, Rosch E (1991) The embodied mind: cognitive science and human experience. MIT Press

Wiener N (1961) Cybernetics or control and communication in the animal and the machine. MIT Press

Woods DD, Hollnagel E (2006) Joint cognitive systems: patterns in cognitive systems engineering. CRC Press

Mechanisms and Correlates

Chapter 5
Perspectives on the Mechanistic Underpinnings of Choice Biases

Konstantinos Tsetsos and Yinan Cao

Abstract Early foundational work in the decision sciences carefully balanced empirical observations and theoretical explanations. Dating back to Daniel Bernoulli, a handful of behavioral regularities observed in theoretical lotteries ignited the refinement of normative theories and the development of new descriptive frameworks of valuation and choice. However, more recent tendencies in behavioral economics and psychology place empirical observations on a pedestal: modern behavioral science has identified more behavioral biases than it has explained. Coupled with replication and reliability crises in experimental psychology, this has resulted in an explanatory gap in the field, in-between the descriptive and predictive levels. Here, we aim to close this explanatory gap by asking how choice biases can emerge from certain decision computations. We demonstrate that biased and irrational choice behavior may arise from multiple, equally viable mechanisms, such as relative value coding and selective information sampling. We posit that this "multiple realizability" problem highlights a broader issue: inferring cognitive mechanisms of complex behavior solely from behavioral measures is often an underdetermined exercise. We propose that using time-resolved neural recordings to track how attention serially parses complex information during multi-attribute, multi-alternative decisions can resolve this "multiple realizability" issue and arbitrate between competing mechanistic explanations of choice biases.

Keywords Choice biases · Irrational decisions · Risky choice · Multiattribute choice · Preference reversal · Information sampling

---

K. Tsetsos (✉)
School of Psychological Science, University of Bristol, Bristol, UK
e-mail: k.tsetsos@bristol.ac.uk

K. Tsetsos · Y. Cao
Department of Neurophysiology and Pathophysiology, University Medical Center Hamburg-Eppendorf, Hamburg, Germany

Y. Cao
Laboratoire de Neurosciences Cognitives et Computationnelles, École Normale Supérieure, Paris, France

U. Ettinger et al. (eds.), *Decision Making*, Studies in Neuroscience, Psychology and Behavioral Economics, https://doi.org/10.1007/978-3-032-00880-0_5

5.1 Introduction

A large body of research within the cognitive and decision sciences has established that human decisions are often influenced by factors that should rationally be ignored (De Martino et al. 2006; Kahneman and Tversky 1984; Summerfield and Tsetsos 2015; Tversky and Kahneman 1981). For example, we tend to stick with an energy plan because it is set as the default option (Baron and Ritov 1994); we prefer volatile stocks when buying but dismiss them when selling (Shafir et al. 1993; Tsetsos et al. 2012a); or we choose salmon fillet over rib-eye steak simply because we noticed that fish fingers are also available (Huber et al. 1982). These examples illustrate that choices are not solely determined by the properties of the available alternatives and the goals of the decision-maker but also by an array of normatively irrelevant factors including: the way the alternatives are presented, the framing of the choice, or the presence of dominated (inferior) alternatives in the choice set (Usher et al. 2019).

The observed sensitivity of human decisions to normatively irrelevant factors (reflected in *choice biases*) has had profound impact on the behavioral sciences, resulting in the distinction between normative (how *should* we decide) and descriptive (how *do* we decide) theories of choice. One of the main goals of descriptive theories of choice is to specify how choice biases come about in the deciding brain. Achieving this goal can subsequently inform deeper multidisciplinary considerations, e.g., on why bias propensity varies among individuals (Aczel et al. 2015; Spektor et al. 2021) or across the lifespan (Parrish et al. 2024; Tentori et al. 2001); or on why biases have persisted despite evolutionary pressure for reward-maximizing choices (Moran and Tsetsos 2018; Tsetsos et al. 2016; Webb et al. 2021). Furthermore, choice biases have taken center stage in applied behavioral science, representing predictable blind spots that can be harnessed in interventions to induce behavioral change (Thaler and Sunstein 2009). Precisely understanding the mechanisms that mediate choice biases can aid the development of targeted approaches that could bolster the limited effectiveness of extant "nudging" interventions (Maier et al. 2022).

Despite the theoretical and applied importance of understanding the computational and neural mechanisms that lead to choice biases, existing theories of choice have not provided definitive insights. This is reflected in the multitude and disparity of frameworks proposed to explain choice biases, ranging from verbally formulated heuristics (Gigerenzer and Gaissmaier 2011; Kahneman and Tversky 1984; Shafir et al. 1993) and algebraic modifications of normative theories (Tversky and Kahneman 1992; Tversky and Simonson 1993) to Bayesian (Bhui and Xiang 2021; Srivastava and Schrater 2012) and dynamical models (Busemeyer et al. 2019). To date, these disparate explanations of choice biases, often cast at different levels of analyses (McClelland 2009), have not been comprehensively related to underlying decision mechanisms. In this chapter, we aim to close this gap by describing how explanations of hallmark choice biases can be situated along the processing stages that occur during decision-making.

We begin by clarifying the notion of choice bias and proceed to show that hallmark choice biases could arise from computations occurring at almost any stage—including relative and non-linear value coding (Louie and De Martino 2014), selective information sampling (Usher et al. 2019), and non-linear accumulation dynamics (Cavanagh et al. 2020). Given this "multiple realizability" issue, we then address how distorting mechanisms along the processing pathway can be better identified. We highlight that during complex decisions, information is sampled partially and serially before reaching the decision formation level; and propose that tracing this information flow with non-invasive high-temporal-resolution neural recordings as decisions unfold can considerably constrain mechanistic inferences. We conclude that a central goal in the cognitive and neural sciences should be to understand the principles that orchestrate information sampling during decision-making.

5.2 Choice Biases: Innocuous and Irrational

From a normative standpoint, an optimal agent should always [or more likely, in the presence of behavioral stochasticity (Loomes and Sugden 1995)] choose the most desirable course of action in any given situation (Summerfield and Tsetsos 2015). Thus, a prerequisite of optimal behavior is the ability to value choice alternatives by transforming their objective properties into "desirability" (or utility) scores based on a set of criteria that represent the goals and needs of the agent at any given moment (Juechems and Summerfield 2019). A longstanding question in the decision sciences is whether human choices comply with this notion of optimality. This has been hard to assess since, in real-life decisions, decision-relevant criteria are inherently subjective and opaque. To circumvent this issue, decision theorists have resorted to two distinct approaches: (i) studying choice behavior in simplified scenarios where decision-relevant criteria can be objectively defined; (ii) specifying a set of rules (or axioms) that utility-maximizing agents must abide by. In the following, we define choice biases in relation to these two approaches.

5.2.1 *Innocuous Biases*

To study human valuation and choice in a tractable way, researchers often rely on laboratory tasks with externally imposed objectives (Summerfield and Tsetsos 2012). This approach has roots in the foundational work on probability theory, where theoretical lotteries were used to benchmark human behavior against statistical optimality (Stearns 2000). In these cases, it can be assumed that the desirability of each lottery option is equivalent to the expected value of the corresponding payout distribution. Thus, statistically optimal agents should always choose the lottery with the largest expected value. However, even in these tractable situations, human choices depart from the statistical ideal. Consider a choice between two prospects A and B. Prospect

A offers £40 for sure while prospect B offers £100 with 50% probability or £0 otherwise. Although prospect B has a higher expected value ($EV_A = £40$ versus $EV_B = £50$, most people in this scenario will choose prospect A (Kahneman and Tversky 1979) (Fig. 5.1a). This example illustrates that human valuation is sensitive to the variances of the payout distributions, which is not relevant for maximizing expected rewards. More broadly, human economic choices disclose several idiosyncrasies, such as risk (Mata et al. 2018) and skewness preferences (Olschewski et al. 2024) or aversion to losses (Novemsky and Kahneman 2005).

Idiosyncratic biases are not specific to theoretical lotteries but can be encountered in any choice task entailing a transparently defined payoff structure. In perceptual choice tasks, participants observe sensory evidence and are asked to make choices based on a criterion defined by the experimenter (e.g., choose the rectangle with

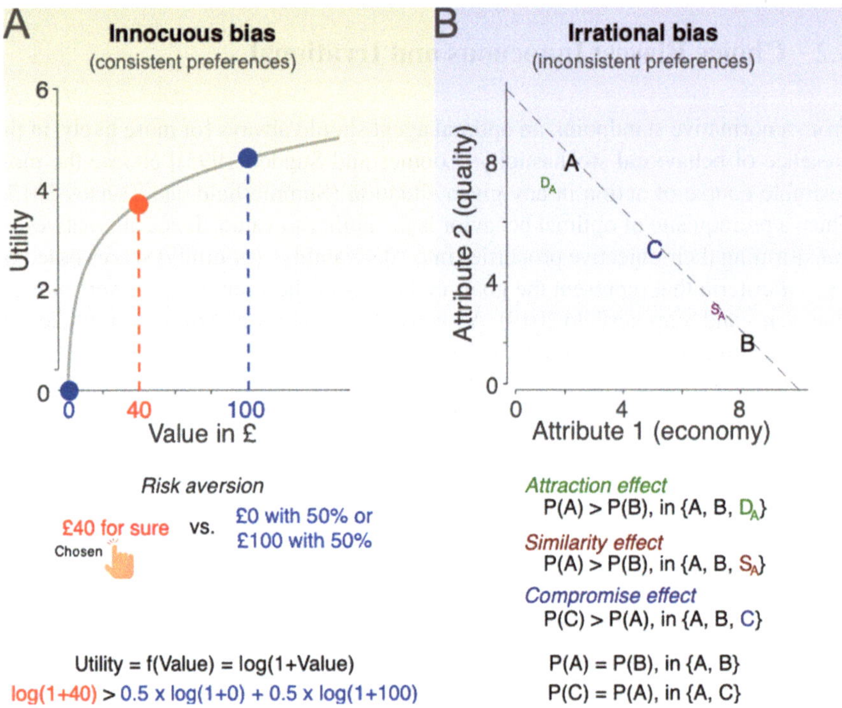

Fig. 5.1 Illustrative examples of innocuous and irrational choice biases. **a** A stable concave (diminishing returns) utility function predicts a risk-aversion bias in a choice between two hypothetical gambles with explicitly described rewards and probabilities. **b** Contextual preference reversal in multiattribute choice. Alternatives vary across two attributes. For illustration purposes, we assume that any two alternatives positioned on the negative diagonal are equally preferred in the respective binary choices. The attraction effect is a choice bias for a target alternative A over a competitor B occurring in ternary choices after an inferior decoy (D_A) is introduced near the target. The similarity effect is a choice bias for the target A after a non-dominated decoy (S_A) is introduced near the competitor B. The compromise effect is a choice bias for the all-average alternative C over A or B, occurring in ternary trials featuring another extreme alternative B

the larger area). In these tasks, humans and other animals exhibit several suboptimal tendencies that prevent them from attaining optimal performance. For instance, across consecutive decisions, they exhibit choice history biases (e.g., repeating or avoiding a previous choice) (Braun et al. 2018; Urai et al. 2019); or, within a decision, they assign larger weight to information presented early in the trial (Tsetsos et al. 2012b). Similarly, in value-learning (bandit) tasks, the objective of maximizing monetary reward is undermined by a list of suboptimal tendencies such as ignoring infrequent rewards or penalties (Hertwig et al. 2004), or overestimating the importance of extreme outcomes (Ludvig et al. 2014).

The presence of this type of choice biases in tasks with clearly defined objectives underlines that "desirability" is an inescapably subjective notion. That is, even when the task dictates maximizing economic value, people will not limit themselves to just that. Instead, in addition to accruing economic value, people seem to be gaining extra utility by avoiding risky prospects and losses, by repeating their previous choices, or by avoiding committing errors in perceptual choices at the expense of dwelling for too long on a given decision (Bogacz et al. 2010). Although these tendencies lead to biased choices and curtail reward accrual, they may satisfy other latent, non-economic metrics. Thus, biased choice behavior can be "rationalized" as maximizing a stable, albeit idiosyncratic, utility function. Because the kind of choice biases we reviewed here do not rule out utility-maximizing behavior under a more liberal definition of utility, we deem them *innocuous.*[1]

5.2.2 *Irrational Biases*

As described above, prominent choice biases can be absorbed into a putative utility function that the agent's choices could be maximizing. Is the premise of utility-maximization even falsifiable? In their influential work on rational choice theory, von Neumann and Morgenstern demonstrated that the premise of utility-maximization is falsifiable: it can hold only if preferences (as revealed by overt choices) satisfy certain rationality axioms (Savage 1972; Von Neumann and Morgenstern 2007). A stable utility function can be defined only when preferences satisfy these axioms. Accordingly, if preferences violate any of these axioms, then the utility-maximizing narrative falls apart. Below we adopt a bird's-eye-view and describe the core intuition underlying the axioms of rational choice theory. More detailed and formal expositions of these axioms can be found elsewhere (Regenwetter and Davis-Stober 2012; Rieskamp et al. 2006).

Rational choice theory axioms are not concerned with the specific preferences of agents but with the internal consistency (or rationality) of those preferences

[1] The point that choice biases reflect idiosyncratic preferences for non-normative aspects is rather theoretical, aiming to emphasise that the presence of certain biases does not falsify an expanded notion of optimal (utility-maximizing) behavior. Practically, certain choice biases likely stem from processing bottlenecks in biological brains rather than from explicit preferences for non-normative aspects.

(Allingham 2002). Thus, they do not prescribe how choice alternatives should be mapped onto utilities; instead, they merely ensure that this mapping does not change due to irrelevant factors. One such irrelevant factor is choice framing (Kahneman and Tversky 1984). Do preferences change when a choice is framed as "select the best" versus when it is framed as "reject the worst"? According to rational choice theory, in both frames, one should consider their needs and wants and assign utility scores to alternatives. In the "selection" frame the alternative with the highest score should be chosen, and in the logically equivalent "rejection" frame, the alternative with the lowest utility score should be eliminated. Thus, when two alternatives are available, selecting *A* coheres with rejecting *B*, and vice versa. However, it has been shown that when alternative *A* is mediocre (e.g., a not-so-expensive and dull holiday destination) and *B* is more extreme (e.g., an expensive but exciting destination) people tend to *both* select and reject the extreme alternative (Shafir 1993; Tsetsos et al. 2012a). This behavioral pattern discloses inconsistent preferences and cannot be reconciled under the maximization of a stable utility function that rational choice theory anticipates.

A second factor that provokes inconsistent choice patterns is the composition of the choice set. Following the rational choice framework, the utility assigned to an alternative should solely be a function of its inherent properties and the goals and needs of the decision-maker. The utility of an alternative should thus be independent of the properties of other alternatives that are available for choice (i.e., the *independence-from-irrelevant-alternatives* axiom). A logical consequence of this schema is that if *A* is preferred over *B* when only these two alternatives are offered, then *A* should still be preferred over *B* when a third alternative *C* is available for choice. However, it has been shown time and again—even in non-primate species including amoebae and bees (Latty and Beekman 2011; Tan et al. 2015)—that preferences change as a function of the choice-set composition (Evangelidis et al. 2024). To illustrate, a preference for an all-inclusive holiday to Berlin (*A*) over an all-inclusive holiday to Rome (*B*) can reverse when a holiday to Rome where you must pay for breakfast (*C*) is introduced in the choice set. In this example, the presence of the inferior alternative (*C*) boosts the desirability of its more similar alternative (*B*). This so-called *attraction effect* (Huber et al. 1982) (Fig. 5.1b) and similar phenomena where preferences change as alternatives are added or removed from the choice set, are collectively referred to as *contextual preference reversals* (Tsetsos et al. 2010). Unlike innocuous biases, the framing and choice-set preference reversals outlined here are puzzling as they cannot be "rationalized" under a utility-maximizing narrative. Therefore, we deem these choice biases *irrational*. An open question, which we will explore in the next sections, is whether innocuous and irrational biases have distinct or common mechanistic underpinnings.

5.2.3 Limitations in Existing Accounts of Choice Biases

Within the judgment and decision-making literature, mainstay accounts of biased behavior consist of ad-hoc formulations that effectively re-describe human behavior without providing deeper explanations. For example, the "take-the-best" heuristic posits that decisions are settled exclusively based on the most important cue or attribute (Gigerenzer and Goldstein 1996). Accordingly, it is assumed that people actively *use* this heuristic rule when making multi-attribute decisions. In this case, the explanation (take-the-best heuristic) and the explanandum (choice bias in favor of the alternative that is better in the most important attribute) are almost indiscernible. This issue is not exclusive to the heuristic and biases framework: in experience-based decisions (Hertwig and Erev 2009), certain behavioral regularities (e.g., recency or ignoring rare events) are ascribed to homologue processing biases (i.e., underweighting early and rare events); or in algebraic models like in Prospect Theory (Kahneman and Tversky 1979), the shape of psychoeconomic functions and the position of the reference point fit the patterns of human behavior but have no a priori motivation. Thus, extant influential accounts of choice biases stay too close to the observed behavioral effects, offering little explanatory depth.

Here, we do not claim that ad-hoc accounts should be dispensed with as they provide a useful and abstract way to summarize how human behavior deviates from normative expectations. It is even conceivable that certain ad-hoc formulations, like explicit loss-aversion, reflect hardwired asymmetries in the way the brain processes information (Tom et al. 2007). Instead, we posit that this level of theorizing cannot readily provide an encompassing answer to the question: how do choice biases occur in the deciding brain? In the next section, we sketch an alternative and deeper level of theorizing about choice biases.

5.3 Mechanistic Explanations of Choice Biases

Here we explore the idea that choice biases *emerge* from dynamic computations during the decision process rather than stemming from a rigid set of ad-hoc rules. This approach can have multiple advantages. First, it can afford precise quantitative predictions, which can be valuable in predicting novel biases or in anticipating how people would respond in different contexts. Second, it can offer a natural interface between behavioral and neural levels (Gold and Shadlen 2007), enabling the understanding of altered decision-making during ageing or in neuropsychiatric disorders. Third, with certain decision computations serving adaptive functions in biological brains (Summerfield and Tsetsos 2020), linking choice biases to these computations can help reconcile the normative-descriptive gap. Finally, a mechanistic framework can, in principle, offer unifying and parsimonious explanations. Multiple choice biases can arise from variations in a single or a minimal set of mechanisms, thereby reducing the dimensionality of the tangled ontology of behavioral biases (Hallsworth 2023).

However, to a large extent, mechanistic inferences are under-constrained by empirical data (Pirrone and Tsetsos 2023), especially in relatively complex behavioral domains (e.g., multi-attribute choices where choice biases abound). As a result, mechanistic models can become over parametrized and arbitrary (Anderson 2013), often ending up as ad-hoc as heuristic formulations. Indeed, several influential multi-attribute models can be criticized for being overly flexible, combining algebraic ad-hoc and dynamic (and biologically grounded) mechanisms to explain preference reversals (Roe et al. 2001; Trueblood et al. 2014; Usher and McClelland 2004). Even though we acknowledge the merit and influence of these more complex models (for a comprehensive review, see Busemeyer et al. 2019), we here consider choice biases within the minimal evidence-accumulation framework developed for simple perceptual decisions. We then expand this framework with a small set of additional mechanisms intended to help agents navigate the rich information involved in multiattribute, multialternative decisions.

5.3.1 Information Processing During Decision-Making

In a very generic description of the decision process, choices arise from first assigning utilities to available alternatives (valuation) and then selecting the alternative that has the highest utility (comparison) (Platt and Plassmann 2014; Vlaev et al. 2011). In most choice theories in psychology and economics, valuation and comparison are formulated in a stylized fashion using algebraic operations and functions [such as weighted summation, (Keeney and Raiffa 1993); or the *softmax* function]. In contrast to these static and stylized formulations, representations and computations in biological brains are shaped by noisy and dynamical processes (Miller et al. 2024). What are the dynamical processes underpinning decision-making?

Akin to biologists using Drosophila as their model organism for studying more complex organisms, psychologists and neuroscientists have used simple sensorimotor decisions to approximate generalizable decision processes (Shadlen and Kiani 2013). Behavioral and neural data (Platt and Glimcher 1999; Ratcliff and McKoon 2008; Yang and Shadlen 2007) indicate that during simple decisions—such as categorizing an ambiguous image as a face or a house or determining the dominant direction of motion in a random dot kinematogram (Heekeren et al. 2008)—noisy information is sampled and accumulated over time into a growing confidence signal, until a criterial degree of confidence (or *boundary*) is reached. Following the law of large numbers, accumulation over time alleviates the detrimental influence of noise and improves decision accuracy. The boundary on confidence controls how long accumulation lasts for, effectively determining how the observer trades off the speed and accuracy of the decision (Bogacz et al. 2010).

Within this *accumulation-to-bound* framework, valuation and comparison are dynamic and temporally multiplexed processes. Valuation arises from *representing*,[2]

[2] In any decision-task, external information needs to be mapped onto decision-relevant information. For example, if the task is to determine the rectangle that has the largest width, then information about the height or color of the rectangles is not relevant and needs to be discarded. Neural representations are subject to non-linear transduction, typically captured by a concave (e.g., logarithmic) transformation of the objective information (see Weber-Fechner law).

sampling[3] and *accumulating* noisy evidence (held in different *accumulators*) in favor of the available choice alternatives (Fig. 5.2a). Choices are made once the activity of one of the accumulators exceeds the decision boundary, thereby implementing the comparison operation (Lo and Wang 2006). The specific implementation of the comparison process can vary across models, particularly in the form and degree of competition among accumulators (Teodorescu and Usher 2013). Having outlined the fundamental mechanisms implicated in simple decisions, we next ask how these mechanisms can lead to choice biases.

5.3.2 *Choice Biases Within the Standard Mechanistic Framework*

Can the standard mechanistic framework outlined above produce innocuous and irrational choice biases? At first glance, this seems like a tall order for accumulation-to-bound models because they essentially stem from the framework of optimal sequential hypothesis testing (Bogacz et al. 2006; Wald 2004; Wald and Wolfowitz 1948). However, as described above, accumulation-to-bound models can vary in their implementational details. Due to such variations, some accumulation-to-bound models can deviate from statistically optimal computations.

One characteristic example of suboptimal computations is the non-uniform temporal weighting of evidence emerging from the accumulation dynamics in competing accumulator models. In the leaky competing accumulator model (Usher and McClelland 2001), the activity of a given accumulator increases with the corresponding incoming evidence and decreases due to self-dissipating activity (or leakage) and lateral inhibition coming from competing accumulators (Fig. 5.2b). When the leakage and inhibition are balanced, the model mimics the diffusion model showing equal sensitivity to early and late evidence (Bogacz et al. 2007). When lateral inhibition is stronger than the leakage, the model operates in an impulsive regime, exhibiting larger sensitivity to early evidence (*primacy*) through strong winner-take-all dynamics. Conversely, when lateral inhibition is weaker than the leakage, the model becomes forgetful or "leaky", being more sensitive to late evidence (*recency*) (Tsetsos et al. 2012b). Similar temporal weighting profiles fall out from variations in the excitation/inhibition ratio in a biophysical cortical circuit model of evidence accumulation (Lam et al. 2022).

Non-uniform temporal weighting can lead to innocuous choice biases when decision-relevant information is processed in a fixed order. Consider the choice between an affordable but dull holiday destination (*A*) and an expensive and

[3] Decision tasks can entail either dynamic or static information. Dynamic information is naturally chunked into monetary samples, which are subject to serial accumulation. In static tasks, such as a face/house discrimination based on an ambiguous photograph, the assumption is that internal sampling turns takes noisy snapshots of the external information, which are then serially accumulated over time.

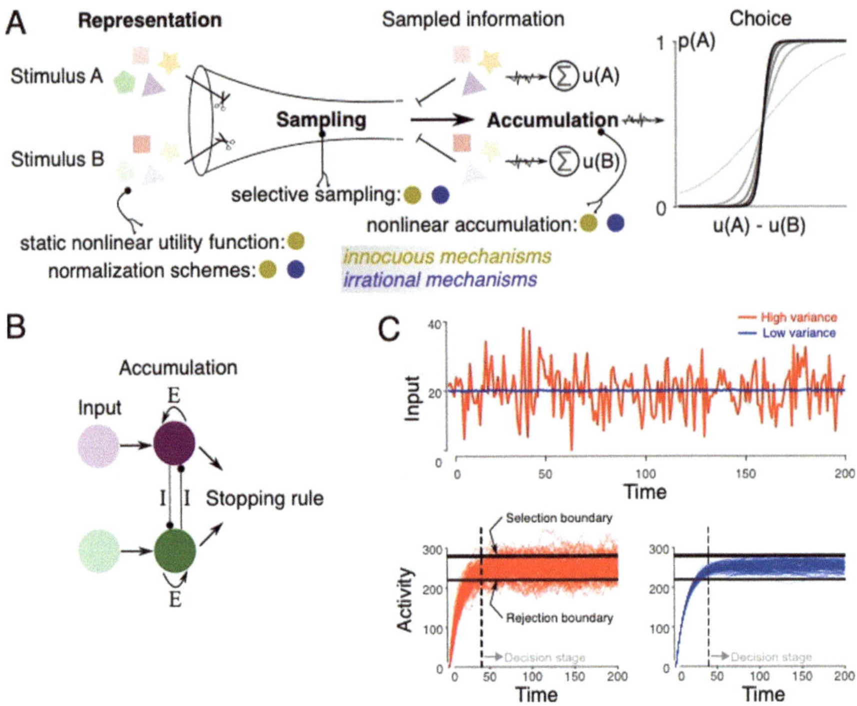

Fig. 5.2 Choice biases within a generalized mechanistic framework. **a** Schematic of the processing hierarchy in decision-making: from stimulus representation to sampling, evidence accumulation, and ultimately, choice. **b** The leaky competing accumulator (LCA) model architecture. "I" denotes lateral-inhibitory connections and "E" self-excitatory connections. With self-excitation being <1, information is subject to dissipation (leakage). The balance between leakage and lateral inhibition strength controls the profile of temporal weighting (see main text). **c** Risk preferences within a bounded accumulation framework. In the example scenario a high-variance (red) and a low-variance (blue) alternative compete for choice. Following an initial pre-decisional period (left relative to the dashed vertical line), evidence is accumulated in two accumulators that are weakly coupled with inhibition. Under the selection (rejection) framing, once an accumulator breaches the upper (lower) boundary a selection choice is made in its favor. The high-variance accumulator has large deflections and thus exceeds both boundaries more often than the low-variance accumulator. This predicts risk-seeking under selection and risk-aversion under rejection

exciting destination (*B*). If the price information is conveyed first, agents with a primacy weighting profile will tend to choose *A*. Interestingly, non-uniform temporal weighting can lead to irrational choice biases if the framing of the task or the choice-set composition alters the order in which information is processed. For example, an agent with a primacy profile will disclose a preference reversal if they first process positive information (*B* is exciting) in the "select the best" framing and negative information (*B* is expensive) in the "reject the worst" framing. Thus, deviations from statistically optimal computations can open the door to both innocuous and irrational choice biases, depending on certain assumptions about the order in which information is considered. Irrespective of assumptions about the order of information processing, how can the standard mechanistic framework produce well-established choice biases?

Next, we consider the innocuous preferences that humans have towards less or more variable alternatives. Empirical findings suggest that people are risk-averse in description-based lotteries but risk (variance)-seeking when value information is experienced sequentially (Tsetsos et al. 2012a). For presentation purposes, we describe both choice biases using a hypothetical choice between two holiday destinations. A preference for the mediocre holiday destination *A* over the more extreme one *B* (risk-aversion) can be explained by a concave transduction function that maps objective values onto internal subjective counterparts (see also footnote 2 and Fig. 5.1a). This is the classical explanation of risk-aversion adopted in expected-utility theory and prospect theory (Kahneman and Tversky 1979). Beyond this rather rigid representational distortion, in Fig. 5.2c we show how the opposite risk-seeking bias (Tsetsos et al. 2012a) naturally emerges within a minimal accumulation-to-bound framework involving two independent (or weakly competing) accumulators racing towards a boundary. Due to its more variable input, the accumulator corresponding to the more extreme alternative *B* shows larger deflections. These deflections translate into an increased likelihood of crossing the decision boundary. Similarly, under a rejection frame—and assuming that elimination happens when an accumulator breaches a lower boundary—*B* will be more frequently eliminated resulting in an irrational (framing) reversal of risk preferences (Shafir 1993).

A preference for the more variable alternative (*B*) can also arise within a biophysical cortical model of evidence accumulation due to convex evidence transduction and non-linear accumulation dynamics (Cavanagh et al. 2020). Although it is not obvious how the biophysical model could produce a framing reversal of this bias, this model can produce a choice-set reversal. In choices among three alternatives that vary in their decision values, increasing the value of the worst alternative improves the relative discrimination accuracy between the two highest-value alternatives (Chau et al. 2014) [but see Cao and Tsetsos (2022) for an alternative interpretation of this effect]. This *positive distractor* effect violates the independence-from-irrelevant-alternatives principle. The biophysical model captures the positive distractor effect because increasing the value of the worst alternative raises the level of the pooled inhibition. This, in turn, adjusts the accumulation dynamics to a regime closer to optimal, approximating the diffusion model that achieves the highest discrimination accuracy (Bogacz et al. 2007).

5.3.3 Choice Biases Within an Extended Mechanistic Framework

We showed that some innocuous and irrational choice biases can emerge within the standard accumulation-to-bound framework. However, this framework cannot readily explain multi-attribute choice-set reversals (Fig. 5.1b). Notably, the standard mechanistic framework was built around simple choice tasks, where the amount of decision-relevant information typically falls within the processing capacity of the cognitive system (Donner et al. 2009; Gold and Shadlen 2007). However, choice biases, especially irrational ones, occur in more complex decision domains involving multiple (more than two) alternatives that often vary in more than one attribute (Busemeyer et al. 2019). In these cases, parsing information in parallel becomes impossible and the cognitive system needs to find ways to efficiently navigate the increased complexity given its processing bottlenecks. Below, we discuss two classes of mechanisms that can help the brain efficiently process large amounts of information. As a byproduct, these mechanisms enable dynamic and context-dependent valuation, and choice biases ensue.

Relative Coding

The first class of mechanisms, collectively referred to as *relative coding* (Summerfield and Tsetsos 2020), entails dynamic and non-linear distortions impacting the representation of decision-relevant information. More specifically, in contrast to static and context-independent transduction non-linearities (see footnote 2), relative coding schemes adjust representations as a function of the temporal (i.e., the recent history of stimulation) or the immediate (i.e., the choice-set) context. The exact form of these adjustments is motivated by the theory of efficient neural coding widely evidenced in sensory systems (Simoncelli 2003). This theory states that neurons minimize redundancy by increasing their representational resolution for the most frequently occurring stimuli (Barlow 1961). Below we provide an overview of prominent relative coding schemes with regards to choice biases.

In the divisive normalization model, the "raw" utility of each alternative is divided by the sum of the raw utilities of all alternatives in the choice set (Louie et al. 2013). Therefore, the model predicts that increasing the utility of the worst alternative reduces the discrimination accuracy between the two high-utility alternatives, a *negative distractor* effect at odds with the independence-from-irrelevant-alternatives axiom (but see Gluth et al. (2020) for a failure to replicate this effect). In the range normalization model (Rustichini et al. 2017), raw utilities are divided by the range of raw utilities (max–min) encountered in the choice-set, that way producing a positive distractor effect (Fig. 5.3). The range normalization principle together with auxiliary assumptions can also explain contextual preference reversals in multi-attribute choice, including the attraction, the similarity, and the compromise effects (Soltani et al. 2012) (Fig. 5.1b). These multi-attribute effects can also be captured by a recurrent version of the divisive normalization model, where the attribute value of an alternative is divided by itself plus the mean choice set values on that attribute (Dumbalska et al. 2020). Practically, relative coding schemes implement context-dependent utility functions thereby generating irrational choice biases. One downside of relative coding schemes is that they afford compressive representations. Thus, these models cannot readily capture the risk-seeking bias obtained in decisions from experience (Tsetsos et al. 2012a).

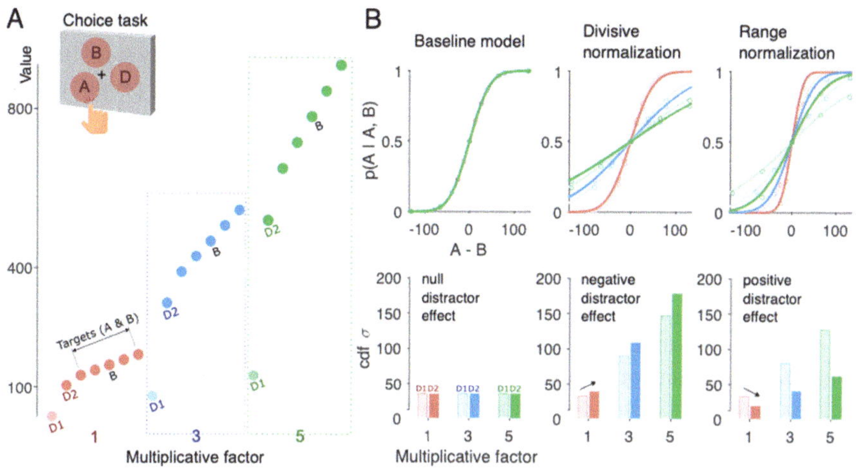

Fig. 5.3 Distractor effects under relative coding schemes. **a** In a ternary choice task, observers choose one alternative among three candidates (*A*, *B*, and a distractor *D*) on each trial. The distractor's value is always lower than the values of targets *A* and *B*, so it should not influence the choice between *A* and *B*, as predicted by the baseline model in panel B. In a thought experiment, observers perform this task in two additional contexts where the option values are multiplicatively scaled up by 3 or 5. This manipulation creates an ideal situation for testing prominent normalization theories. The distractor effect is defined as the change in sensitivity (slope measure of a Gaussian cumulative-density-function fit to choice probabilities) to the value difference between *A* and *B* (fixed), modulated by changes in the distractor value. **b** Divisive normalization assumes the option value is transformed into the mean firing rate $\mu_i = KV_i/(\sigma_h + \Sigma wV_j)$, where V_i is the raw value of the option under consideration, $K > 0$, $\sigma_h > 0$, and $w > 0$ represent gain, semi-saturation, and weight, respectively (Louie et al. 2013). When $w = 0$, the model reduces to the baseline version, meaning the value coding is independent of other options in the choice-set. Range normalization assumes $\mu_i = KV_i/(\sigma_h + w(\max(V_j) - \min(V_j)))$, meaning the denominator involves the range of the values rather than the summation. Both normalization models predict an increase in the distractor effect as the multiplicative factor increases, but they predict the distractor effect in opposite directions. These very specific predictions can be contrasted with the predictions of other non-normalization models of distractor effects (e.g., decision-by-sampling). Code for reproducing this figure can be found at: https://github.com/YinanCao/bookchapter

Selective Sampling

The second class of mechanisms we review operate downstream from the early distortions that relative coding schemes induce. These mechanisms govern selective information sampling by determining, at each moment, which aspects of the available information should be passed on for accumulation. Selective information sampling is a pragmatic solution to the challenge posed by the rich information involved in multi-alternative and multi-attribute decisions. Indeed, it is commonly observed that during complex decisions, attention serially traverses across attributes and alternatives (Fiedler and Glöckner 2012; Russo and Dosher 1983), driving the online construction of preferences (Slovic 1995). Recent work has incorporated attentional fluctuations into the accumulation-to-bound framework by positing that the gain of processing increases for attended items (Krajbich et al. 2010). However, the principles that orchestrate these attentional fluctuations, and the reasons why sampling can end up being partial and biased, remain unknown. Below we review proposals that infer principles of information sampling through explaining contextual preference reversals.

Various models of multi-attribute choice assume that only one attribute can be processed at a time, with attention stochastically fluctuating across attributes over time (Roe et al. 2001; Turner et al. 2018; Tversky 1972). More recent models additionally assume that within each attended attribute, only a subset of the available alternatives is considered at a given instance (Wollschläger and Diederich 2012). In the decision-by-sampling multi-attribute model (Noguchi and Stewart 2018), two alternatives are compared within each sampled attribute, with more similar alternatives forming comparison pairs more frequently. The binary outcome of the comparison updates the "counting" accumulator of the winner. Due to these principles and some auxiliary assumptions, the model can explain the attraction, similarity, and compromise effects (Fig. 5.1b). A recent adaptation of the decision-by-sampling framework can also explain a positive distractor effect in single-attribute decisions (Tohidi-Moghaddam and Tsetsos 2024). Another sampling model, the selective integration model, assumes that within a focused attribute, attentional selection prioritizes the processing (i.e., assigns a larger gain) of high-valued alternatives at the expense of low-valued alternatives, a principle that leads to the attraction and other contextual preference reversal effects (Tsetsos 2012; Tsetsos et al. 2012a) including violations of transitivity (Tsetsos et al. 2016). The selective integration model also predicts risk-seeking that reverses under a rejection frame where observers prioritize the processing of lower values (Usher et al. 2019). Taken together, in the models described above, selective sampling leads to innocuous and irrational choice biases by prioritizing the processing of certain choice aspects at the expense of others.

5.4 Identifying the Mechanisms Underpinning Choice Biases

The previous section highlights that, even within a limited search space of mechanisms, there are multiple and equally viable[4] explanations for choice biases situated at all stages of the processing pathway (Table 5.1). This is a stark reminder that the cognitive and neural mechanisms of more complex behaviors are grossly underdetermined by empirical data (Pirrone and Tsetsos 2023). How can the mechanisms underlying the various choice biases be securely identified?

[4] Here, arbitrating competing mechanisms based on auxiliary criteria, like biological plausibility or normative justification, seems fruitless. For instance, given efficient codes in the brain, two competing schemes, divisive and range normalization, seem equally biologically plausible. Similarly, while relative coding schemes can maximize information transfer while keeping metabolic

Table 5.1 Summary of the standalone predictions of various mechanisms in relation to innocuous and irrational choice biases

Mechanism		Choice biases					
		Risk-seeking (under selection)	Risk-aversion (under selection)	Framing risk-reversal	Negative distractor	Positive distractor	Multi-attribute preference reversals
Stable non-linear function (R)							
	Convex	Yes	No	No	No	No	No
	Concave	No	Yes	No	No	No	No
Relative value coding (R)							
	Divisive normalization	No	Yes	?	Yes	?	No
	Range normalization	No	Yes	?	No	Yes	Yes
	Recurrent divisive normalization	No	Yes	?	?	?	Yes
Selective sampling (S)							
	Selective integration	Yes	No	Yes	?	?	Yes
	Decision-by-sampling	?	?	?	No	Yes	Yes
Non-linear accumulation (A)							
	Bounded accumulator	Yes	No	Yes	No	No	No
	Biophysical cortical circuit	Yes	No	?	No	Yes	No

"R" stands for representation, "S" for sampling, and "A" for accumulation. "?" is inserted when certain effects have not been adequately explored in relation to certain choice biases

Identifying the mechanisms that underlie a cognitive process requires the ability to experimentally manipulate or observe the process input while simultaneously observing the output at the behavioral and neural levels. As we alluded to in previous sections, this approach is exemplified in the study of sensorimotor decisions, where the use of well-controlled psychophysical stimuli have uncovered the neural underpinnings of perceptual decision-making (Gold and Shadlen 2007). However, in multi-attribute choice problems—which entail information across different attributes and require comparisons between multiple alternatives—the input to the decision process is not under experimental control. Instead, due to the rich information exceeding the parallel processing capacities of the cognitive system, attention can voluntarily fluctuate over time in numerous different ways (i.e., examining one alternative on all attributes at a given time, dividing attention between two alternatives, focusing on all alternatives in one attribute etc.). Observing these fluctuations is necessary to uncover the flow of the input that drives the decision-making process. In turn, knowing the dynamical input of the decision process can radically constrain mechanistic inferences and answer how choice biases come about. Simply put, understanding *how* the brain processes information requires knowing *what* information it processes.

Mainstay approaches have attempted to empirically characterize information sampling using eye-tracking techniques (Ettinger and Klein 2016; Russo and Dosher 1983). However, interpreting eye movements and fixations is not straightforward in the context of decision-making tasks. Eye movements can also correspond to operations unrelated to the decision-making process (i.e., merely reading the information), which cannot be dissociated from decision-related oculomotor activity (i.e., accumulating the fixated information). Instead, it is often asserted that only fixations in the middle part of the deliberation are decision-related, while early fixations correspond to the "scanning" stage and late fixations to the "validation" stage (Bettman and Kakkar 1977). Importantly, even if decision-related fixations could be dissociated from unrelated operations, knowing where people fixate does not say much about what they think. Does looking at the price of holiday destination A guarantees that the price of holiday B is not covertly considered at the same time in a comparative manner?

Thus, eye-tracking techniques are limited to overt attention, while the computational role of the tracked measures is unclear. What has been missing is a way to track both the locus of attention (overt and covert) and at the same time the state of the accumulators as complex decisions take shape. Recent work from our lab (Siems et al. 2023) and others (Mostert et al. 2018) has offered new possibilities for continuously tracking the locus of covert spatial attention using non-invasive neural recording techniques that have high temporal resolution [magneto/electro-encephalography (M/EEG)]. This relies on a dedicated functional localizer task and an encoding model that is inverted to estimate the locus of attentional allocation on a *single trial* basis from the pattern of M/EEG signals across the scalp in another task, such as multi-alternative decision-making or a classic bandit task in reinforcement

costs bounded, the selective integration can maximize a different metric, namely decision accuracy in the presence of late noise (Tsetsos et al. 2016).

learning (Fig. 5.4a). The key technical advantage of this approach is the clear dissociation of covert processes from oculomotor "contaminations", with the spatial location decoding of attention being robust across multiple locations and unaffected by eye gaze shifts, no matter how minuscule they are (Fig. 5.4b). Combining the tracking of covert attention with well-established M/EEG signals that track the state of decision accumulators (e.g., the beta-band lateralization in parietal and pre-motor cortices) (Donner et al. 2009; O'connell et al. 2012) can reveal regularities and biases in information sampling (Siems et al. 2023), decisively constraining mechanistic inferences about the general decisions processes as well as those that generate innocuous and irrational choice biases.

5.5 Conclusions

Research in behavioral economics and psychology has identified several choice biases that still lack conclusive mechanistic explanations. In this chapter, we explored how choice biases can be mapped onto the neural and computational mechanisms underlying decision-making. Using as our starting point the standard accumulation-to-bound framework developed for simple decisions, we described how choice biases can emerge from decision computations throughout the processing pathway: from representing, to sampling, to accumulating decision-relevant information. While non-linear accumulation dynamics within the standard accumulation-to-bound framework can explain some choice biases, a more complete explanation of irrational preference reversals requires invoking relative coding at the representation level or selective information sampling. Given that relative coding and selective sampling are both descriptively adequate, biologically plausible, and normatively motivated, distinguishing between these two types of mechanisms seems impossible. This resonates with the broader issue that identifying the mechanisms underlying complex decision-making is underdetermined, given that the effective input to the decision process is opaque and intractable with conventional process tracing techniques. We outlined how using non-invasive time-resolved neural recordings can track attentional fluctuations during decision-making, that way measuring the effective decision input feeding to downstream decision computations.

We argue that observing how information is being sampled during complex decisions can unlock the mechanistic understanding of puzzling behavioral regularities. As we described in this chapter, non-linear accumulation dynamics can lead to non-uniform temporal weighting of information (Tsetsos et al. 2012b). Across these lines, more recent findings suggest that evidence accumulation is not simply a feed-forward process, with the state of downstream accumulators biasing the way incoming evidence is weighted (Talluri et al. 2018). Thus, with information being unequally weighted over time, the order in which information is being parsed can be the major determinant of choice. If choice-set or framing manipulations do not systematically alter the order of information sampling, then relative coding computations would appear necessary to explain irrational choice biases. However, if

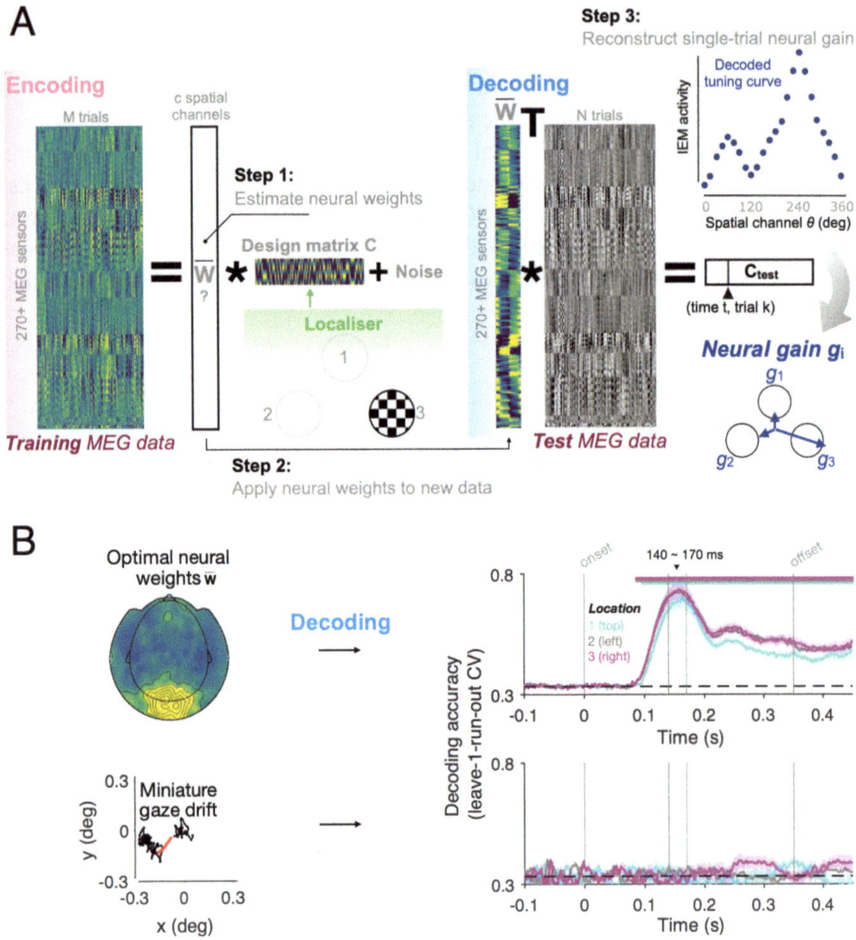

Fig. 5.4 Decoding the locus of spatial attention using M/EEG. **a** In the localizer task, a checkerboard patch appears at one location on each trial while participants fixate centrally and press a button as soon as the fixation cue changes color (a rare event: ~10% of trials). The encoding–decoding model (IEM) involves three main steps. First, optimal neural weights are estimated using a linear model that maps multi-channel M/EEG activities onto hypothetical spatial channels (rectified sinusoids). The M/EEG amplitude measured at each sensor during the localizer task ("Training data") is modelled with spatial channels, each selectively tuned to a different angular position. Next, the encoding model is inverted to estimate the channel responses from the pattern of M/EEG signals across the scalp in another task ("Test data"), such as multi-alternative decision-making or a classic bandit task in reinforcement learning. Finally, neural gains projected spatially across the stimulus space are reconstructed for each single trial. **b** Decoding performance revealed by training IEM on neural data versus concurrent eye-gaze data in the spatial localizer task. Upper panel: Topography shown as PCA loadings that maximally differentiate between the three spatial stimulus locations, indicating where the neural activation patterns across the scalp best diversify across these locations. Cross-validation shows that the IEM trained on neural signals exhibits excellent and highly reliable decoding performance, peaking around 140–170 ms after stimulus onset, and is consistent across individual locations. In stark contrast, because the localizer task prohibits eye movement, IEM trained on eye positions contains no decodable information about stimulus locations (lower panel). This demonstrates that the method reliably captures covert neural processes and neural gain modulation in the visual-parietal regions

framing and choice-set manipulations systematically alter the order of information processing, then puzzling choice biases could simply result from a combination of non-linear accumulation dynamics and specific patterns of information sampling (presumably consistent with extant sampling proposals, such as selective integration and decision-by-sampling). In this latter case, better understanding the principles and mechanisms that orchestrate information sampling during decision-making would become a critical new goal in the cognitive and decision sciences.

Acknowledgements This work was supported by the EU Horizon 2020 Research and Innovation Program (ERC starting grant no. 802905) to K.T.

References

Aczel B, Bago B, Szollosi A, Foldes A, Lukacs B (2015) Measuring individual differences in decision biases: methodological considerations. Front Psychol 6:163396

Allingham M (2002) Choice theory: a very short introduction. OUP Oxford, Oxford

Anderson JR (2013) The adaptive character of thought. Psychology Press, London

Barlow HB (1961) Possible principles underlying the transformation of sensory messages. Sens Commun 1(01):217–233

Baron J, Ritov I (1994) Reference points and omission bias. Organ Behav Hum Dec Process 59(3):475–498. https://doi.org/10.1006/obhd.1994.1070

Bettman JR, Kakkar P (1977) Effects of information presentation format on consumer information acquisition strategies. J Consum Res 3(4):233–240

Bhui R, Xiang Y (2021) A rational account of the repulsion effect

Bogacz R, Brown E, Moehlis J, Holmes P, Cohen JD (2006) The physics of optimal decision making: a formal analysis of models of performance in two-alternative forced-choice tasks. Psychol Rev 113(4):700–765. https://doi.org/10.1037/0033-295X.113.4.700

Bogacz R, Usher M, Zhang J, McClelland JL (2007) Extending a biologically inspired model of choice: multi-alternatives, nonlinearity and value-based multidimensional choice. Philos Trans R Soc B Biol Sci 362(1485):1655–1670

Bogacz R, Hu PT, Holmes PJ, Cohen JD (2010) Do humans produce the speed–accuracy trade-off that maximizes reward rate? Quart J Exp Psychol 63(5):863–891

Braun A, Urai AE, Donner TH (2018) Adaptive history biases result from confidence-weighted accumulation of past choices. J Neurosci 38(10):2418–2429

Busemeyer JR, Gluth S, Rieskamp J, Turner BM (2019) Cognitive and neural bases of multi-attribute, multi-alternative, value-based decisions. Trends Cogn Sci 23(3):251–263. https://doi.org/10.1016/j.tics.2018.12.003

Cao Y, Tsetsos K (2022) Clarifying the role of an unavailable distractor in human multiattribute choice. Elife 11:316. https://doi.org/10.7554/eLife.83316

Cavanagh SE, Lam NH, Murray JD, Hunt LT, Kennerley SW (2020) A circuit mechanism for decision-making biases and NMDA receptor hypofunction. Elife 9:664. https://doi.org/10.7554/eLife.53664

Chau BK, Kolling N, Hunt LT, Walton ME, Rushworth MF (2014) A neural mechanism underlying failure of optimal choice with multiple alternatives. Nat Neurosci 17(3):463–470. https://doi.org/10.1038/nn.3649

De Martino B, Kumaran D, Seymour B, Dolan RJ (2006) Frames, biases, and rational decision-making in the human brain. Science 313(5787):684–687. https://doi.org/10.1126/science.1128356

Donner TH, Siegel M, Fries P, Engel AK (2009) Buildup of choice-predictive activity in human motor cortex during perceptual decision making. Curr Biol 19(18):1581–1585. https://doi.org/10.1016/j.cub.2009.07.066

Dumbalska T, Li V, Tsetsos K, Summerfield C (2020) A map of decoy influence in human multialternative choice. Proc Natl Acad Sci USA 117(40):25169–25178. https://doi.org/10.1073/pnas.2005058117

Ettinger U, Klein C (2016) Eye movements. Neuroeconomics 31:481–502

Evangelidis I, Bhatia S, Levav J, Simonson I (2024) 50 years of context effects: merging the behavioral and quantitative perspectives. J Consum Res 51(1):19–28

Fiedler S, Glöckner A (2012) The dynamics of decision making in risky choice: an eye-tracking analysis. Front Psychol 3:335

Gigerenzer G, Gaissmaier W (2011) Heuristic decision making. Ann Rev Psychol 62:451–482

Gigerenzer G, Goldstein DG (1996) Reasoning the fast and frugal way: models of bounded rationality. Psychol Rev 103(4):650

Gluth S, Kern N, Kortmann M, Vitali CL (2020) Value-based attention but not divisive normalization influences decisions with multiple alternatives. Nat Hum Behav 4(6):634–645

Gold JI, Shadlen MN (2007) The neural basis of decision making. Ann Rev Neurosci 30:535–574. https://doi.org/10.1146/annurev.neuro.29.051605.113038

Hallsworth M (2023) A manifesto for applying behavioural science. Nat Hum Behav 7(3):310–322. https://doi.org/10.1038/s41562-023-01555-3

Heekeren HR, Marrett S, Ungerleider LG (2008) The neural systems that mediate human perceptual decision making. Nat Rev Neurosci 9(6):467–479

Hertwig R, Erev I (2009) The description-experience gap in risky choice. Trends Cognit Sci 13(12):517–523. https://doi.org/10.1016/j.tics.2009.09.004

Hertwig R, Barron G, Weber EU, Erev I (2004) Decisions from experience and the effect of rare events in risky choice. Psychol Sci 15(8):534–539

https://bris.on.worldcat.org/atoztitles/link?sid=ProQ:&issn=&volume=&issue=&title=Information+integration+in+perceptual+and+value-based+decisions&spage=&date=2012-01-01&atitle=Information+integration+in+perceptual+and+value-based+decisions&au=Tsetsos%2C+K.&id=doi:

Huber J, Payne JW, Puto C (1982) Adding asymmetrically dominated alternatives: violations of regularity and the similarity hypothesis. J Consum Res 9(1):90–98

Juechems K, Summerfield C (2019) Where does value come from? Trends Cogn Sci 23(10):836–850. https://doi.org/10.1016/j.tics.2019.07.012

Kahneman D, Tversky A (1979) Prospect theory: analysis of decision under risk. Econometrica 47(2):263–291. https://doi.org/10.2307/1914185

Kahneman D, Tversky A (1984) Choices, values, and frames. Am Psychol 39(4):341–350. https://doi.org/10.1037/0003-066x.39.4.341

Keeney RL, Raiffa H (1993) Decisions with multiple objectives: preferences and value trade-offs. Cambridge University Press, Cambridge

Krajbich I, Armel C, Rangel A (2010) Visual fixations and the computation and comparison of value in simple choice. Nat Neurosci 13(10):1292–1298. https://doi.org/10.1038/nn.2635

Lam NH, Borduqui T, Hallak J, Roque A, Anticevic A, Krystal JH, Wang X-J, Murray JD (2022) Effects of altered excitation-inhibition balance on decision making in a cortical circuit model. J Neurosci 42(6):1035–1053

Latty T, Beekman M (2011) Irrational decision-making in an amoeboid organism: transitivity and context-dependent preferences. Proceed R Soc B Biol Sci 278(1703):307–312

Lo C-C, Wang X-J (2006) Cortico–basal ganglia circuit mechanism for a decision threshold in reaction time tasks. Nat Neurosci 9(7):956–963

Loomes G, Sugden R (1995) Incorporating a stochastic element into decision theories. Eur Econ Rev 39(3–4):641–648

Louie K, De Martino B (2014) The neurobiology of context-dependent valuation and choice. Neuroeconomics. Elsevier, Amsterdam, pp 455–476

Louie K, Khaw MW, Glimcher PW (2013) Normalization is a general neural mechanism for context-dependent decision making. Proc Natl Acad Sci USA 110(15):6139–6144. https://doi.org/10.1073/pnas.1217854110

Ludvig EA, Madan CR, Spetch ML (2014) Extreme outcomes sway risky decisions from experience. J Behav Dec Making 27(2):146–156

Maier M, Bartos F, Stanley TD, Shanks DR, Harris AJL, Wagenmakers EJ (2022) No evidence for nudging after adjusting for publication bias. Proc Natl Acad Sci USA 119(31):e2200300119. https://doi.org/10.1073/pnas.2200300119

Mata R, Frey R, Richter D, Schupp J, Hertwig R (2018) Risk preference: a view from psychology. J Econ Perspect 32(2):155–172

McClelland JL (2009) The place of modeling in cognitive science. Top Cognit Sci 1(1):11–38

Miller EK, Brincat SL, Roy JE (2024) Cognition is an emergent property. Curr Opin Behav Sci 57:101388

Moran R, Tsetsos K (2018) The standard Bayesian model is normatively invalid for biological brains. Behav Brain Sci 41:e237. https://doi.org/10.1017/S0140525X18001449

Mostert P, Albers AM, Brinkman L, Todorova L, Kok P, de Lange FP (2018) Eye movement-related confounds in neural decoding of visual working memory representations. eNeuro 5(4):17. https://doi.org/10.1523/ENEURO.0401-17.2018

Noguchi T, Stewart N (2018) Multialternative decision by sampling: a model of decision making constrained by process data. Psychol Rev 125(4):512

Novemsky N, Kahneman D (2005) The boundaries of loss aversion. J Market Res 42(2):119–128

O'connell RG, Dockree PM, Kelly SP (2012) A supramodal accumulation-to-bound signal that determines perceptual decisions in humans. Nat Neurosci 15(12):1729–1735

Olschewski S, Spektor MS, Le Mens G (2024) Frequent winners explain apparent skewness preferences in experience-based decisions. Proceed Natl Acad Sci 121(12):e2317751121

Parrish AE, Dawes J, Thompson HL (2024) Exploring the impact of decoys on decision-making by young children. J Behav Dec Mak 37(3):e2385

Pirrone A, Tsetsos K (2023) Toward an atlas of canonical cognitive mechanisms. Cogn Sci 47(2):e13243. https://doi.org/10.1111/cogs.13243

Platt ML, Glimcher PW (1999) Neural correlates of decision variables in parietal cortex. Nature 400(6741):233–238

Platt ML, Plassmann H (2014) Multistage valuation signals and common neural currencies. Neuroeconomics 34:237–258

Ratcliff R, McKoon G (2008) The diffusion decision model: theory and data for two-choice decision tasks. Neural Comput 20(4):873–922

Regenwetter M, Davis-Stober CP (2012) Behavioral variability of choices versus structural inconsistency of preferences. Psychol Rev 119(2):408

Rieskamp J, Busemeyer JR, Mellers BA (2006) Extending the bounds of rationality: evidence and theories of preferential choice. J Econ Literat 44(3):631–661

Roe RM, Busemeyer JR, Townsend JT (2001) Multialternative decision field theory: a dynamic connections model of decision making. Psychol Rev 108(2):370

Russo JE, Dosher BA (1983) Strategies for multiattribute binary choice. J Exp Psychol Learn Mem Cogn 9(4):676–696

Rustichini A, Conen KE, Cai X, Padoa-Schioppa C (2017) Optimal coding and neuronal adaptation in economic decisions. Nat Commun 8(1):1208

Savage LJ (1972) The foundations of statistics. Courier Corporation

Shadlen MN, Kiani R (2013) Decision making as a window on cognition. Neuron 80(3):791–806

Shafir E (1993) Choosing versus rejecting: why some options are both better and worse than others. Memory Cognit 21(4):546–556

Shafir E, Simonson I, Tversky A (1993) Reason-based choice. Cognition 49(1–2):11–36. https://doi.org/10.1016/0010-0277(93)90034-S

Siems M, Cao Y, Tohidi-Moghaddam M, Donner TH, Tsetsos K (2023) Rhythmic sampling of multiple decision alternatives in the human brain. bioRxiv. https://doi.org/10.1101/2023.12.08.570734
Simoncelli EP (2003) Vision and the statistics of the visual environment. Curr Opin Neurobiol 13(2):144–149
Slovic P (1995) The construction of preference. Am Psychol 50(5):364–371. https://doi.org/10.1037/0003-066x.50.5.364
Soltani A, De Martino B, Camerer C (2012) A range-normalization model of context-dependent choice: a new model and evidence. PLoS Comput Biol 8(7):e1002607
Spektor MS, Bhatia S, Gluth S (2021) The elusiveness of context effects in decision making. Trends Cognit Sci 25(10):843–854
Srivastava N, Schrater PR (2012) Rational inference of relative preferences. Adv Neural Inform Process Syst 25:1134
Stearns SC (2000) Daniel Bernoulli (1738): evolution and economics under risk. J Biosci 25(3):221–228
Summerfield C, Tsetsos K (2012) Building bridges between perceptual and economic decision-making: neural and computational mechanisms. Front Neurosci 6:70. https://doi.org/10.3389/fnins.2012.00070
Summerfield C, Tsetsos K (2015) Do humans make good decisions? Trends Cogn Sci 19(1):27–34. https://doi.org/10.1016/j.tics.2014.11.005
Summerfield C, Tsetsos K (2020) Rationality and efficiency in human decision-making. Cognit Neurosci 6:427–439
Talluri BC, Urai AE, Tsetsos K, Usher M, Donner TH (2018) Confirmation bias through selective overweighting of choice-consistent evidence. Curr Biol 28(19):3128–3135. https://doi.org/10.1016/j.cub.2018.07.052
Tan K, Dong S, Liu X, Chen W, Wang Y, Oldroyd BP, Latty T (2015) Phantom alternatives influence food preferences in the eastern honeybee A pis cerana. J Anim Ecol 84(2):509–517
Tentori K, Osherson D, Hasher L, May C (2001) Wisdom and aging: irrational preferences in college students but not older adults. Cognition 81(3):B87–B96
Teodorescu AR, Usher M (2013) Disentangling decision models: from independence to competition. Psychol Rev 120(1):1
Thaler RH, Sunstein CR (2009) Nudge: improving decisions about health, wealth, and happiness. Penguin, New York
Tohidi-Moghaddam M, Tsetsos K (2024) The timescale and functional form of context-dependence during human value-learning. bioRxiv
Tom SM, Fox CR, Trepel C, Poldrack RA (2007) The neural basis of loss aversion in decision-making under risk. Science 315(5811):515–518
Trueblood JS, Brown SD, Heathcote A (2014) The multiattribute linear ballistic accumulator model of context effects in multialternative choice. Psychol Rev 121(2):179–205. https://doi.org/10.1037/a0036137
Tsetsos K, Usher M, Chater N (2010) Preference reversal in multiattribute choice. Psychol Rev 117(4):1275–1293. https://doi.org/10.1037/a0020580
Tsetsos K, Chater N, Usher M (2012a) Salience driven value integration explains decision biases and preference reversal. Proc Natl Acad Sci USA 109(24):9659–9664. https://doi.org/10.1073/pnas.1119569109
Tsetsos K, Gao J, McClelland JL, Usher M (2012b) Using time-varying evidence to test models of decision dynamics: bounded diffusion versus the leaky competing accumulator model. Front Neurosci 6:79. https://doi.org/10.3389/fnins.2012.00079
Tsetsos K, Moran R, Moreland J, Chater N, Usher M, Summerfield C (2016) Economic irrationality is optimal during noisy decision making. Proc Natl Acad Sci USA 113(11):3102–3107. https://doi.org/10.1073/pnas.1519157113

Tsetsos K (2012) Information integration in perceptual and value-based decisions (Publication Number U600209) [Ph.D., University of London, University College London (United Kingdom)]. ProQuest Dissertations and Theses Global. England
Turner BM, Schley DR, Muller C, Tsetsos K (2018) Competing theories of multialternative, multiattribute preferential choice. Psychol Rev 125(3):329–362. https://doi.org/10.1037/rev0000089
Tversky A (1972) Elimination by aspects: a theory of choice. Psychol Rev 79(4):281
Tversky A, Kahneman D (1981) The framing of decisions and the psychology of choice. Science 211(4481):453–458. https://doi.org/10.1126/science.7455683
Tversky A, Kahneman D (1992) Advances in prospect theory: cumulative representation of uncertainty. J Risk Uncert 5:297–323
Tversky A, Simonson I (1993) Context-dependent preferences. Manag Sci 39(10):1179–1189
Urai AE, de Gee JW, Tsetsos K, Donner TH (2019) Choice history biases subsequent evidence accumulation. Elife 8:331. https://doi.org/10.7554/eLife.46331
Usher M, McClelland JL (2001) The time course of perceptual choice: the leaky, competing accumulator model. Psychol Rev 108(3):550
Usher M, McClelland JL (2004) Loss aversion and inhibition in dynamical models of multialternative choice. Psychol Rev 111(3):757–769. https://doi.org/10.1037/0033-295X.111.3.757
Usher M, Tsetsos K, Glickman M, Chater N (2019) Selective integration: an attentional theory of choice biases and adaptive choice. Curr Direct Psychol Sci 28(6):552–559
Vlaev I, Chater N, Stewart N, Brown GD (2011) Does the brain calculate value? Trends Cogn Sci 15(11):546–554. https://doi.org/10.1016/j.tics.2011.09.008
Von Neumann J, Morgenstern O (2007) Theory of games and economic behavior: 60th anniversary commemorative edition. In: Theory of games and economic behavior. Princeton University Press, Princeton
Wald A, Wolfowitz J (1948) Optimum character of the sequential probability ratio test. Ann Math Stat 31:326–339
Wald A (2004) Sequential analysis. Courier Corporation
Webb R, Glimcher PW, Louie K (2021) The normalization of consumer valuations: context-dependent preferences from neurobiological constraints. Manag Sci 67(1):93–125
Wollschläger LM, Diederich A (2012) The 2 N-ary choice tree model for N-alternative preferential choice. Front Psychol 3:189
Yang T, Shadlen MN (2007) Probabilistic reasoning by neurons. Nature 447(7148):1075–1080

Chapter 6
Testing a Formal Model of Coherence-Based Information Search in Decision-Making

Arndt Bröder and Sophie E. Scharf

Abstract To make informed decisions, people must retrieve decision-relevant information either from memory or from sources available in the environment. Formalized theories of the information-search process are scarce. In this chapter, we review evidence for the spreading-activation coherence-based model iCodes which predicts the so-called "attraction search effect" (ASE) and several moderating variables for this effect. The ASE addresses the search direction based on the current attractiveness differences between options, which has not yet been tackled by other theories. While the ASE seems ubiquitous, and several predicted moderators and new predictions have been confirmed, we also report and discuss prediction failures of the model. These improvement opportunities notwithstanding, the ASE poses an explanatory challenge for *every* new theory of pre-decisional information search.

Keywords Information search · Cognitive modeling · Parallel constraint satisfaction

6.1 Introduction

Imagine you have moved to a new neighborhood, and you want to try the best Italian restaurant there. You learn that there are two restaurants around the corner, "La Cipolla" and "Da Mario", and internet recommendations inform you that the food quality of La Cipolla is excellent (4.8 out of 5 stars). However, there are other relevant attributes to making a decision, such as price, ambiance, and friendliness of staff. Knowing that La Cipolla's food is excellent, which information will you seek next? Will you first check out the food quality of Da Mario or require more information

A. Bröder (✉) · S. E. Scharf
School of Social Sciences, Chair of Experimental Psychology, University of Mannheim, Mannheim, Germany
e-mail: broeder@uni-mannheim.de

S. E. Scharf
e-mail: sophie.scharf@uni-mannheim.de

U. Ettinger et al. (eds.), *Decision Making*, Studies in Neuroscience, Psychology and Behavioral Economics, https://doi.org/10.1007/978-3-032-00880-0_6

about La Cipolla? Would your search strategy be different if you first learned that La Cipolla's food quality is mediocre (1.7 out of 5 stars)?

Although it is natural to view information acquisition as an integral part of the decision process in everyday decision-making, theorizing about decisions in psychology and economics has traditionally focused on the information *integration* process. Gigerenzer et al. (2012, p. 243) bemoaned that "in judgment and decision-making research, most approaches [...] do not model information search". In fact, early psychological research on decision-making was very much inspired by mathematical foundations of rational choice in economic game theory (Edwards 1954; Von Neumann and Morgenstern 1947), and so the 1950s to 1960s were dominated by testing "structural models" of decisions (i.e., the coherence of preferences as assumed in Expected Utility Theory), often using lotteries or urns-and-beads tasks. Starting with Amos Tversky's and Daniel Kahneman's research in the 1970s, psychologists began to look closer into psychological processes involved in decision-making, but the typical experimental paradigm used vignettes and textbook problems (such as the famous Linda problem or medical inference tasks) in which all decision-relevant information was presented and did not require further search. Thereafter, beginning in the late 1980s, decision-making began to be viewed as a genuine part of cognitive psychology, and efforts were made to apply its theoretical language to analyze choices as well. A seminal paper by Johnson and Payne (1985) utilized the concept of *elementary information processes* for analyzing various decision heuristics that contained information acquisition as integral components. This led to various methodological improvements such as information boards, the famous *MouseLab* (Payne et al. 1988), eye-tracking, and other experimental techniques to monitor information acquisition during decision-making (see Bröder 2019; Schulte-Mecklenbeck et al. 2019, for overviews). In fact, this evolving experimental focus on information search has been criticized by some as potentially misleading if used to test hypotheses on information integration (Bröder 2000; Glöckner and Betsch 2008b).

In this chapter, we will summarize the empirical evidence hitherto collected concerning a coherence-based decision model named iCodes (**i**ntegrated **co**herence-based model for **de**cisions and **s**earch) that was proposed by Jekel et al. (2018). The current chapter is structured as follows: First, we will very briefly discuss several perspectives of information acquisition in judgment and decision-making (JDM) research. Second, we will describe the formalized iCodes model and its intellectual roots in a non-formal way. We will also describe the intuition behind the "attraction search effect" (ASE) which is the hallmark prediction of the model. In the third section, we will present and discuss the empirical evidence that has been provided for the effect and its moderators. Finally, we will distinguish the effect from related concepts and discuss potential future research questions related to the model.

6.2 Perspectives on Information Acquisition in JDM

Without any claims of completeness, we will briefly sketch and discuss three influential perspectives on information acquisition in JDM: the normative perspective, the sampling perspective, and the heuristic toolbox perspective.

Normative Perspective and Sampling Models

Early normative approaches formulated models of optimal stopping thresholds in information acquisition that trade off the expected informativeness of further information against acquisition costs (Edwards 1965; Rapoport and Burkheimer 1971). Experimental paradigms used sequential choice tasks like the "secretary problem" or sequential sampling, and optimal stopping rules have been used as benchmarks to evaluate actual behavior. In these tasks, the informativeness or diagnosticity of information is not always assessed accurately by decision-makers (Snapper and Peterson 1971). Participants sometimes even acquire *pseudodiagnostic* information (Doherty et al. 1979), and they have been shown to be too conservative in probability updating (Phillips and Edwards 1966). Nonetheless, participants are sensitive to variations in the payoff structure of environments, acquiring less information when its utility is small relative to its costs (Bröder 2003), and they can adjust their stopping even to varying optimization criteria (Meder and Nelson 2012). Furthermore, sequential sampling of outcome information in risky decision tasks may explain well-known violations of Expected Utility Theory that have been found in the context of traditional lottery problems in which all information is readily available and does not have to be acquired (Bhatia 2014; Busemeyer et al. 2019; Hertwig et al. 2006).

While normative models are largely mute about specific underlying processes, various sampling models have been formulated that explicitly specify the decision process as evidence accumulation until a subjective threshold is reached. They come in different flavors, such as *drift diffusion models* (Voss et al. 2013), *Decision Field Theory* (DFT) for risky choice (Busemeyer and Townsend 1993), or *Leaky Competing Accumulator* models (Usher and McClelland 2001). Forstmann et al. (2016) provide a comprehensive overview. Assuming noisy information intake, the models can account for various context effects observed in the choice literature (Busemeyer et al. 2019). An important contribution of sampling models is the introduction of process measures by linking the weighting of information to attentional processes as measured by eye movements, for example (e.g., Krajbich 2019; Yang and Krajbich 2023).

Hence, normative models yield valuable insights for *stopping*, and sampling models provide a rationale for understanding speed-accuracy tradeoffs. But both are often tied to a small range of well-defined task-structures which do not allow predictions of search direction. Further, if the underlying paradigm allows for assessing search direction, the models will often make only vague predictions lacking predictive power.[1]

Decision Heuristics and the Adaptive Toolbox

One of the most influential ideas in cognitive decision research is the assumption that we possess a cognitive "toolbox" filled with various decision heuristics and strategies that we can flexibly apply when appropriate (Gigerenzer et al. 1999; Payne et al. 1993; Svenson 1979; Thorngate 1980). Information search is a fundamental and constituting element of these heuristics, and many step-by-step descriptions of various heuristics explicitly contain predictions about search order and search direction, suggesting process-tracing methods as a useful tool to investigate the strategies empirically (see Payne 1976, for an early example). As briefly described above, by decomposing the strategies into elementary information processes, Johnson and Payne (1985) were able to quantify the "cognitive costs" associated with complex versus simple strategies which provided a metric to compare them against their respective success in terms of accuracy. This gave a new impetus to the cost–benefit model of strategy selection (Beach and Mitchell 1978), and Payne et al. (1993) developed this idea further into a cognitive framework.

In this view and in the adaptive toolbox framework promoted by Gigerenzer et al. (1999), an important distinction is made between *compensatory* and *noncompensatory* strategies. The former strategies essentially involve acquiring all available decision-relevant information and weighing the attributes against each other. Hence, good values on less important attributes may compensate for bad values on other more important attributes. Examples of these strategies are the "Weighted Additive Strategy" (WADD), "Multi-Attribute-Utility Theory" (MAUT), or simple tallying of pros and cons (Equal Weight Strategy, EQW). Noncompensatory strategies, on the other hand, focus on the most important cues or attributes and may ignore some of the available information. Examples are Tversky's (1969) idea of a lexicographic semiorder, the strict lexicographic strategy (LEX; called take-the-best—TTB—for inference decisions with binary attributes), or "elimination by aspects" (EBA). Using LEX, for example, people look up the most important attribute. If the values of the options differ to a sufficient degree, a choice will be made, disregarding other less important attributes. If the most important attribute does not differentiate, the next most important cue is sought, and so on.

The conceptualization of these heuristic strategies therefore includes stepwise information acquisition, and the direction of the search is determined, particularly for the non-compensatory heuristics, as *attribute-wise* and from the most to the least important attribute. In the pizza place example above, learning about food quality in La Cipolla, searching attribute-wise would mean we acquire information about the food quality at Da Mario next. In contrast, compensatory strategies are often associated with an *option-wise* search, so we would try to learn more information

[1] This statement is somewhat simplified, since DFT assumes a stochastic search process that is guided by attribute importance. However, no straightforward mechanism is postulated that would allow for the prediction of the attraction search effect described below.

about La Cipolla next before switching to Da Mario.[2] Hence, the heuristics in the toolbox make predictions about search order and thus include information acquisition as an integral part. Importantly, if decision-makers choose to use a certain strategy (LEX or WADD, for example), they are assumed to follow its predetermined search order, irrespective of the information found. Coming back to the pizza place example, these strategies assume that you will go on with the same kind of search process, regardless of whether you find out that La Cipolla serves high- or low-quality food. But is it plausible that the nature of the information you found would not influence your subsequent search?

6.3 Coherence-Based Information Acquisition

The concept of coherence, or consistency, has a long tradition in psychology, dating back at least to the Gestalt psychologists who described the tendency to organize perceptual (Wertheimer 2017) or conceptual (Duncker 1935) inputs in a parsimonious and coherent manner. The idea gained popularity also in early social cognition research with Kurt Lewin's Field Theory eventually inspiring influential models like Heider's (1958) Balance Theory or Festinger's (1957) Dissonance Theory. The fundamental assumption of this view is that our cognitive apparatus strives for coherent representations which can be achieved by eliminating or downplaying contradictory information. This view has been influential, but the theories remained somewhat imprecise until the ideas became formalized in connectionist networks in the 1980s (Rumelhart and McClelland 1986; Simon and Holyoak 2002; Thagard 1989). There are different variants of connectionist networks, but they have in common the ideas of (1) interconnected nodes representing cognitive units at different levels, and (2) spreading activation between nodes proportional to the "strength" of the interconnections. If the network (or one node in it) is activated, the activation spreads through the network, resulting in a dynamic flow of activation that will eventually settle in a stable "low energy" state, characterized by a final pattern of activation across the nodes in the network. The network structure (nodes and connection strengths) determines the information flow in all parts of the network simultaneously, and the processing thus follows *parallel constraint satisfaction* (PCS). A comprehensive review of coherence-based cognition and its formalization as PCS networks is provided by Simon and Read (2023).

Glöckner and Betsch (2008a) suggested formalized PCS models as an alternative to the sequential strategies that populate the "adaptive toolbox", claiming that the human mind is capable of processing more information in parallel than suggested by scholars proposing "simple heuristics" as the main mode of decision-making. Glöckner and Betsch (2008a) assumed that decision situations can be represented

[2] This has been criticized for methodological reasons since several compensatory strategies can also be implemented with an attribute-wise search (Bröder 2000), but this does not affect the general argument made here.

as networks in which cues or attributes as well as decision options are nodes, interconnected by links of varying strength (determined, for example, by attribute importance). The options are connected by inhibitory links which represent a contrast-enhancing mechanism: The activation of one option leads to a decrease in activation of competing options. The available information pieces affect the network simultaneously and generate a dynamic flow of activation that results in a stable state after several iterations. The final state is characterized by an activation pattern across options that predicts the choice of the most highly activated option with a confidence determined by the activation difference to the less highly activated option(s). The decision time, on the other hand, can be predicted by the number of iterations the network needs to achieve the stable end state. Experiments testing the choice, confidence, and reaction time predictions of the formalized PCS-DM model[3] have generally favored the model over the sequential competitors from the toolbox (Glöckner and Betsch 2012; Glöckner et al. 2014). One unique prediction of these models is *coherence shifts,* defined as shifts in the perceived importance of attributes due to coherence processes. These shifts are hard to explain by decision heuristics, and they have been reported repeatedly (Glöckner et al. 2010; Lee and Holyoak 2021; Simon et al. 2001).

One shortcoming of the PCS-DM model, however, is the lack of predictions for information search because the spreading activation process describes the information integration only in situations where all information is readily available. Marewski (2010) bemoaned that this fact makes it hard to compare the explanatory success of heuristic models to PCS-DM since the empirical domains are not identical. Furthermore, Söllner et al. (2013) showed that PCS-DM fails to describe behavior in situations involving even the most modest search costs. Hence, PCS-DM and the toolbox metaphor are not on an equal footing when describing decisions, and it is hard to test them against each other empirically. Therefore, Jekel et al. (2018) extended the original PCS-DM model to include search and developed the *iCodes* model. Conceptually, this involved representing cues with *hidden information* as an additional layer of nodes in the network and assuming that these information pieces are acquired stepwise in order of decreasing activation. Importantly, like in the original PCS-DM model, nodes representing searchable information receive "bottom-up" activation according to their importance as well as "top-down" activation from activated option nodes. To illustrate the general idea, Fig. 6.1 shows a simplified network representation.

In the depicted decision situation, it is assumed that the decision-maker can obtain two pieces of information about each of both options (here: the restaurants), namely food quality and price. These cues and their respective cue values (i.e., the concrete characteristics such as the rating of the food quality of La Cipolla) are connected to the options. The cue nodes are connected to a source node. This node has no psychological meaning in itself, but it starts the activation spread by pumping activation into the network. In this first step [marked by (1) in Fig. 6.1a], activation spreads from the source node to the cues and their cue values proportional to cue importance

[3] "DM" stands for decision making.

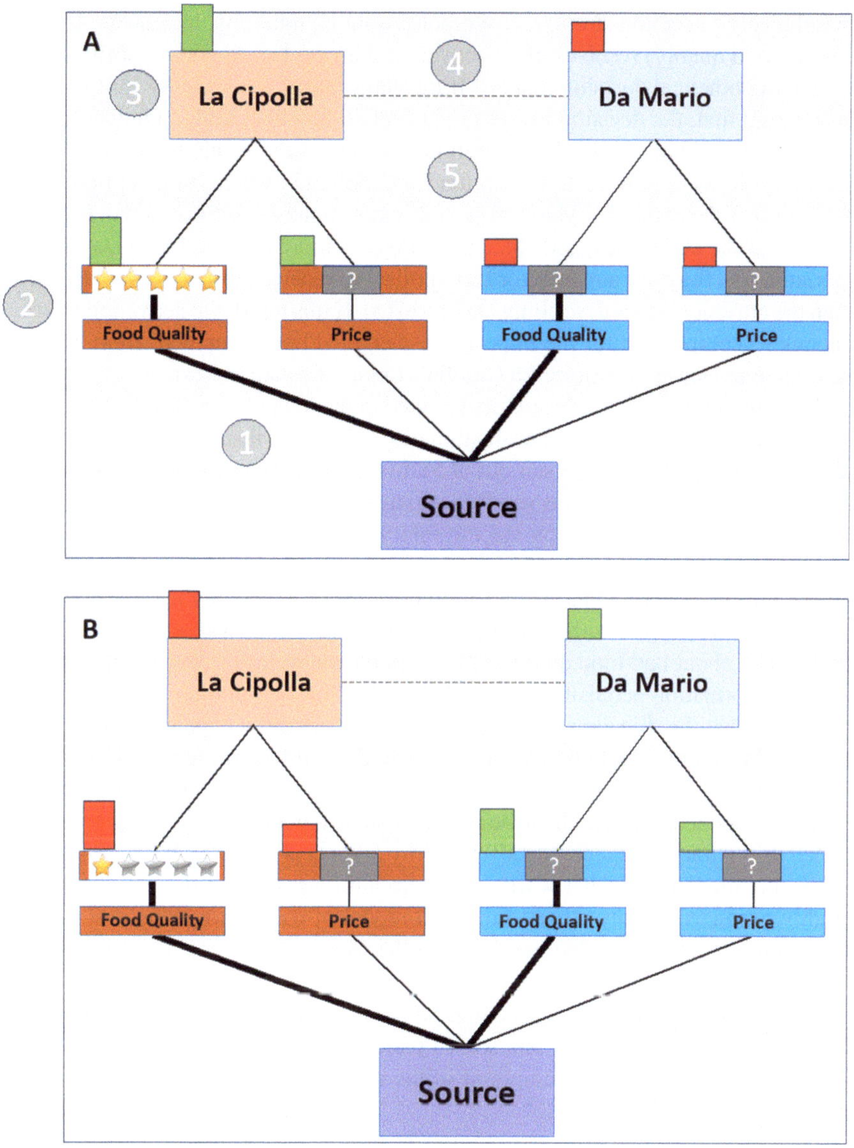

Fig. 6.1 Outline of two simple decision situations modeled by iCodes. In Panel **a**, the decision maker first finds positive information about food quality for Option "La Cipolla", in Panel **b**, the information found is negative. This leads to specific patterns of positive (green bars) or negative (red bars) activation as explained in the text

or validity. Connection strength is symbolized in the figure by thicker lines. Hence, Cue 1 (food quality) receives more activation than the less important attribute price for the hypothetical decision maker.[4] If the cue value of *food quality* for La Cipolla is acquired first, the decision maker might find an excellent evaluation like in Panel A of Fig. 6.1. This will lead to a preliminary activation of this option by positive activation flowing upward to the option La Cipolla (3) which, in turn, inhibits the activation of the option Da Mario by the inhibitory link (4). Activation will flow back "down" to the cue values, and since La Cipolla is more activated than Da Mario in this situation, the price attribute for La Cipolla will receive more top-down activation than the attribute values describing Da Mario (5). If this top-down activation is strong enough to compensate for the importance-weighted bottom-up activation, this will result in a tendency to acquire La Cipolla's price information next.

Now, imagine a different scenario depicted in the lower Panel B of Fig. 6.1. Here, the decision maker finds a bad evaluation of La Cipolla when acquiring information about its food quality (e.g., 1 star out of 5). In this case, Steps (3)–(5) would develop differently: La Cipolla would receive negative activation (3) which in turn would increase Da Mario's activation via the inhibitory link (4). With Da Mario having higher activation than La Cipolla, more activation flows back to the cue values associated with this option than for La Cipolla (5), resulting in a higher activation level of the nodes associated with Da Mario than those associated with La Cipolla. Hence, finding out about bad food quality in La Cipolla would lead to switching the direction of information acquisition to Da Mario. To sum up, the model predicts (as a tendency) that the direction of information search critically depends on the content of the information you find. This temporarily changes the perceived attractiveness of the options, which in turn, changes the activation of the remaining information.

This dependency of search on the current relative attractiveness of options has been called the "attraction search effect" (ASE) by Jekel et al. (2018).[5] What is surprising about this prediction is that while it is quite intuitive, it is (to our knowledge) not readily predicted by any other formalized model of decision-making that includes information acquisition. The heuristics in the toolbox assume fixed search orders, regardless of the information found, whereas sampling models of decision-making would have to be enriched with additional assumptions to produce the same effect. In contrast, the prediction of this search pattern is an inherent consequence of the structural and processing assumptions of the iCodes model.

---

[4] Of course, subjective attribute importance may differ between persons.

[5] In fact, the effect has empirically been described before by Carlson and Guha (2011) who termed it "leader-focused-search". We were unaware of this work when developing iCodes. We also learned later that DeKay (2015) had proposed a spreading-activation network as a conceptual idea to explain the effect, although without a mathematical formalization of the idea which can be found in Jekel et al. (2018).

6.4 Empirical Tests of the Predictions

The ASE is the simplest and most obvious prediction of iCodes. Hence, the first empirical tests focused on this effect. Another prediction directly derived from the model without additional assumptions concerns response times in search, a variable that has—to our knowledge—never been investigated in decision research. How long will a person take to open the next piece of information in a MouseLab setting? Simulating different information constellations with iCodes, one can compare their respective numbers of iterations needed to reach a stable state in the network. If the coherence process has a psychological reality, one would expect search times to correlate with these iterations. The allure of these predictions is their qualitative nature, i.e., they can be tested without complicated procedures of model-fitting that might fall prey to many hidden and often nontransparent assumptions. For more qualitative predictions of the model, auxiliary assumptions are necessary about the connections between model parameters to other cognitive variables, such as attention allocation, salience effects, and a mental focus on attractiveness.

The Attraction Search Effect

Jekel et al. (2018) report two experiments testing the attraction search effect with a total of $N = 144$ participants. In each of these, an abstract stock-market game was used in which participants saw recommendations of fictitious brokers with different predictive validities recommending investments in fictitious stocks. In each trial, an *information pattern* was shown in which either Option A or Option B was more attractive, while some pieces of information were still hidden. Participants could acquire additional information. Depending on the experimental condition, only one piece of information could be acquired or more information, either with associated costs, or for free. Table 6.1 shows three exemplary information patterns in their two versions each.

A simple index to measure conformity of search behavior with the ASE is to subtract the probability of acquiring information about Option A in Version A of a

Table 6.1 Three exemplary information patterns used by Jekel et al. (2018)

	Pattern 1				Pattern 2				Pattern 3			
	Version A		Version B		Version A		Version B		Version A		Version B	
	A	B	A	B	A	B	A	B	A	B	A	B
Cue 1	+	?	–	?	+	–	–	+	X	–	X	–
Cue 2	+	?	–	?	?	?	?	?	X	–	X	+
Cue 3	?	?	?	?	+	–	+	–	?	+	?	+
Cue 4	?	–	?	?	–	?	–	?	–	?	–	?

In version A of each pattern, option A is more attractive, in version B, the more attractive option is B. Cues are ordered according to decreasing validities

Note "+" = positive cue value, "–" = negative cue value, "?" = hidden, but available cue value, "X" = hidden, unavailable cue value

pattern minus the probability of acquiring Option A information in Version B. The index can take values between 1 and −1, with positive values indicating an ASE, while negative values indicate the opposite of iCodes's prediction. An index of 0 means that there is no relation between search direction and pattern version (as would be predicted by most heuristics). Jekel et al. (2018) found positive values of the index, with large effects according to Cohen's (1988) conventions, ranging from $d = 0.86$ for cost-free unrestricted search and larger effects for cost-free unrestricted and restricted search ($d = 1.5$ and $d = 2.15$, respectively). Hence, the experiments clearly supported the ASE prediction. To eliminate the possibility that the information patterns had been chosen in a way to favor the effect, Jekel et al. (2018) also reanalyzed five MouseLab studies with completely closed information matrices, the latest of which had been published 12 years before. Here, the ASE score was defined as the probability of a switch search to another option after finding a negative cue value first minus the switch probability after first finding a positive value. Again, the ASE score was significantly positive in all five studies (Cohen's d ranging from 1.07 to 1.75).

The initial results including the re-analyses of five old studies therefore match iCodes' prediction of an ASE. However, critics might object that all studies used (1) a hypothetical simplified stock market scenario and (2) a matrix presentation of options by attributes. One concern was that these features of the experimental design may have evoked the effect. Scharf et al. (2019) report three experiments in which they used 12 different semantic embeddings (e.g., choosing a hotel for vacation, a hairdresser, or predicting the weather) in Experiments 1 and 2, and they even dispensed with the usual attribute-by-option-matrix in Experiment 3. In this experiment, a simulated online shop was used, and participants had to choose offers of clothing from wholesalers receiving information about the fabric, ease of care, variety of sizes, and fit. All three experiments revealed considerable ASE scores on average, thus showing that the effect was not produced by peculiarities of the presentation format (matrix) or the semantic embedding (stock market). Rather, the effect seems generalizable, thus lending support to iCodes or at least to the principle of coherence-based search.

Predicting Search Times

To our knowledge, no researcher to date has investigated the time until acquiring the next piece of information as a potentially interesting variable. Although traditional MouseLab studies often report the overall time spent for information acquisition or the percentage of time spent on the most important attribute (e.g., Payne et al. 1988), the search time for individual acquisition choices has hitherto been neglected. This is probably because no theory to date has made predictions about such a variable. With iCodes, however, one can simulate different information layouts and assess the number of iterations until a stable state is reached. These serve as a proxy for predicting search times. As an intuitive illustration, one can imagine that more coherent information patterns (most cues speak for A, some against B) settle more quickly than more ambiguous patterns (e.g., some cues favor A, others favor B). Scharf et al. (2024a, b) assessed the iterations needed for each of the 16 patterns used in Jekel et al.'s (2018) experiments via simulation. Then, they correlated each

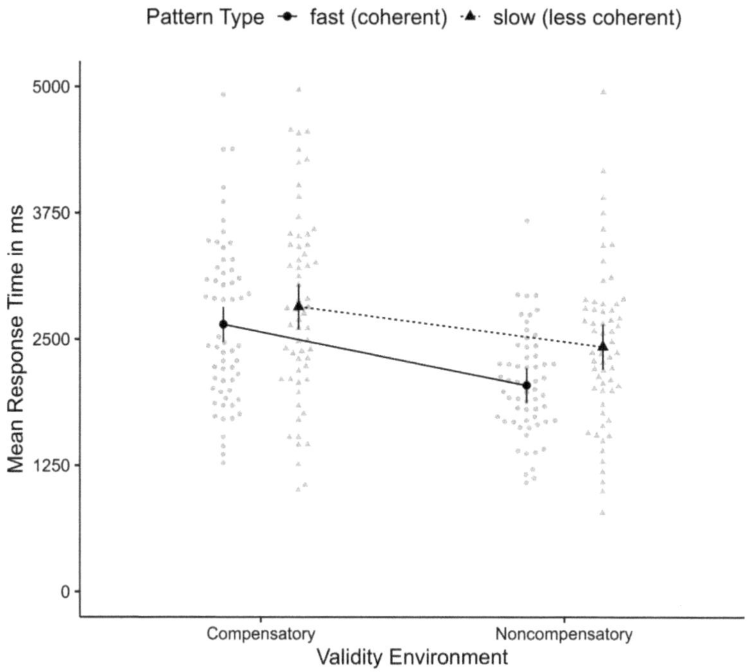

Fig. 6.2 Search times in the new experiment by Scharf et al. (2024a, b) dependent on environment and pattern type

participant's median response time[6] for these patterns with the predictions from the simulations. The correlations were significantly positive in all experimental conditions, ranging from $r = 0.12$ to $r = 0.25$. These correlations may seem small at first, but note that response times are particularly noisy, especially if no speed instruction is given.

In a confirmatory experiment, the authors strove to back up the reanalysis with predicted and experimentally manipulated effects. They generated coherent (fast) and less coherent (slow) cue patterns, using the original stock-market task. In addition, two different information environments were tested: In the compensatory environment, cues had a small range of validities (0.53–0.62), whereas the validities varied considerably stronger in the non-compensatory environment (0.56–0.80). Simulations with iCodes suggested that search times should be faster in the latter than in the former environment. Testing 120 participants in a 2 × 2 mixed design, both predicted main effects on search times were supported in addition to a non-predicted interaction (see Fig. 6.2).

Whereas the ASE seems psychologically obvious, the prediction of search times is much more subtle. To the best of our knowledge, no formal decision model other

[6] As reaction times became considerably shorter over the course of the experiment, Scharf et al. (2024a, b) partialled out the trial order before aggregating response times.

than iCodes can predict these search times. Hence, we consider the results as strong evidence in favor of coherence-based search, and we see this as a viable challenge to competing theories of information search in decision-making.

Effect of an Attractiveness Focus

The ASE and search time predictions follow from iCodes as a PCS extension without further assumptions. However, it is possible to extend the model's scope by making some psychologically plausible auxiliary assumptions. For example, the ASE depends on the amount of "top–down" activation flowing back from the attractive option to the associated cue values. Scharf et al. (2022) made the additional assumption that an awareness focus on option attractiveness should strengthen this top–down activation flow and hence increase the size of the ASE. In a two-group design, they had $N = 202$ participants make choices in the stock-market paradigm of Jekel et al. (2018). Approximately half of the participants made a relative attractiveness judgment of the two stocks when presented with the cue patterns. Only after this attractiveness assessment, they were able to acquire additional information. As predicted, the ASE was much larger in this group compared to the control group without attractiveness focus. Furthermore, the iCodes model contains a parameter γ which determines the relative weight of the bottom-up connections between source node and cues and the top–down bidirectional connections between options and cues. Whereas this parameter is typically fixed for simulations, one can also fit the model to individual data and estimate the parameter values. Scharf et al. (2022) did this and found increased mean parameter values in the attractiveness focus group. Hence, the results lend further credibility to the processes formulated in iCodes and the parameters representing them.

Salience Influences on Cue Integration

The bottom-up influence activating cue values is conceptualized as validity-driven (or importance-driven) in the original iCodes version. One may think of other factors as well, for example, the visual salience of information which may direct attention to respective parts of the display, thereby increasing their activation. Salience effects on decision strategies have been reported repeatedly (Bryant 2014). Following this line of reasoning, Bröder et al. (2021) varied the visual salience of cue values by highlighting certain cues on the display. Three experiments with a total sample size of $N = 170$ followed the same design and varied only in detail. Importantly, iCodes can simulate what happens if additional activation of a cue due to salience is assumed: A straightforward prediction are *direct* salience effects that increase the ASE when a hidden cue value of the more attractive option is salient, whereas the ASE should decrease when the salient cue value is part of the less attractive option. In addition to these rather obvious direct effects, iCodes also predicts subtle interaction effects when positive or negative (rather than unknown) cue values are highlighted. The ASE should increase when a positive value is salient on the attractive option or a negative one is salient on the less attractive option (match). Likewise, the ASE should decrease under opposite conditions (mismatch of cue-value valence and option attractiveness). The results were clear: Although the ASE per se as well as the *direct* salience contrasts were found in all three experiments as predicted, none of the experiments (nor the pooled data of all three) confirmed the existence of the indirect effects. Bröder et al. (2021) were able to reconcile the latter prediction failure by implementing a different auxiliary assumption: If the attention due to the higher salience of a cue value encompasses the whole option rather than the specific salient attribute, iCodes produces predictions very similar to the observed effects. This re-interpretation of the data is of course post hoc, and the changed auxiliary hypothesis remains to be validated in further studies.

Attention Allocation as Measured with Eye-Tracking

The previously described studies all utilized a MouseLab task to assess the information-search process. In many decision situations, however, all information is readily available to the decision-maker and information search is better represented by a visual process. To predict this more subtle form of information search with iCodes, an auxiliary assumption is necessary, namely that the activation level at the cue-value nodes determines the relative attention to the respective information in the decision task. With this auxiliary assumption, several predictions regarding attention allocation can be made, most notably that new information on the attractive option should be attended to first (i.e., the visual counterpart to the ASE) and that coherent information should receive more attention than incoherent information. In two eye-tracking experiments with a total of $N = 107$, Scharf, Bröder et al. (2024a, b) investigated whether informational coherence and option attractiveness also influence the arguably less structured visual search process, using a stock-market task similar to the previously described experiments. The results of the experiments showed that new information was generally more likely to be fixated on the attractive option first, supporting the prediction of an Attraction Attention Effect. However, this oculomotor pattern was attenuated when task constraints on the visual search process decreased. Concerning the coherence influence on attention allocation, the experiments supported a preference for coherent information on the aggregate level. Yet, a closer look at the data revealed that participants only preferred coherent information that is positive and characterizes the attractive option. The more subtle prediction of a preference for (the also coherent) negative information on the unattractive option was not supported. The results of these experiments therefore support the relevance of option attractiveness for the information-search process also in the visual domain. However, they also show that additional auxiliary assumptions might be necessary for iCodes to adequately predict attention allocation and that the role of coherence in attention allocation still needs to be clarified.

6.5 Summary and Discussion

ICodes is a mathematical model implementing the ideas of coherence-based decision-making and information search in a spreading-activation network structure. Although the model is relatively complex, it makes *qualitative* and testable predictions. We have summarized about a dozen experiments testing those predictions. The hallmark prediction of an ASE was confirmed by each study as well as by a re-analysis of data collected in five older studies. Hence, we consider the principle of coherence-based search and the influence of option attractiveness to be well-established. The generalization tests by Scharf et al. (2019) are particularly important for demonstrating that the effect is not restricted to specific tasks and presentation formats.[7] In addition, iCodes predicted effects concerning a hitherto unexplored behavioral variable, namely search times for new information. To our knowledge, iCodes is the only formalized decision model that generates predictions about this variable. The ability to successfully predict not only new data patterns, but also new *kinds* of data is in our view a very strong argument for the model, and it can be seen as a challenge for competing theories to accommodate this new kind of evidence.

The success of predictions with auxiliary assumptions was mixed: Whereas a focus on attractiveness judgments or the salience of attributes enhanced the ASE, more subtle predictions like indirect salience effects or those on attention allocation were less successful. Unfortunately, we are caught in the Duhem-Quine-problem here, not knowing whether central tenets of iCodes fail, or whether only auxiliary assumptions like the transfer to eye-movements are questionable.

Construction Site Issues of iCodes

Currently, iCodes lacks certain features that limit its applicability in general: First, only binary attributes have been modeled, yet. This is a severe limitation currently implemented for simplicity. Extensions to continuous or multi-valued attributes are desirable and possible in principle. Furthermore, these extensions would be necessary to put iCodes on equal footing with evidence accumulation models that can deal with this kind of information. This also concerns applications to risky choice or sequential choice. Extensions to the applications are under construction, and the current version with its successes is an encouraging starting point.

A further unsolved issue of iCodes is the nature of the stopping process: Whereas evidence accumulation models have a built-in stopping threshold allowing to model speed-accuracy-tradeoffs, a psychologically plausible stopping criterion for iCodes still remains to be established. Candidates are certain activation differences between options (i.e., choice confidence) or minimum activation of hidden attribute values. Which mechanism is most promising awaits future modeling and experimentation.

While iCodes is the only formalized model to our knowledge predicting the ASE, similar or even the same search patterns have been published before (e.g., leader-focused search, Carlson and Guha, 2011; confirmation bias, see Nickerson, 1998, for a review; positive test strategies, Klayman and Ha, 1987, for example). Currently, it is not clear whether all these phenomena stem from the same underlying information-search process as described by iCodes or whether different cognitive (or motivational) processes can result in similar search patterns across experimental paradigms. A task

[7] Of course, the generalizations were all within certain boundaries, such as binary choices and "1-click-information-acquisition", and generalizations beyond these parameters would be desirable.

for future research is to compare the various phenomena of attractiveness-biased search and to clarify iCodes's contribution to this literature.

6.6 Conclusion

Independent of the question of whether iCodes as a specific model or whether its extensions to other domains will survive future empirical tests, the ubiquity of the ASE has clearly been established by our experiments, and it supports coherence-based search in principle. Certainly, the effect must be accepted as an explanandum for any theory that claims to explain pre-decisional information search.

Acknowledgements Most experiments discussed here were funded by the German Research Foundation (DFG), grants GL 632/5-1 (Andreas Glöckner), and BR 2130/13-1 (Arndt Bröder).

References

Beach LR, Mitchell TR (1978) A contingency model for the selection of decision strategies. Acad Manag Rev 3(3):439–449. https://doi.org/10.5465/amr.1978.4305717

Bhatia S (2014) Sequential sampling and paradoxes of risky choice. Psych Bull Rev 21(5):1095–1111. https://doi.org/10.3758/s13423-014-0650-1

Bröder A (2000) A methodological comment on behavioral decision research. Psychol Beiträge 42(4):645–662

Bröder A (2003) Decision making with the "adaptive toolbox": influence of environmental structure, intelligence, and working memory load. J Exper Psychol Learn Memory Cogni 29(4):611–625. https://doi.org/10.1037/0278-7393.29.4.611

Bröder A (2019) Methods for studying human thought. In: Sternberg RE, Funke J (eds) The psychology of human thought: an introduction. Heidelberg University Publishing, New York, pp 27–53

Bröder A, Scharf S, Jekel M, Glöckner A, Franke N (2021) Salience effects in information acquisition: no evidence for a top-down coherence influence. Memory Cognit 49(8):1537–1554. https://doi.org/10.3758/s13421-021-01188-9

Bryant DJ (2014) Strategy selection in cue-based decision making. Can J Exper Psychol 68(2):97–110. https://doi.org/10.1037/cep0000020

Busemeyer JR, Townsend JT (1993) Decision field theory: a dynamic-cognitive approach to decision making in an uncertain environment. Psychol Rev 100(3):432–459. https://doi.org/10.1037/0033-295X.100.3.432

Busemeyer JR, Gluth S, Rieskamp J, Turner BM (2019) Cognitive and neural bases of multi-attribute, multi-alternative, value-based decisions. Trends Cognit Sci 23(3):251–263

Carlson KA, Guha A (2011) Leader-focused search: the impact of an emerging preference on information search. Organ Behav Hum Dec Process 115(1):133–141. https://doi.org/10.1016/j.obhdp.2010.12.002

Cohen J (1988) Statistical power analysis for the behavioral sciences, 2nd edn. Lawrence Erlbaum Associates

DeKay ML (2015) Predecisional information distortion and the self-fulfilling prophecy of early preferences in choice. Curr Direct Psychol Sci 24(5):405–411. https://doi.org/10.1177/0963721415587876

Doherty ME, Mynatt CR, Tweney RD, Schiavo MD (1979) Pseudodiagnosticity. Acta Psychol 43(2):111–121. https://doi.org/10.1016/0001-6918(79)90017-9

Duncker K (1935) Zur Psychologie des produktiven Denkens = the psychology of productive thought. Springer, New York

Edwards W (1954) The theory of decision making. Psychol Bull 51(4):380–417. https://doi.org/10.1037/h0053870

Edwards W (1965) Optimal strategies for seeking information: models for statistics, choice reaction times, and human information processing. J Math Psychol 2(2):312–329. https://doi.org/10.1016/0022-2496(65)90007-6

Festinger L (1957) A theory of cognitive dissonance. Stanford University Press

Forstmann BU, Ratcliff R, Wagenmakers E-J (2016) Sequential sampling models in cognitive neuroscience: advantages, applications, and extensions. Annu Rev Psychol 67(1):641–666. https://doi.org/10.1146/annurev-psych-122414-033645

Gigerenzer G, Dieckmann A, Gaissmaier W (2012) Efficient cognition through limited search. Ecological rationality: intelligence in the world. Oxford University Press, Oxford, pp 241–273

Gigerenzer G, Todd PM, The ABC Research Group (1999) Simple heuristics that make us smart. Oxford University Press

Glöckner A, Betsch T (2008a) Modeling option and strategy choices with connectionist networks: towards an integrative model of automatic and deliberate decision making. Judgment Dec Making 3(3):215–228. https://doi.org/10.1017/S1930297500002424

Glöckner A, Betsch T (2008b) Multiple-reason decision making based on automatic processing. J Exper Psychol Learn Memory Cognit 34(5):1055–1075. https://doi.org/10.1037/0278-7393.34.5.1055

Glöckner A, Betsch T (2012) Decisions beyond boundaries: when more information is processed faster than less. Acta Psychol 139(3):532–542. https://doi.org/10.1016/j.actpsy.2012.01.009

Glöckner A, Betsch T, Schindler N (2010) Coherence shifts in probabilistic inference tasks. J Behav Dec Making 23(5):439–462. https://doi.org/10.1002/bdm.668

Glöckner A, Hilbig BE, Jekel M (2014) What is adaptive about adaptive decision making? A parallel constraint satisfaction account. Cognition 133(3):641–666. https://doi.org/10.1016/j.cognition.2014.08.017

Heider F (1958) The psychology of interpersonal relations. Wiley, Amsterdam

Hertwig R, Barron G, Weber EU, Erev I (2006) The role of information sampling in risky choice. Information sampling and adaptive cognition. Cambridge University Press, Cambridge, pp 72–91

Jekel M, Glöckner A, Bröder A (2018) A new and unique prediction for cue-search in a parallel-constraint satisfaction network model: the attraction search effect. Psychol Rev 125(5):744–768. https://doi.org/10.1037/rev0000107

Johnson EJ, Payne JW (1985) Effort and accuracy in choice. Manag Sci 31(4):395–414

Klayman J, Ha Y-W (1987) Confirmation, disconfirmation, and information in hypothesis testing. Psychol Rev 94(2):211–228. https://doi.org/10.1037/0033-295X.94.2.211

Krajbich I (2019) Accounting for attention in sequential sampling models of decision making. Curr Opin Psychol 29:6–11. https://doi.org/10.1016/j.copsyc.2018.10.008

Lee DG, Holyoak KJ (2021) Coherence shifts in attribute evaluations. Decision 8(4):257–276. https://doi.org/10.1037/dec0000151

Marewski JN (2010) On the theoretical precision and strategy selection problem of a single-strategy approach: a comment on Glöckner, Betsch, and Schindler (2010). J Behav Dec Making 23(5):463–467. https://doi.org/10.1002/bdm.680

Meder B, Nelson JD (2012) Information search with situation-specific reward functions. Judgment Dec Making 7(2):119–148. https://doi.org/10.1017/S1930297500002977

Nickerson RS (1998) Confirmation bias: a ubiquitous phenomenon in many guises. Rev Gen Psychol 2(2):175–220. https://doi.org/10.1037/1089-2680.2.2.175

Payne JW (1976) Task complexity and contingent processing in decision making: an information search and protocol analysis. Organ Behav Hum Perform 16(2):366–387. https://doi.org/10.1016/0030-5073(76)90022-2

Payne JW, Bettman JR, Johnson EJ (1988) Adaptive strategy selection in decision making. J Exper Psychol Learn Memory Cognit 14(3):534–552. https://doi.org/10.1037/0278-7393.14.3.534

Payne JW, Bettman JR, Johnson EJ (1993) The adaptive decision maker. Cambridge University Press, Cambridge

Phillips LD, Edwards W (1966) Conservatism in a simple probability inference task. J Exper Psychol 72(3):346–354. https://doi.org/10.1037/h0023653

Rapoport A, Burkheimer GJ (1971) Models for deferred decision making. J Math Psychol 8(4):508–538. https://doi.org/10.1016/0022-2496(71)90005-8

Rumelhart DE, McClelland JL (eds) (1986) Parallel distributed processing: explorations in the microstructure of cognition, vol 1. Foundations. MIT Press Bradford

Scharf SE, Wiegelmann M, Bröder A (2019) Information search in everyday decisions: the generalizability of the attraction search effect. Judgment Dec Making 14(4):488–512

Scharf SE, Jekel M, Glöckner A (2022) Awareness of option attractiveness increases the attraction search effect: modeling the awareness effect in an extended iCodes model. Decision 9(1):43–59. https://doi.org/10.1037/dec0000162

Scharf SE, Bröder A, Jekel M, Glöckner A (2024a) The influence of coherence and option attractiveness on attention allocation in decisions with open information displays. Manuscript in preparation

Scharf SE, Jekel M, Bröder A (2024b) Predicting information search times with iCodes. Manuscript in preparation

Schulte-Mecklenbeck M, Kühberger A, Johnson JG (2019) A handbook of process tracing methods, 2nd edn. Routlegde, Taylor & Francis Group. https://doi.org/10.4324/9781315160559

Simon D, Holyoak KJ (2002) Structural dynamics of cognition: from consistency theories to constraint satisfaction. Person Soc Psychol Rev 6(4):283–294. https://doi.org/10.1207/S15327957PSPR0604_03

Simon D, Read SJ (2023) Toward a general framework of biased reasoning: coherence-based reasoning. Perspect Psychol Sci 12:579. https://doi.org/10.1177/17456916231204579

Simon D, Pham LB, Le QA, Holyoak KJ (2001) The emergence of coherence over the course of decision making. J Exper Psychol Learn Memory Cognit 27(5):1250–1260. https://doi.org/10.1037/0278-7393.27.5.1250

Snapper KJ, Peterson CR (1971) Information seeking and data diagnosticity. J Exper Psychol 87(3):429–433. https://doi.org/10.1037/h0030557

Söllner A, Bröder A, Hilbig BE (2013) Deliberation versus automaticity in decision making: which presentation format features facilitate automatic decision making? Judgment Dec Making 8(3):278–298. https://doi.org/10.1017/S1930297500005982

Svenson O (1979) Process descriptions of decision making. Organ Behav Hum Perform 23(1):86–112

Thagard P (1989) Explanatory coherence. Behav Brain Sci 12(3):435–467. https://doi.org/10.1017/S0140525X00057046

Thorngate W (1980) Efficient decision heuristics. Behav Sci 25(3):219–225. https://doi.org/10.1002/bs.3830250306

Tversky A (1969) Intransitivity of preferences. Psychol Rev 76(1):31–48. https://doi.org/10.1037/h0026750

Usher M, McClelland JL (2001) The time course of perceptual choice: the leaky, competing accumulator model. Psychol Rev 108(3):550–592. https://doi.org/10.1037/0033-295X.108.3.550

Von Neumann J, Morgenstern O (1947) Theory of games and economic behavior, 2nd edn. Princeton University Press, Princeton

Voss A, Nagler M, Lerche V (2013) Diffusion models in experimental psychology. Exper Psychol 60(6):385–402. https://doi.org/10.1027/1618-3169/a000218

Wertheimer M (2017) Untersuchungen zur Lehre von der Gestalt: I. Prinzipielle Bemerkungen. Gestalt Theory 39(1):79–89

Yang X, Krajbich I (2023) A dynamic computational model of gaze and choice in multi-attribute decisions. Psychol Rev 130(1):52–70. https://doi.org/10.1037/rev0000350

Chapter 7
LATER Models of Reaction Times and Decisions in Neurophysiology

Imran Noorani and Andrew Anderson

Abstract Choosing between options in the face of uncertainty in the information presented from the outside world reflects a crucial function of the brain. A simple conceptual framework for understanding such decision-making is the LATER (Linear Approach to Threshold with Ergodic Rate) model. LATER accounts for random variability in reaction times by postulating that neural activity rates can vary randomly between one decision and the next. The model has been developed to account for reaction times in not only simple decision tasks, but also more advanced ones particularly in oculomotor neurophysiology, such as the anti-saccade task that is popular in clinical research for neurological conditions. This chapter discusses the history and evolution of the LATER model, and future applications.

Keywords Decision-making · LATER model · Rise-to-threshold model · Saccade · Reaction times

7.1 Introduction

How does one choose between multiple options for action when presented with uncertainty in the information and in the potential consequences of each action? Governed by Bayesian probability principles, the LATER (Linear Approach to Threshold with Ergodic Rate) model has emerged as a useful conceptual tool for predicting response outcomes, and reaction times (latencies), in choice tasks. Originally conceived for explaining decision times in simple dichotomous choice tasks in the laboratory, the LATER model—with its few free parameters and ability to predict complete reaction time distributions—has been applied to increasingly complex decision-making tasks, enabling the prediction of errors and the timing and frequency of outcomes in tasks

I. Noorani (✉)
Department of Neurosurgery, National Hospital for Neurology and Neurosurgery, University College London, London, UK
e-mail: imran.noorani@cantab.net

A. Anderson
Department of Optometry and Vision Sciences, University of Melbourne, Parkville, Australia

U. Ettinger et al. (eds.), *Decision Making*, Studies in Neuroscience, Psychology and Behavioral Economics, https://doi.org/10.1007/978-3-032-00880-0_7

with multiple potential responses. As such, the model has become more widely used for clinical research to assess progress of neurological diseases, such as dementia and movement disorders (Antoniades et al. 2012; Burrell et al. 2012). In this chapter, we discuss the importance of understanding reaction or decision-making times, the use of saccadic tasks for studying decision times, the history behind the LATER model and its application to such decision tasks, and how such models can be helpful in the clinic for potentially quantifying deficits in neurological disorders.

7.2 Reaction Times as a Measure of Decision

The saccade is an eye movement we make two or three times each second while we are awake; even more frequently than our heartbeat. Yet, surprisingly, it takes nearly 200 ms on average after a visual stimulus is presented for us to start making this saccadic eye movement. Why are we so slow to initiate such a movement in response to a given sensory stimulus? This question has intrigued physiologists for many years. Even by accounting for the physiological processes required for an eye movement to occur in response to a visual stimulus, it becomes clear that simple sensory and motor delays do not account for this long latency prior to moving. For example, it takes approximately 40 ms for sensory information to travel from the eye to the superior colliculus in the brainstem, and then another 20 ms for the motor outputs to travel from the superior colliculus to the ocular muscles to generate an eye movement. Clearly, this duration of 60 ms is far less than the average 200 ms needed to initiate an eye movement. This ocular motor procrastination, where there seems to be a deliberate delay in initiating a response, is known to reflect the brain's decision-making period during which neural processes are at play evaluating incoming sensory evidence to make a decision on the best course of action. The LATER model was introduced over four decades ago to precisely explain this phenomenon in quantitative detail (Carpenter 1981b, a; Noorani 2014; Noorani and Carpenter 2016). It is worth noting that shorter latency saccades (70–100 ms or so), alternatively called 'early' saccades, do constitute a subpopulation of our saccades; the physiology behind these has been reviewed elsewhere.

In addition to the surprisingly long duration of reaction times, another interesting feature is their great variability. Even for simple eye movement tasks—such as the step task, where an observer moves their eye to an abruptly appearing target—for the same individual we can observe that from one trial to the next there is substantial variability in reaction times for the same response. When we plot reaction time (T) distributions as histograms, we can observe that there is universally a positive skew towards longer reaction times. However, when histograms are plotted as a function of the *reciprocal* of reaction time (1/T), there is a remarkable change—the distributions lose their skewness and instead followed a normal or Gaussian distribution (Carpenter 1981b, a; Noorani and Carpenter 2016). Moreover, if we use a nonlinear probit scale for frequency on the vertical Y axis, then these normal distributions will generate a straight line. Such plotting is called a reciprobit plot.

From a mathematical perspective, each distribution can then be described with two parameters alone: the mean or median (μ) and the slope of the line (equivalent to the standard deviation, σ). Full reaction-time distributions closely follow straight lines when plotted on reciprobit plots not only for eye movement tasks, but also for example for auditory tasks and those requiring manual responses to stimuli, suggesting that this is a more universal approach to interpreting decision reaction times.

7.3 LATER Model

From an empirical point of view, the fact that many reaction time distributions can be described using only two parameters (μ and σ) in reciprobit space suggests the basis of a simple model for neural decisions. This model is called the Linear Approach to Threshold with Ergodic Rate model, or LATER in short, with the model's name also serving as a reminder that neural procrastination is key to the process of decision-making. The model assumes that there is a decision signal or process (S) that rises towards a threshold (S_T) in response to the presentation of an outside stimulus (Fig. 7.1). Once this signal reaches the threshold, a response is generated. If the rate of rise varies randomly between trials, then 1/T will also be distributed normally and give rise to the skewed distribution of T that is observed in practice. Why such random variation would be deliberately introduced into a decision-making model will be discussed later in this chapter.

Although needing only a small number of parameters to explain full reaction time distributions is clearly of benefit for scientific analysis, it must be borne in mind that

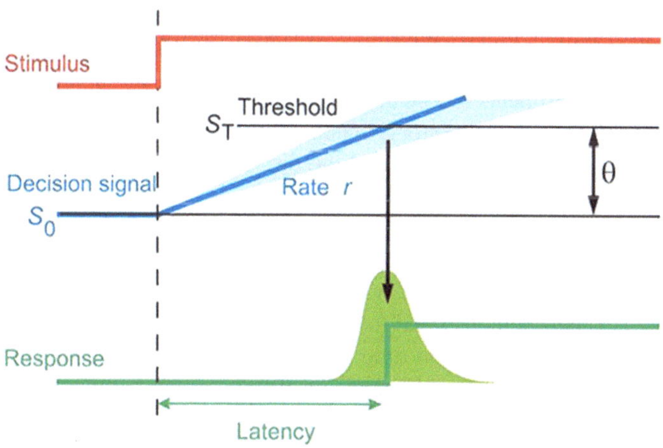

Fig. 7.1 A schematic diagram of the LATER model. In this model, once a stimulus is presented to the subject a decision signal begins to rise linearly towards a threshold at which point a decision is made. The rate of rise of the signal varies randomly from trial to trial, although the mean rate, r, is constant

better fitting of such data can be done if more free parameters are added. Indeed, this is the case for alternative but similar models to LATER, such as the linear ballistic accumulator (LBA) model and the extended LATER model (Nakahara et al. 2006; Brown and Heathcote 2008). In these models not only is the rate of rise allowed to vary from trial to trial, but also the distance to threshold varies.

So how does LATER function as a decision-making model and what are its underlying tenets? An ideal decision-making system needs to be able to make decisions about the existence of stimuli in the world based on noisy incoming sensory signals. To start the decision process, the decision signal S has the value S_0 that represents the log prior odds that this particular response will occur in a trial: for example, the probability of a visual stimulus being on the right as opposed to the left, and so requiring the decision to look right. When the visual stimulus is then presented, S will begin to rise steadily at a rate r as more incoming sensory information is provided for the stimulus on the right. Once the log odds reach the threshold evidence level S_T, this leads to a decision taking place which, in this instance, will be to look towards the stimulus on the right. Further details into the mathematical basis of this Bayesian approach to decision-making in LATER can be found in Carpenter and Noorani (2023).

7.4 Testing the Variables in LATER

An important feature of any scientific hypothesis or model is that there are testable predictions that can be made from it. In the case of LATER, altering the distance to threshold will cause the distribution to swivel around the intercept K (Fig. 7.2), whereas changing the mean rate of rise to threshold will cause a horizontal parallel shift in the distribution. Finally, altering σ will cause the distribution to rotate around its median.

Fortunately, experiments can be done to test the validity of this model by manipulating its variables in the laboratory. Specifically, one can manipulate the Bayesian factors of the prior probability (S_0) and the threshold level S_T. Changing the level of stimulus expectation (S_0) is simple: for example, we can make a stimulus appear more often on the right than on the left, thereby altering the prior probability for stimulus existence on the right. In this scenario, by increasing S_0, the reaction times will decrease, and we should see a swivel of the reciprobit plots around a constant intercept K. This was demonstrated to be the case in human subjects, with a reduction of 80 ms in median reaction time per log unit of prior probability (Carpenter and Williams 1995). Similarly, the decision threshold can be altered by changing the level of urgency in a decision task. Here, the subject is told to either respond as quickly as possible and not worry about making mistakes (high urgency) or, on the contrary, to take their time and avoid making any errors (low urgency). Clearly, this reflects a trade-off between speed and accuracy. Results show that if the subject is told to respond as fast as possible with high urgency, then the reciprocal plot of reaction times swivels around a fixed intercept with a reduction in latency, as would

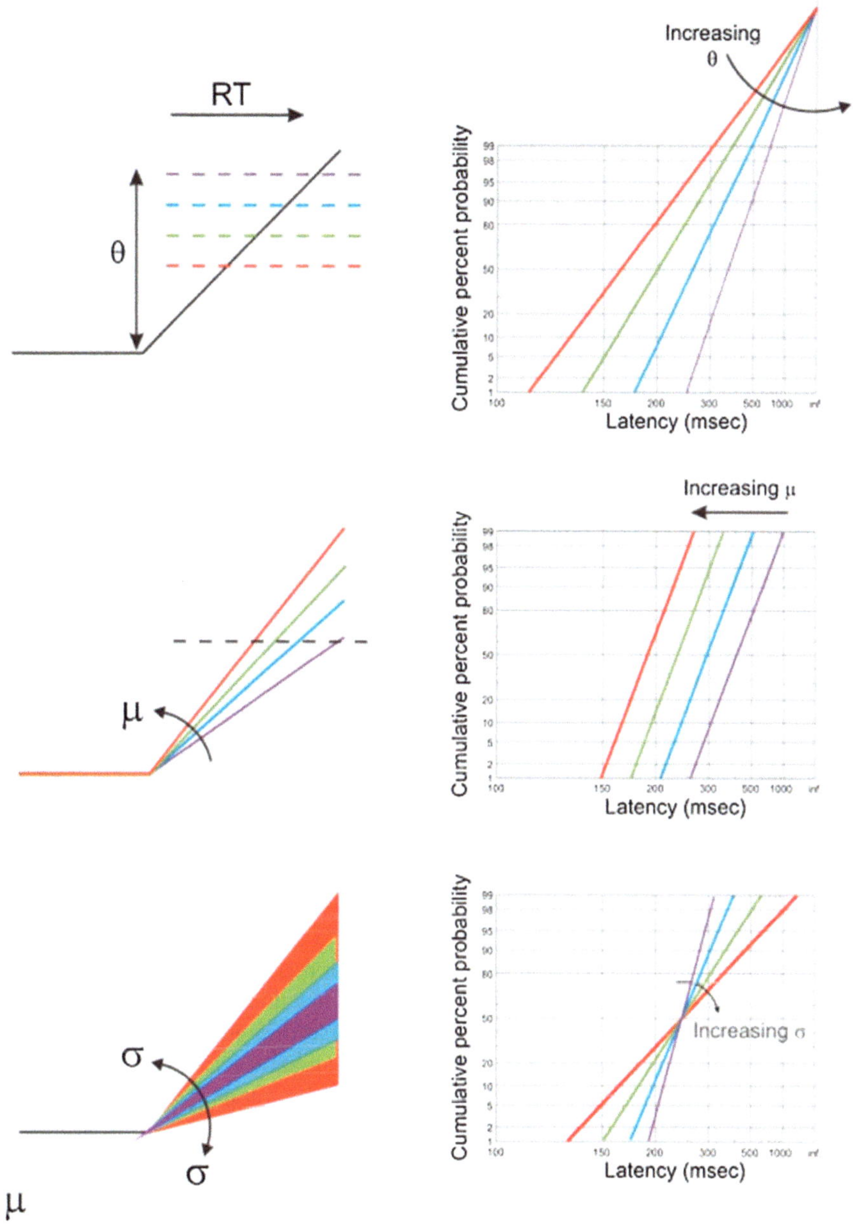

Fig. 7.2 There is a predictable relationship between the parameters of the LATER model and the parameters of the reciprobit plot. In the top panel, we can see that variation in the distance to threshold leads to a swivel of the plot around the infinite intercept; in the second panel, we observe that changes in the mean rate rise (μ) this leads to a shift or a parallel translation of the plot; in the lower panel, we show that changes in the variance of LATER results in a change in the slope of the distribution without a change in the median latency

be predicted from a reduction in the threshold S_T in the LATER model (Reddi and Carpenter 2000). The rate of rise towards threshold r is governed by the rate of incoming information in support of a particular response, and can also be manipulated. A useful task for doing this in the laboratory is the random dot kinetogram (RDK) in which continually moving random dots on the screen contain a proportion of dots moving in one direction (the signal) and the rest moving randomly. The subject's task is to determine in which direction the signal dots move (Britten et al. 1993).

If we alter the proportion of signal dots in the RDK task, we are correspondingly changing the rate of information supply r to the subject, which would be the equivalent of increasing the rate of rise towards threshold in the LATER model. This would be predicted to cause a shift in the median latency towards a lower time rather than a swivel of the reciprobit plot. Indeed, such results are verified in experiments, further adding credence to the LATER model representing a simple way of explaining Bayesian based decision-making (Carpenter and Williams 1995).

7.5 Complex Decision Tasks and LATER

It is useful to see how we may conceptualize more complex decision-making tasks, such as when a previously planned action needs to be cancelled because of newly presented information and perhaps an alternative action performed (Carpenter and Noorani 2017; Noorani 2017; Noorani and Carpenter 2017). For example, in a countermanding task (a "stopping" task), a subject typically makes a saccade towards a peripheral target as for the step task above. However, in randomly interleaved trials, an extra stimulus is presented instructing the subject not to make the saccade –the "stop" signal (Hanes and Schall 1995; Hanes and Carpenter 1999). The probability of a successful cancellation of a saccade is dependent on the time between the presentation of the original stimulus and the stop signal (time D): with a large D, there is a low chance of a successful cancellation or stop (Logan et al. 1984). We can model the behaviour in this task, including the full reaction time distributions, using a LATER model incorporating a go LATER unit that responds to the stimulus with the saccade, and an additional stop LATER unit that is activated by the stop stimulus. These two LATER units are racing towards threshold in parallel, and if the stop unit reaches threshold first, it will disable the go unit and enable successful cancellation of the upcoming saccade (Hanes and Carpenter 1999). The randomness in rate of rise towards threshold between trials of these two units is what generates the randomness of successful cancellation of a saccade in a given trial. The μ and σ parameters of the go and stop unit are independent from one another.

The Go/No Go task is related to the countermanding task, in the sense that a subject must respond to a certain stimulus but not another. For example, they must look towards a blue stimulus but not towards a red one. This task has a similarity with countermanding in that colour information takes longer to arrive in the relevant cortical brain regions than information regarding the existence of a visual stimulus (akin to the D value in countermanding). If we study the reaction time distributions in the Go/No Go task, we can see that the earliest saccades have a similar latency distribution regardless of which colour stimulus is being responded to. Only at later times, do we see the reaction times for a blue stimulus and a red stimulus begin to diverge dramatically, and very few responses to the wrong colour are made. This can be quantitatively modelled by a stop unit being activated by the delayed arrival of colour information which can cancel planned responses to the wrong color and activate a go unit that appropriately responds to the right colour (Noorani and Carpenter 2011; Noorani et al. 2011).

A particularly popular complex decision task that is widely used in both clinical and basic research laboratories is the anti-saccade task (Hallett and Adams 1980; Antoniades et al. 2013). In this task, the subject is presented with a visual stimulus just like in the step task, except that here the subject must try to suppress their response towards the target and instead must try to make a saccade in the opposite direction. In a considerable proportion of trials, the subject makes errors and cannot help but look towards the target (a pro-saccade). However, a similarity between the anti-saccade task and the countermanding and Go/No Go tasks described above is the need for a stop mechanism to prevent errors. Therefore, the anti-saccade task can similarly be conceptualized as—and, indeed, quantitatively modeled by—two Go units (one driving the pro-saccade and the other driving the anti-saccade) and a stop unit all racing against one another towards threshold, with the winner of the race determining the outcome. Note that the anti-saccade unit is activated after a time delay corresponding to the extra time needed for the brain to transform the direction of the stimulus to the mirror image goal. This model is able to predict in fine detail the reaction time distributions of both the errors and anti-saccades (Fig. 7.3), in addition to the corrective anti-saccades that subjects make after making errors (Noorani and Carpenter 2013; Noorani and Carpenter 2014a, b). There is some correlative neural recording evidence in animal models that there are competing basal ganglia pathways representing such a race to threshold underlying these responses (Schmidt, Leventhal et al. 2013; Noorani and Carpenter 2014a, b). There are of course other models of the anti-saccade task (such as (Cutsuridis, Smyrnis et al. 2007), though a review of these models is beyond the scope of this Chapter.

In summary, the findings above suggest that various implementations of LATER can model reaction times across a range of tasks. It would, however, be almost inconceivable that truly distinct instantiations of LATER exists for each task. What is more plausible is that rise-to-threshold LATER units are able to be flexibly deployed as appropriate to the task at hand. There is some behavioural evidence to support this (Anderson et al. 2021). Quite different tasks can contain some trials in which the stimulus appearance—and the required response—are the same: for example, the presence of pro-saccadic trials (i.e., trials made directly to a solitary appearing target) in step and countermanding tasks. Fitting reaction times to these trials suggests at least some LATER parameters are unchanged across tasks, which might be expected if the same LATER units were deployed within each task (Anderson et al. 2021).

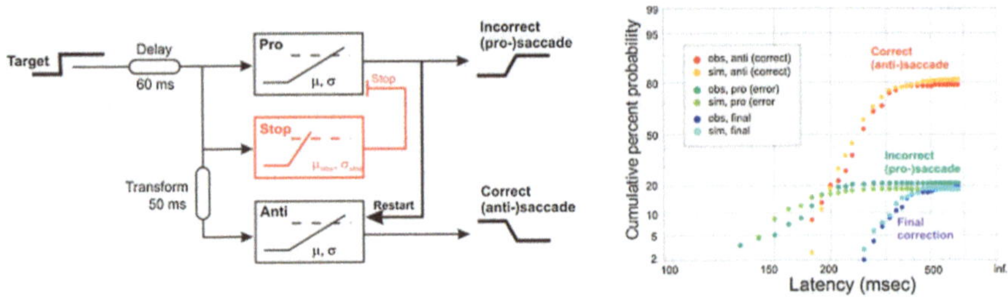

Fig. 7.3 The antisaccade task. In the left panel, a LATER model contains two identical go units (Pro and Anti), with the latter unit only being activated after the time delay; there is also a stop unit that cancels the incorrect pro-saccade. In the case of an error being made, the anti saccade unit is restarted and generates the final corrective saccade. In the right panel, we can see that the model simulates the full distributions of all types of saccade made by an individual subject in this task

7.6 Other Reaction Time Models

Whilst LATER can explain reaction time variability, there are other types of decision-making models that do likewise. Arguably the most common class of decision-making models are *diffusion* and *random walk* models (Ratcliff et al. 2016). Here, a sensory signal is integrated until some threshold boundary is reached.[1] For example, imagine that a person's task is to decide whether an experimental stimulus has been increased or decreases in intensity. Because sensory signals are noisy, moment-to-moment variation in the sensory information encoding stimulus intensity causes the integrated signal to wander: one moment it increases, the next moment it may decrease. However, if there is a bias in the sensory signal—that is, on average, the signal has increased—then the meandering integrated signal will tend to rise and eventually cross a threshold corresponding to the decision that the stimulus has indeed increased. Random variation in the sensory signal mean that the integrated signal will reach the the decision boundary at differing times. Hence, reaction times will vary on a trial-by-trial basis. Further variability can occur through between-trial random variation of model parameters, although it has been argued that such variations may not be critical to the overall behaviour of diffusion models (Tillman et al. 2020). Whilst any number of competing diffusion signals can be set racing against each other to decide between competing outcomes, a typical approach is for the integrated signal to have upper and lower decision boundaries representing two competing outcomes, with the assumption that sensory evidence in support of one outcome is evidence against the other (Ratcliff and Rouder 1998). The experimental situation of deciding between an increase or decrease in stimulus intensity would be an example of such a situation. Both diffusion models and LATER and can produce similar reaction time distributions (Ratcliff 2001), and large datasets and an analysis

[1] The key difference between diffusion and random walk models is that in the former, integration is continuous over time, whereas in the latter, integration occurs in discrete steps Ratcliff, R. and J. N. Rouder (1998). "Modeling response times for two-choice decisions." Psychological Science 9(5): 347–356.

of complete reaction time distributions may be needed to distinguish between them (Noorani and Carpenter 2011). The models do, however, have clear philosophical differences. In the former, randomness is the result of noisy sensory information. In the latter, randomization is a deliberately introduced strategy.

Which model is correct? It is possible that both are. It has been suggested that responding to a stimulus requires two stages—detection of the stimulus elements, followed by a decision stage (Ratcliff 2001; Ludwig et al. 2005; Carpenter, Reddi et al. 2009). This allows for the idea that certain low-level stimulus features may be common to competing hypotheses (for example, the presence of a vertical stem when deciding whether the next letter is a "k" or an "h") making it inappropriate for the detection of such features to be influenced by factors like prior probability, that may differ between hypotheses (Carpenter et al. 2009). In a particular model instantiating this two-stage idea, the initial detection stage operates using a random walk process, with the subsequent decision stage operating as a LATER process (Carpenter et al. 2009). When sensory stimuli are close to threshold, a greater time is spent in the detection phase and so reaction time behaviour is dominated by the random walk element. For stimuli well above threshold, detection is rapid, and reaction time behaviour is dominated by the subsequent LATER decision mechanism. Such a model helps explains the observation that visual information appears to be fully integrated more quickly, and less variably, than would be predicted by the saccadic reaction times in response to this visual information (Ludwig et al. 2005). It also explains why the effects of contrast and the prior probability of a stimulus appear to add linearly and independently (Carpenter 2004; Taylor et al. 2006), and why reaction time distributions over a wide range of stimulus saliencies cannot be predicted based on a random-walk or LATER unit alone (Carpenter et al. 2009). Neurophysiological evidence also supports the idea that the timing and variability of response times to salient stimuli is related to processing stages subsequent to target discrimination when stimuli are salient (Thompson et al. 1996), but not when target discrimination is challenging (Sato et al. 2001). Behavioural experiments have confirmed the idea of a two-stage model to turn sensory evidence into action (Thura et al. 2012; Ruter et al. 2013) and neurophysiological models have also been developed that embody two sequential stages (Schall et al. 2011; Sun and Landy 2016; Servant et al. 2019). It may be that the highly discrete stages outlined in models are neurally implemented within integrated, dynamic systems, however (Thura et al. 2022).

7.7 The Need for Randomization

As noted above, inherent in the LATER model is the idea that variability in decision-making time is not a result of unavoidable noise within biological sensory systems, but rather a deliberately introduced process. This variability means that there is variability in which LATER unit wins the race to threshold, and so our behaviours show a level of randomization on a trial-by-trial basis. Randomization is potentially useful for breaking the deadlock inherent in the classic "Buridan's Ass" decision-making paradox (Rescher 2006). A hungry ass, placed equidistant between two identical bales of hay, cannot decide between these perfectly balanced options, and so starves to death. However, only a small amount of variability would be required here, and so solving this highly unusual situation seems a poor justification for the substantial variation—in the order of hundreds of milliseconds, even for very salient stimuli—observed for reaction times. Furthermore, the random variation inherent in diffusion models would be sufficient, without the need to introduce variability deliberately.

While substantial randomness of timing per se may not matter very much, it does mean that there will be randomization of the final choice not merely for the situation where outcomes are *perfectly* balanced (as with Buridan's ass), but also for a range of choices where only an approximate balance exists. At first sight, such randomisation appears wrong-headed and to undermine the "optimal" nature of a deterministic Bayesian process. However, it is necessary to take a wider view of what may be "optimal" for an organism, and the possible benefits of randomization.

Randomization can have an impact an organism's survival. For prey, it is important that decisions whilst fleeing be unpredictable, and not fully determined by sensory input (most of which is, of course, equally available to the pursuing predator!). This is also true for humans, where game theory says that randomising your choice is the optimal strategy for outwitting opponents in certain circumstances. Even in situations where Bayesian approaches might be theoretically optimal given unconstrained processing time and/or resources, incorporating randomisation may be optimal in practice when there are limits on the amount of memory available for calculating decisions (Icard 2021).

Less ruthlessly than outwitting opponents and predators, introducing randomness into decisions means sometimes we do things that are contrary to what prior evidence suggests is best, or what current sensory evidence most supports. This may produce terrible mistakes, but sometimes it will produce novel, beneficial responses that are better than those currently in our repertoire. A striking example of these potential benefits was demonstrated in an analysis of people's use of public transport, as revealed by tracking electronic ticketing data (Larcom et al. 2015). A two-day strike closed certain portions of London's underground train network, forcing regular commuters to seek alternative routes to work. Remarkably, a significant proportion of commuters failed to return to their regular work route after the strike ceased: the "forced experimentation" produced by the strike had shown these commuters that their previous travel route was not, in fact, optimal. Having variability deliberately injected into our decision-making similarly provides a way to gently "force" exploration and discovery. On a smaller scale, the benefits of exploration are demonstrated in work showing that motor variability—rather than being unwanted noise to be suppressed—facilitates the rapid learning of new actions (Wu et al. 2014). Of course, getting the correct balance versus exploring for new knowledge (the "exploration-exploitation dilemma") is potentially challenging (Berger-Tal et al. 2014).

Whilst randomness of choices may be useful for surprising opponents, it raises an interesting problem: if our actions are the result of a random process like LATER, shouldn't they be just as surprising to ourselves as they are to others (Carpenter 1984)? Undoubtedly, we are occasionally surprised by our actions, although in general we have the distinct impression that our actions are consciously controlled by us, and not random at all. A potential solution to this problem is that we are in a unique position to witness both the rise-to-threshold and the outcome of a race-to-threshold prior to any resultant action. Such foresight could lead to the impression that we consciously willed an action—that we have free will—even if the decision was ultimately made by mechanisms outside our control. Willed actions therefore become those we have seen being prepared (Carpenter 1984).

7.8 Fitting Data Using the LATER Model

Whilst the LATER model is simple when compared with other decision-making models, visualising and fitting cumulative latency distributions in LATER's reci-normal space may be challenging for some. Previously, the computer program "SPIC"—developed by LATER's author, the late Professor Roger Carpenter—allowed data to be visualised and modelled without any computer programming experience (Carpenter 1994). This program is no longer freely available, however. Recently, an open-source *R Shiny* application "LATER fitting" for fitting data with the LATER model has been developed (https://later.researchsoftware.unimelb.edu.au) to replicate some of SPIC's core functionality. This new application operates through a graphical user interface, and so requires no programming. However, its functionality is ultimately derived from an open-source R package "LATERmodel", whose functions can be utilized directly by those familiar with programming in R (available through GitHub: https://github.com/unimelbmdap/LATERmodel/, and CRAN: https://cran.r-project.org/web/packages/LATERmodel/index.html).

7.9 Conclusions and Future Directions

At its core, the LATER model represents arguably the simplest model for reaction time distributions. Despite this simplicity, LATER units can be flexibly arranged to explain behaviour across a range of potentially complicated tasks. A key philosophical distinction between LATER and many other decision-making models is its reconceptualization of what random components in decision-making models mean. Rather than being an undesirable source of noise to be minimised, randomisation in LATER is deliberately injected—even for highly salient stimuli—to ensure a degree of unpredictability in behaviour.

The LATER model was originally conceived from an analysis of empirical distributions of reaction times (Carpenter 1981b, a). Subsequently, there have been many human behavioural experiments designed to test specific predictions of LATER, as described above. However, it has long been noted that aspects of LATER are at least consistent with what is known from electrophysiological experiments in monkeys (Reddi 2001). In future, electrophysiological experiments that specifically address key aspects of the LATER model would increase the rigour with which the model is tested and give further insights into how precisely LATER could be instantiated in the brain. In addition, the use of high-temporal resolution neuroimaging offers the possibility of conducting similarly targeted testing in human observers (Smyrnis et al. 2012).

References

Anderson AJ, Smyrnis N, Noorani I, Carpenter RHS (2021) Modelling prosaccade latencies across multiple decision-making tasks. Neuroscience 452:345–353

Antoniades C, Ettinger U, Gaymard B, Gilchrist I, Kristjansson A, Kennard C, John Leigh R, Noorani I, Pouget P, Smyrnis N, Tarnowski A, Zee DS, Carpenter RH (2013) An internationally standardised antisaccade protocol. Vision Res 84:1–5

Antoniades CA, Carpenter RH, Temel Y (2012) Deep brain stimulation of the subthalamic nucleus in Parkinson's disease: similar improvements in saccadic and manual responses. NeuroReport 23(3):179–183

Berger-Tal O, Nathan J, Meron E, Saltz D (2014) The exploration-exploitation dilemma: a multidisciplinary framework. PLoS ONE 9(4):e95693

Britten KH, Shadlen MN, Newsome WT, Movshon JA (1993) Responses of neurons in macaque MT to stochastic motion signals. Vis Neurosci 10(6):1157–1169

Brown SD, Heathcote A (2008) The simplest complete model of choice response time: linear ballistic accumulation. Cogn Psychol 57(3):153–178

Burrell JR, Hornberger M, Carpenter RH, Kiernan MC, Hodges JR (2012) Saccadic abnormalities in frontotemporal dementia. Neurology 78(23):1816–1823

Carpenter R (1981) Oculomotor procrastination. In: Fisher DF, Monty RA, Senders JW (eds) Eye movements: cognition and visual perception Lawrence Erlbaum, Hillsdale, pp 237–246

Carpenter R, Noorani I (2017) Movement suppression: brain mechanisms for stopping and stillness. Philos Trans R Soc Lond B Biol Sci 372(1718)

Carpenter R, Noorani I (2023) LATER the neurophysiology of decision-making. Cambridge University Press

Carpenter RH, Williams ML (1995) Neural computation of log likelihood in control of saccadic eye movements. Nature 377(6544):59–62

Carpenter RHS (1981) Oculomotor procrastination. In: Fisher DF, Monty RA, Senders JW (eds) Eye movements: cognition and visual perception. Lawrence Erlbaum Associates, Hillsdale, NJ, pp 237–246

Carpenter RHS (1984) Neurophysiology. Edward Arnold, London

Carpenter RHS (1994) SPIC: a PC-based system for rapid measurements of saccadic responses. J Physiol (Proc) 480(4P)

Carpenter RHS (2004) Contrast, probability and saccadic latency: evidence for independence of detection and decision. Curr Biol 14(17):1576–1580

Carpenter RHS, Reddi BAJ, Anderson AJ (2009) A simple two-stage model predicts response time distributions. J Physiol 587(16):4051–4062

Cutsuridis V, Smyrnis N, Evdokimidis I, Perantonis S (2007) A neural model of decision-making by the superior colicullus in an antisaccade task. Neural Netw 20(6):690–704

Hallett PE, Adams BD (1980) The predictability of saccadic latency in a novel voluntary oculomotor task. Vision Res 20(4):329–339
Hanes DP, Carpenter RH (1999) Countermanding saccades in humans. Vision Res 39(16):2777–2791
Hanes DP, Schall JD (1995) Countermanding saccades in macaque. Vis Neurosci 12(5):929–937
Icard T (2021) Why be random? Mind 130(517):111–139
Larcom S, Rauch F, Willems T (2015) The benefits of forced experimentation: striking evidence from the London Underground network. Department of Economics Discussion Paper Series. Manor Road Building, Oxford OX1 3UQ, University of Oxford, pp 1–35
Logan GD, Cowan WB, Davis KA (1984) On the ability to inhibit simple and choice reaction time responses: a model and a method. J Exp Psychol Hum Percept Perform 10(2):276–291
Ludwig CJH, Gilchrist ID, McSorley E, Baddeley RJ (2005) The temporal impulse response underlying saccadic decisions. J Neurosci 25(43):9907–9912
Nakahara H, Nakamura K, Hikosaka O (2006) Extended LATER model can account for trial-by-trial variability of both pre- and post-processes. Neural Netw 19(8):1027–1046
Noorani I (2014) LATER models of neural decision behavior in choice tasks. Front Integr Neurosci 8:67
Noorani I (2017) Towards a unifying mechanism for cancelling movements. Philos Trans R Soc Lond B Biol Sci 372(1718)
Noorani I, Carpenter RH (2011) Full reaction time distributions reveal the complexity of neural decision-making. Eur J Neurosci 33(11):1948–1951
Noorani I, Carpenter RH (2013) Antisaccades as decisions: LATER model predicts latency distributions and error responses. Eur J Neurosci 37(2):330–338
Noorani I, Carpenter RH (2014a) Basal ganglia: racing to say no. Trends Neurosci 37(9):467–469
Noorani I, Carpenter RH (2014b) Re-starting a neural race: anti-saccade correction. Eur J Neurosci 39(1):159–164
Noorani I, Carpenter RH (2016) The LATER model of reaction time and decision. Neurosci Biobehav Rev 64:229–251
Noorani I, Carpenter RH (2017) Not moving: the fundamental but neglected motor function. Philos Trans R Soc Lond B Biol Sci 372(1718)
Noorani I, Gao MJ, Pearson BC, Carpenter RH (2011) Predicting the timing of wrong decisions with LATER. Exp Brain Res 209(4):587–598
Ratcliff R (2001) Putting noise into neurophysiological models of simple decision making. Nat Neurosci 4:336–337
Ratcliff R, Rouder JN (1998) Modeling response times for two-choice decisions. Psychol Sci 9(5):347–356
Ratcliff R, Smith PL, Brown SD, McKoon G (2016) Diffusion decision model: current issues and history. Trends Cogn Sci 20(4):260–281
Reddi BA (2001) Decision making: the two stages of neuronal judgement. Curr Biol 11:R603–R606
Reddi BA, Carpenter RH (2000) The influence of urgency on decision time. Nat Neurosci 3(8):827–830
Rescher N (2006) Choice without preference: the problem of "Buridan's ass." Stud History Logic, De Gruyter 10:91–139
Ruter J, Sprekeler H, Gerstner W, Herzog MH (2013) The silent period of evidence integration in fast decision making. PLoS ONE 8(1):e46525
Sato T, Murthy A, Thompson KG, Schall JD (2001) Search efficiency but not response interference affects visual selection in frontal eye field. Neuron 30:583–591
Schall JD, Purcell BA, Heitz RP, Logan GD, Palmeri TJ (2011) Neural mechanisms for saccade target selection: gated accumulator model of the visual-motor cascade. Eur J Neurosci 33:1991–2002
Schmidt R, Leventhal DK, Mallet N, Chen F, Berke JD (2013) Canceling actions involves a race between basal ganglia pathways. Nat Neurosci 16(8):1118–1124

Servant M, Tillman G, Schall JD, Logan GD, Palmeri TJ (2019) Neurally constrained modeling of speed-accuracy tradeoff during visual search: gated accumulation of modulated evidence. J Neurophysiol 121(4):1300–1314

Smyrnis N, Mylonas DS, Rezaie R, Siettos CI, Ventouras E, Ktonas PY, Evdokimidis I, Papanicolaou AC (2012) Single-trial magnetoencephalography signals encoded as an unfolding decision process. Neuroimage 59:3604–3610

Sun P, Landy MS (2016) A two-stage process model of sensory discrimination: an alternative to drift-diffusion. J Neurosci 36(44):11259–11274

Taylor MJ, Carpenter RH, Anderson AJ (2006) A noisy transform predicts saccadic and manual reaction times to changes in contrast. J Physiol 15(573):741–751

Thompson KG, Hanes DP, Bichot NP, Schall JD (1996) Perceptual and motor processing stages identified in the activity of macaque frontal eye field neurons during visual search. J Neurophysiol 76(6):4040–4055

Thura D, Beauregard-Racine J, Fradet CW, Cisek P (2012) Decision making by urgency gating: theory and experimental support. J Neurophysiol 108(11):2912–2930

Thura D, Cabana JF, Feghaly A, Cisek P (2022) Integrated neural dynamics of sensorimotor decisions and actions. PLoS Biol 20(12):e3001861

Tillman G, Van Zandt T, Logan GD (2020) Sequential sampling models without random between-trial variability: the racing diffusion model of speeded decision making. Psychon Bull Rev 27(5):911–936

Wu HG, Miyamoto YR, Gonzalez Castro LN, Olveczky BP, Smith MA (2014) Temporal structure of motor variability is dynamically regulated and predicts motor learning ability. Nat Neurosci 17(2):312–321

Chapter 8
Testing a Dynamic Dual-Process Model of Intertemporal Choice: Serial and Quasi-Parallel Processing

Adele Diederich

Abstract Dual process theories of decision making describe choice as the result of an automatic System 1, which is quick to activate but behaves impulsively, and a deliberative System 2, which is slower to activate but makes decisions in a rational and controlled manner. However, most existent dual process theories are verbal descriptions and do not generate testable qualitative and quantitative predictions. In this paper, I describe a formalized dynamic dual process model framework of intertemporal choice that allows for precise, experimentally testable predictions regarding choice probability and response time distributions. The framework is based on two-stage stochastic process models to account for the two postulated systems and to capture the dynamics and uncertainty involved in decision making. As a proof of principle, I show how an existing static-deterministic model on intertemporal choice (Loewenstein et al. 2015) can be implemented in the framework allowing for testable predictions. The focus of the present paper is on the architecture of the two postulated systems: serial versus parallel processing. One serial and six quasi-parallel model versions are tested on data from Zhao et al. (2019).

Keywords Intertemporal choice · Dual process models · Two-stage stochastic process · Serial processing · Parallel processing · Model comparison and data

8.1 Introduction

Intertemporal choices describe a situation which involves tradeoffs between current and future outcomes. Spend the money now or invest in retirement fonds? Accept the offered job or finish the university degree? Skip the health checkup or make the appointment for it? Buy the car now or wait for a future model? Many personal decisions and almost all policy issues, for instance on energy, environment, household spendings include an intertemporal component.

A. Diederich (✉)
Carl von Ossietzky University of Oldenburg, Oldenburg, Germany
e-mail: adele.diederich@uni-oldenburg.de

U. Ettinger et al. (eds.), *Decision Making*, Studies in Neuroscience, Psychology and Behavioral Economics, https://doi.org/10.1007/978-3-032-00880-0_8

In experiments on intertemporal choice, participants are typically asked to choose between (two) monetary choice options which vary regarding the amount of money offered and the time when it is delivered. The sooner or immediate reward R_I received at time t_1 is smaller than the later or delayed reward R_D available at a later time, $t_2 > t_1$. Empirical results have shown that participants tend to choose immediate rewards over delayed rewards which suggests that they discount the subjective value of future outcomes. Particularly, with increasing delaying time the future rewards seem to decrease even more in subjective value (e.g., Ballard et al. 2023, for list of studies).

A multitude of different models in various disciplines have been proposed to account for those observations referred to as temporal, time, or delay discounting (see comprehensive reviews in (behavioral) economics (Ericson and Laibson 2019) and psychology (Ballard et al. 2023; Dai and Busemeyer 2014)). More recently, dual-process approaches have been discussed in this context (e.g., Figner et al. 2010; Hare et al. 2009; Peters and Büchel 2011).

Dual-process approaches assume that (choice) behavior is a product of two interacting, antagonistic cognitive systems (Evans 2008; Gawronski and Creighton 2013; Grayot 2020). The specific descriptions and labels of the two processes depend on the research area and focus (e.g., automatic versus controlled (Schneider and Shiffrin 1977), intuitive versus analytic (Hammond 1996), heuristic versus analytic (Evans 2006), affective versus deliberate (Loewenstein et al. 2015; Stanovich and West 2000), intuitive versus reflective (Gürçay and Baron 2017), system 1 versus system 2 (Kahneman and Frederick 2002) and many more, see e.g., (DeNeys 2021; Evans 2008; Milli et al. 2021 for overviews and nomenclature). For simplicity, I will use System 1 and System 2 throughout for referring to the two types of processes.

Despite the growing number of perspectives advocating for the existence of two distinct systems, there is a notable scarcity of formal mathematical models capable of capturing the dynamics of their interaction and facilitating precise predictions. The majority of dual-process models are primarily verbal in nature, lacking testable quantitative predictions (see Diederich and Trueblood 2018). The assignment of observations to either system often occurs post hoc and is frequently ad hoc. Additionally, there are ongoing debates regarding the nature of the interaction between the two systems. Many theories propose a sequential occurrence, with the slower System 2 modifying the response of the faster System 1 (e.g., DeNeys 2021; Evans 2008; Milli et al. 2021, for reviews). Alternatively, a few approaches endorse competitive-parallel processing (DeNeys 2021; Evans and Stanovich 2013; Keren and Schul 2009; Milli et al. 2021), offering discussion and criticism but without presenting a mechanism for testing different architectures. Formalizing assumptions about dual processes to enable predictions and rigorous testing would address some of the concerns held by both proponents and opponents of this approach.

Instances where formal dual-process models have been proposed are limited, such as those for self-control (Fudenberg and Levine 2006), probabilistic reasoning (Klauer et al. 2010), risky decision-making (Loewenstein et al. 2015; Mukherjee 2010), and intertemporal choice (Fudenberg and Levine 2006; Loewenstein et al. 2015; Zhao et al. 2019). However, many of these models blend the attributes of the postulated systems, often combining values independently attributed to System 1 and 2, raising doubts about the validity of the two-system concept (e.g., Alós-Ferrer 2018; Keren and Schul 2009; Osman 2004).

More crucially, these models do not address the temporal aspect of the two systems, a fundamental element in verbal models based on dual process theories. For instance, the assumption that System 1 operates rapidly leads to associating quicker decisions with intuition. Similarly, the perception that System 2 functions slowly results in the belief that slower responses reflect rationality. However, as demonstrated in meticulous detail by Krajbich et al. (2015), relying on such reverse inference is problematic because it overlooks additional factors in the data that can impact decision speed, such as choice difficulty. Moreover, even when a processing architecture is mentioned (e.g. Evans 2008; Gürçay and Baron 2017; Kahneman 2011; Sloman 1996, for an overview), none of them integrates these factors into the model.

Diederich and Trueblood (2018) introduced a dynamic stochastic modeling framework for dual process theory, offering insights into the timing and interaction of the two systems. Drawing from multistage sequential sampling models investigated in prior research (e.g., Diederich 2016), the framework was originally employed to analyze risky choices within gain and loss frames (Diederich and Trueblood 2018 and intertemporal decision-making scenarios (Diederich and Zhao 2019), assuming a serial processing approach. In a recent paper, Diederich (2024) later extended the model framework to include parallel processing and conducted a comparative analysis against predictions made under serial processing assumptions.

The present study seeks to test the framework on intertemporal decision-making data published in Zhao et al. (2019) with a focus on evaluating different processing architectures, i.e. serial versus parallel.

8.2 Model Framework Description

First, I provide a very brief summary of the basic assumptions. For details and derivations refer to (e.g., Diederich 1997, 2024; Diederich and Busemeyer 2003; Diederich and Oswald 2014; Diederich and Trueblood 2018).

The processes underlying each system are stochastic in nature. The model parameters for interpretation are grounded in Decision Field Theory (DFT, Busemeyer and Townsend 1993) and its extension, Multiattribute Decision Field Theory (MDFT, Diederich 1997). In this context, decisions rely on a preference state denoted as $P(t)$, indicating the relative preference strength for choosing one option (in this case, the delayed reward, R_D) over another option (in this case, the immediate reward, R_I) at time t. A positive value of $P(t)$ signifies a momentary preference favoring option R_D, while a negative value represents a momentary preference favoring option R_I. The preference strength is updated from one moment, t, to the next, $(t + \tau)$, by an input valence, $V(t)$, which reflects the momentary comparison of consequences resulting from imagining the choice of either option R_D or R_I. The valence is computed as $V(t) = V_{R_D}(t) - V_{R_I}(t)$, where $V_{R_D}(t)$ represents the momentary valence for the delayed reward and $V_{R_I}(t)$ for the immediate reward. The valence fluctuates due to the decision maker's attention switching back and forth to anticipated consequences. For dual processes, it is assumed that the input valence also depends on the system i, where $i = 1, 2$, in which the decision maker (DM) is operating.

The preference dynamics are captured by the following equation:

$$P(t+\tau) = P(t) + V_i(t+\tau), \tag{8.1}$$

where τ represents a small-time increment, and $V_i(t)$ denotes the valence input specific to the system. The mean valence within system i, when assessing choice options over time t, is expressed as $E[V_i(t)] = \mu_i t$. Here, μ_i, referred to as the *drift rate*, signifies the direction and intensity of the preference towards selecting either option R_D or R_I.

Specifically, when $\mu_i > 0$, the preference leans towards choosing option R_D, while for $\mu_i < 0$, it inclines towards option R_I. The magnitude of μ_i directly correlates with the strength of preference for the given choice option. In the case of $\mu = 0$, the decision-maker (DM) remains indifferent between the two choices. The preference process concludes when the preference state surpasses a predefined decision threshold or boundary. If $P(t) < \theta_{R_I} < 0$, the DM opts for the immediate reward R_I, and if $P(t) > \theta_{R_D} > 0$, the DM selects the delayed reward R_D.

It is assumed that the decision threshold is established by the DM prior to the task and depends on various factors, including the time the DM has for making the decision. Under tight time constraints, the boundaries are assumed to be smaller compared to situations without time constraints. Consequently, the likelihood of choosing the delayed reward R_D over the immediate reward R_I is contingent on $P(t)$ reaching the positive threshold before reaching the negative threshold.

8.2.1 Valence Construction for Intertemporal Choice Situations

The modeling centers around the valance $V_i(t)$ in Eq. 8.1 and its corresponding drift rate μ_i. In System 1, the valence is defined as

$$V_1(t) = V_{1_{R_D}}(t) - V_{1_{R_I}}(t) \tag{8.2}$$

with drift rate μ_1, while in System 2, it is expressed as

$$V_2(t) = V_{2_{R_D}}(t) - V_{2_{R_I}}(t) \tag{8.3}$$

with μ_2.

The formulation of valences involves assumptions about how decision-makers (DM) evaluate alternatives. For instance, delayed rewards may be weighted using a discounting function, considering the delay time (see, e.g., Dai and Busemeyer 2014; Ericson and Laibson 2019, for a summary of various functions). Regardless of the specific functional form, the resulting value is eventually mapped onto the mean valence (drift rate).

Thus the magnitude of drift rates in System 1 relate to the following preferences:

$$\mu_1 \begin{cases} > 0 & \text{in favor of choosing } R_D \\ = 0 & \text{indifferent between choosing } R_D \text{ and } R_I \\ < 0 & \text{in favor of choosing } R_I \end{cases} \tag{8.4}$$

The same holds for System 2:

$$\mu_2 \begin{cases} > 0 & \text{in favor of choosing } R_D \\ = 0 & \text{indifferent between choosing } R_D \text{ and } R_I \\ < 0 & \text{in favor of choosing } R_I \end{cases} \tag{8.5}$$

As an example for how to construct the valences take the model proposed by Loewenstein et al. (2015).

8.2.1.1 The Loewenstein et al. (2015) Model

The model is a static, deterministic dual-process model with two processes: affective (System 1) and deliberate (System 2). System 1 preferences are constructed according to a motivational function $M(R, a)$ with variable a capturing the intensity of affective motivations (p. 59 Loewenstein et al. 2015). It is weighted by a function $h(W, \sigma)$ reflecting the willpower strength W and cognitive demands σ. For two rewards, one immediate and one delayed, the affective motivational function (System 1) is defined as

$$M(R, a) = R_I + \exp(-\delta_A(a) \cdot T_D) \cdot R_D \tag{8.6}$$

where $\delta_A(a)$ is the discounting factor of an exponential discounting function for the affective system. Increased affective intensity a implies a smaller $\delta_A(a)$ (p. 62 Loewenstein et al. 2015). $M(R, a)$ is weighted by $h(W, \sigma)$. The exact form of this function is not specified in Loewenstein et al. (2015) but is meant to be decreasing in W and increasing in σ.

Preferences in System 2 are defined as

$$U(R) = R_I + \exp(-\delta_D \cdot T_D) \cdot R_D \tag{8.7}$$

where δ_D is the discounting factor of an exponential discounting function for the deliberate system with $\delta_A(a) < \delta_D$.

Note that $M(R, a)$ and $U(R)$ have the same functional form. Bringing both processes together Loewenstein et al. (2015) assumed that they operate simultaneously, and the subjective value $\mathcal{V}$ of an option (R) is the sum of both processes, $M(R, a)$ weighted by $h(W, \sigma)$:

$$\mathcal{V}(R) = U(R) + h(W, \sigma) \cdot M(R, a). \tag{8.8}$$

8.2.1.2 Implementation

One way to implement the static-deterministic dual process model (Eq. 8.8) into the dynamic-stochastic dual process framework (Eqs. 8.2 and 8.3) is as follows.

Preference evolves in System 1 (affective) and System 2 (deliberate) by comparing the consequences when choosing the immediate reward R_I and the delayed reward R_D option, producing valence values $V_{R_I}(t)$ and $V_{R_D}(t)$, respectively, at any moment in time. The expected values of the valences in System 1 are $E[V1_{R_I}(t)] = h(W, \sigma) \cdot R_I \cdot t$ for the immediate reward and $E[V1_{R_D}(t)] = h(W, \sigma) \cdot \exp(-\delta_1 \cdot T_D) \cdot R_D \cdot t$ for the delayed reward, resulting in a mean difference in valence (which defines the drift rate) of

$$\mu_1 = h(W, \sigma) \cdot [\exp(-\delta_1 \cdot T_D) \cdot R_D - R_I], \tag{8.9}$$

where $\delta_A(a)$ is relabeled to δ_1 referring to the discounting factor for System 1. Similarly operating in System 2 produces a mean difference in valence

$$\mu_2 = \exp(-\delta_2 \cdot T_D) \cdot R_D - R_I. \tag{8.10}$$

with δ_D relabeled to δ_2 referring to the discounting factor for System 2.

This is closest to the ideas of DFT and MDFT. An alternative is to define the drift rates as in (Loewenstein et al. 2015) with $\mu_1 = h(W, \sigma) \cdot M(R, a)$ and $\mu_2 = U(R)$, i.e., the sum of the input instead of a comparison expressed as difference value. Or another model version may assume that only the immediate reward is considered in System 1 and a comparison of rewards occurs only in System 2. Or the discounting functions could be defined differently. Investigating these interesting topics is beyond the scope of this chapter. The focus here is on testing different architectures of the dynamic-stochastic dual process framework on published data. Before I present the model accounts I briefly summarize the ideas of serial and parallel processing in this context.

8.3 Dynamic-Stochastic Dual Process Framework: Architectures

Two different architectures of dual-process models have been proposed. The "default-interventionist" approach (Evans 2008, p. 266) assumes a sequential ordering of the two systems with System 1 starting first and System 2 intervening later. This is the prevalent assumption in reasoning and preferential choice research with System 1 driven by heuristics and mainly responsible for biases (e.g. Evans 2008; Kahneman 2011, for an overview).

The "parallel-competitive" approach (Evans 2008, p. 266) assumes that the systems operate in parallel and are competitive (Sloman 1996) (see Baron and Gürcay 2017; Gürçay and Baron 2017, for moral judgments).

Neither of these views has proposed a formal model to test the assumptions rigorously. In the following I illustrate the basic ideas of a serial dual process model, called "two-stage dynamic-stochastic dual model" and a parallel version, called "parallel dynamic-stochastic dual process model".

8.3.1 *Two-Stage Dynamic-Stochastic Dual Model*

Fig. 8.1 demonstrates the underlying ideas of the two-stage dynamic-stochastic dual model: System 1 (intuitive, affective) takes the initial lead before the reins are handed over to System 2 (rational, deliberate). The vertical dashed line in each panel indicates the point in time (t_1) when operating in System 1 switches to operating in System 2. Note that for demonstration purposes t_1 is fixed here but can follow a distribution (Diederich and Oswald 2014). The sequence of processes implies a transition in decision-making from predominantly relying on System 1 to predominantly engaging with System 2 at some point in time.

Figure 8.1 also shows μ combinations related to Eqs. 8.4 and 8.5. The model makes several qualitative and quantitative predictions for choice probabilities and mean choice response times patterns depending on the relation between μ_1 and μ_2 (Diederich and Trueblood 2018; Diederich and Zhao 2019). For instance, if μ_1 and μ_2 are either both positive or both negative and $|\mu_1| > |\mu_2|$ (Fig. 8.1: A, C, J, L), the model predicts faster mean response times for the more frequently chosen alternative, and if $|\mu_1| < |\mu_2|$ (Fig. 8.1: B, E, G, K), the model predicts faster mean response times for the less frequently chosen alternative. For $\mu_1 > 0 > \mu_2$ and $\mu_1 < 0 < \mu_2$ (Fig. 8.1: D, H) the response time patterns are more complex, however, the patterns remain consistent (Diederich and Oswald 2014; Diederich and Zhao 2019).

More important in this context, however, is how the amount of time operating in System 1 affects possibles choice probabilities. Following the labeling of Loewenstein et al. (2015), a preference for the immediate reward is called *myopic behavior* and the magnitude thereof a *myopic effect*.

Figure 8.2 illustrates the two-stage dynamic-stochastic dual model's predictions of the size of the myopic effect due to the dynamics of the two systems. Each colored trajectory in each panel indicates the accumulation process of one trial. Once a trajectory hits a criterion θ_{R_D} or θ_{R_I} a decision is made in favor of the alternative associated with the boundary. Across the panels the amount of time the decision maker operates in System 1 increases, switching later to System 2 (from A to D). In panel A, the DM switches to System 2 early on in the preference accumulation process, and the decision is mainly based on System 2, which favors the delayed

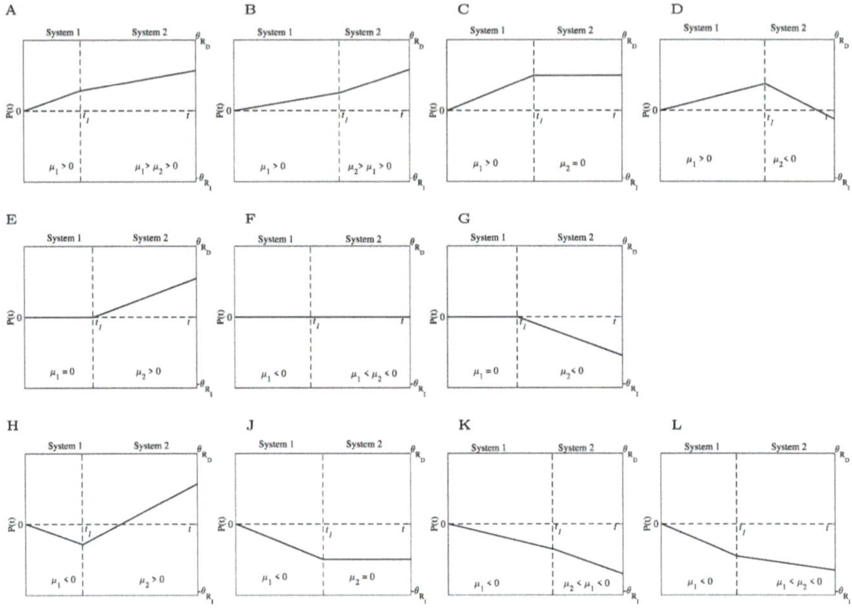

Fig. 8.1 Serial processing. A subset of drift rate combinations for systems 1 and 2 according to Eqs. 8.9 and 8.10. For the serial model version System 1 precedes System 2. Operating time in System 1 lasts t_1 time units before it switches to operating in System 2. Note that operating time may be a random variable T

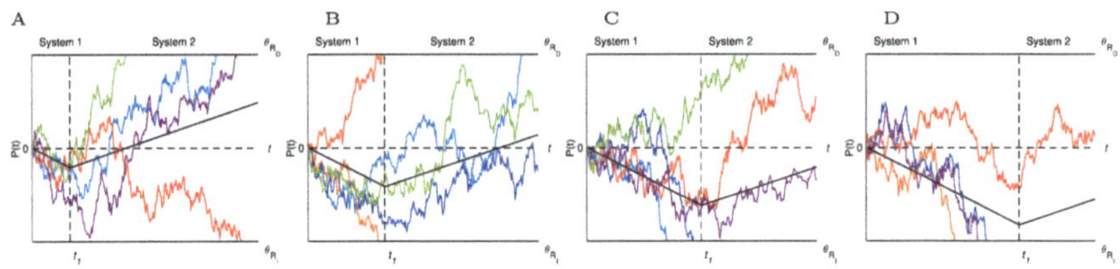

Fig. 8.2 Four different operating time durations for System 1 with drift rate $\mu_1 < 0$ in the direction of choosing the immediate reward R_I and drift rate $\mu_2 > 0$ in the direction of choosing the delayed reward R_D. Each colored trajectory indicates the preference accumulation process of one trial. The black line is the drift rate, i.e., the expected value of all possible colored trajectories. Once a trajectory hits a boundary, the process stops, and a response is initiated in favor of the associated alternative

reward, and therefore, the myopic effect is small. That is, no decision is made solely based on System 1. The longer the DM operates in System 1, the stronger the myopic effect is because more and more decisions are made relying only on System 1.

Note that operations in System 1 indirectly influence System 2, with System 2's preference process starting where System 1's process ends. Additionally, System 1 may influence System 2 by allowing the equations defining drift rate μ_2 to be a function of drift rate μ_1 (see examples in Diederich and Trueblood 2018).

8.3.2 Parallel Dynamic-Stochastic Dual Process Model

Diederich (2024) proposed several ways to model a “parallel-competitive” approach (Evans 2008, p. 266). The most straightforward approach is an independent race model, i.e. a race takes place between the processes in System 1 and System 2, and the winner of the race determines the response. However, the description of System 1 being much faster than System 2 (e.g.,Kahneman 2011) entails that System 1 would win the race in almost all cases. Consequently, all responses are determined by processes operating in the intuitive System 1, making System 2 redundant (Diederich and Trueblood 2018) (see also Alós-Ferrer 2018, for a critical evaluation of race models in this context).

An alternative is to give up the stochastic independence between the racers. This requires a two-dimensional stochastic process departing from the one-dimensional approach employed thus far. Recently, N-dimensional Wiener processes have been suggested for situations involving multiple alternatives (Mallahi-Karai and Diederich 2019, 2021). Dependencies between these processes can be modeled through covariances. Identifying specific dependencies, such as positive or negative correlations between the racers, is a task not addressed in this context.

Another alternative is to presume rapid alternation between the two systems, resulting in a pseudo-parallel processing with dependency. This approach is explored further in the following discussion, and Fig. 8.3 visually outlines the fundamental concepts. Assuming the drift rate μ_1 in System 1 to be smaller than the drift rate μ_2 in System 2 in absolute values ($|\mu_1| < |\mu_2|$) as shown in panels A and A′, the overall preference moves towards the boundary for choosing the delayed reward R_D. In contrast, assuming the drift rate μ_1 in System 1 to be larger than the drift rate μ_2 in System 2 in absolute values ($|\mu_1| > |\mu_2|$) as in panels B and B′, the overall preference moves towards the boundary for choosing the immediate reward R_I.

For illustration purposes the times spent operating in each system shown Fig. 8.3 are constant with longer times in each system in panels A and B and shorter times in panels A′ and B′. Similar to the two-stage dual process model the duration may follow a distribution; the underlying distributions and their expected values are referred to as *time schedule* (Diederich and Oswald 2014). Assuming identical operation time distributions in both systems, the switching frequency prominently hinges on the expected values of these distributionsâŁ"the mean time for operation in each system. Smaller expected values lead to higher switching frequencies (panels A′ and B′), while larger values result in lower frequencies (panels A and B).

Panels B″ and B‴ show the situation when operating times in both systems differ. In fact, the significance of one system in the decision process can be linked to the assumed distributions and/or their expected values. The μ values in panels B″ and B‴ are identical to the ones for B (and B′) but in panel B″ the operating time in System 1 is much shorter than in System 2, and the overall preference is directed towards choosing the delayed option – in contrast to the situation described in panel B. In panel B‴ the operating time in System 1 is much longer than in System 2, and the overall preference is directed towards choosing the immediate reward but

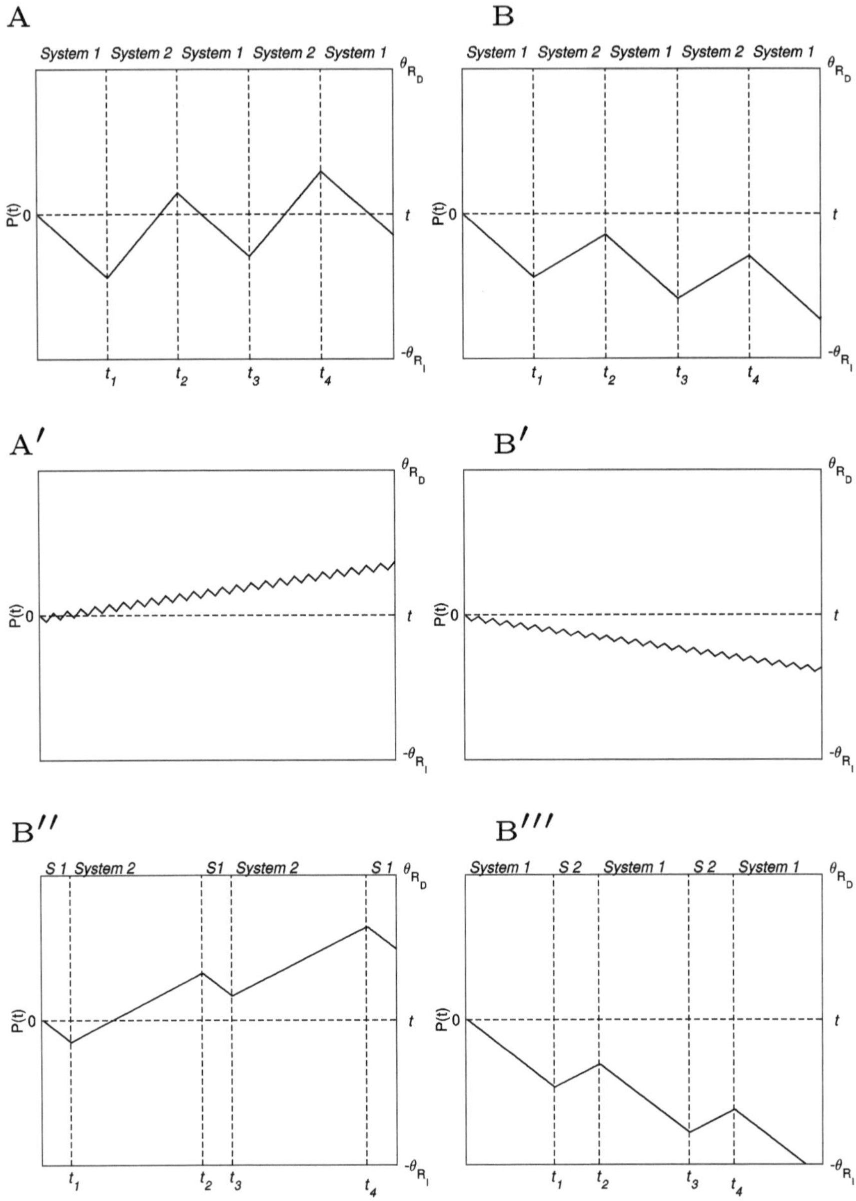

Fig. 8.3 Quasi-parallel processing. A subset of drift rate combinations for systems 1 and 2 according to Eqs. 8.9 and 8.10 with $\mu_1 < 0 < \mu_2$. Operating switches back and forth between the two systems. In the cartoons, System 1 takes the lead which, however, is to the default situation. See text for details

reaches a decision sooner than for the situation described in panel B. However, one can argue that a model closely resembling a parallel model should maintain identical distributions and expected values in both systems, as depicted in Fig. 8.3, A, B, A′ and B′.

The "default-interventionist" approach and therefore the two-stage dual process models holds that System 1 activation precedes the transition to System 2 engagement switching to operating in System 2. In the context of a parallel architecture, System 1 may initiate the process first or alternatively, System 2 may take the lead. It is also conceivable that both systems have an equal (or defined) probability (distribution) of initiating the process. The sequence in which the systems are considered is called *order schedule* (Diederich and Oswald 2014). Both time and order schedules are parameters of the model.

Arguably, a model closely aligned with a parallel structure should assign an equal probability for each system to initiate the process. That is, the probability to start with System 1 and with System 2, respectively, is both 0.5. Furthermore, the operating times in each system should have the same distributions with identical parameters, here referred to as "basic quasi-parallel" model. In this case, the operating times in System 1 and System 2 should have no or little effect on the size of the myopic effect, in contrast to the two-stage dual process model. Furthermore, the absolute larger drift rate determines the direction of the preference process towards the choice alternatives. Therefore, no preference reversals should occur as a function of operating times in both systems. It is more difficult to make qualitative predictions for the remaining time and order schedules (Diederich 2024)

Next, we test the model versions on data.

8.4 Testing the Models

The models are tested on a set of data from (Zhao et al. 2019). Participants (n = 51) made decisions between immediate and delayed rewards. The values of the immediate rewards were \$3, \$5.5, \$ 8.5, and \$11; the time delays T_D were 3, 7, 15, and 30 days; and multipliers for the delayed reward, R_D/R_I, were 1.1, 1.3, 1.7, and 2.5, producing 64 different choice pairs. Each choice pair was presented 10 times to each participant. Each choice proportion is based on about 504 observations (standard deviation 5.3).

The following four parameters apply to all model versions: The discounts factors δ_1 and δ_2 with $\delta_1 < \delta_2$ (Loewenstein et al. 2015) are coefficients as part of the drift rates μ_1 (Eq. 8.9) for System 1 and μ_2 (Eq. 8.10) for System 2. Because W and σ and the functional form of $h(W, \sigma)$ are not specified (Loewenstein et al. 2015), $h(W, \sigma) := h$ is set here to a constant parameter. The boundaries $\theta_{R_D} = -\theta_{R_I}$ are symmetric; no a priori bias for any of the choice options is assumed. For the operating times in a system, I assume a geometric distribution with mean $E(T) = 1/r$ (see Diederich and Oswald 2014, for further distributions).

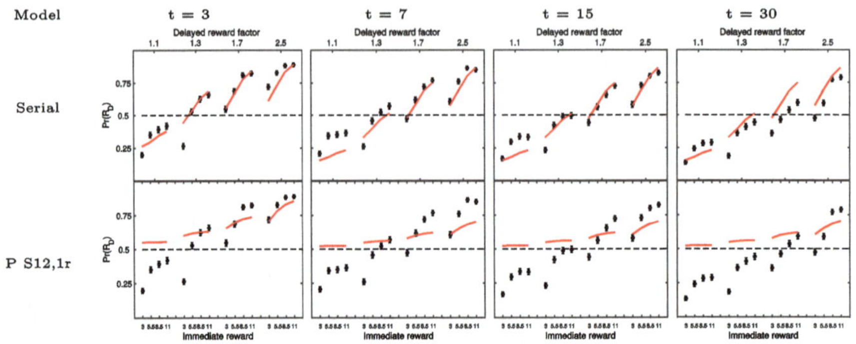

Fig. 8.4 Observed (black symbols) choice proportions and predicted (red lines) choice probabilities of all 64 conditions (4 immediate reward × 4 delayed reward factors × 4 time delays) for the serial model version (upper row) and the basic quasi-parallel model version, i.e., assuming an equal lead for any of the two systems (S12) and the same distribution and expected value $E(T) = 1/r_2 = 1/r_2$ of operating time in both systems (1r)

For the serial version we have the following parameters to be estimated from 64 observations (choice proportions): $\delta_1, \delta_2, h, \theta$ and, $1/r_1$, where by default processing starts with System 1 and only one switch occurs from System 1 to System 2.

For the quasi-parallel dual process model the order and time schedules need to be specified. With respect to the order schedule, either System 1 comes first, or System 2 comes first, or both systems have a specified probability of operating first. This adds one more model parameter. For the time schedule, we either assume identical operating time distributions in both systems with expected values $E(T) = 1/r_1 = 1/r_2$, adding another parameter, or non-identical operating time distributions, or the same distributions but with different parameters, $1/r_1$ and $1/r_2$, adding 2 more parameters. Thus, we have either 6 or seven parameters to be estimated from the data. Details are found in the supplement.

Figure 8.4 shows the observed choice frequencies (black circles with error bars) and the model accounts for the serial version (upper row, red lines) and the quasi-parallel model version (lower row, red lines) closest to a parallel model (same operation distributions and means for both systems and fixing the probability to 0.5 for starting with any system, abbreviated with P S12, r1: P for parallel, S12 for starting with any of the 2 systems and 1r for one parameters for the operating times). The immediate reward values are on the lower x-axes; the delayed reward factors are on the upper x-axes (the immediate reward multiplied by the factor yields the delayed reward; e.g., $3 \times 1.1 = 3.3$; $5.5 \times 1.1 = 6.05$; etc.). The time delays t are stated on top of each column. The probability of choosing the delayed reward $\Pr(R_D)$ is on the y-axis. The myopic effect occurs when $\Pr(R_D) < 0.5 = \Pr(R_I) > 0.5$, i.e. choice proportions are shown below the 0.5 line in each panel. When the differences between the immediate reward and the delayed reward are small (delayed reward factor 1.1), we observe a strong myopic effect, which is intensified even with increasing time delay t. With increasing delayed reward factors, the preference increases towards

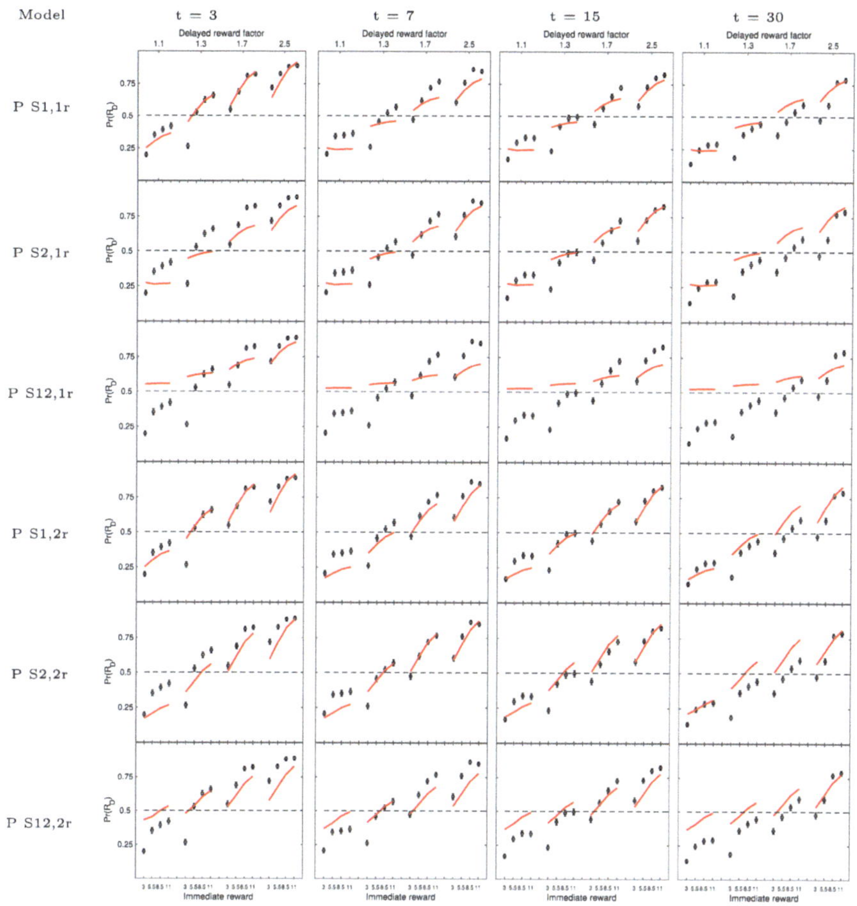

Fig. 8.5 Observed (black symbols) choice proportions and predicted (red lines) choice probabilities of all 64 conditions (4 immediate reward × 4 delayed reward factors × 4 time delays) for six quasi-parallel model versions. S1 refers to System1 takes the lead, S2 to System 2 takes the lead, S12 to any of the systems comes first (here the probability is 0.5). 1r and r2 refers to same or different expected times in each system, respectively. For simplicity, the distribution is the same in both systems but with different expected values

the delayed reward option. However, the increase is attenuated with increasing time delay. Indeed, for $t = 30$, most choices are made in favor of the immediate reward option. The two-stage dual process model can account for these patterns. Note that it predicts preferences reversals (from below 0.5 to above 05), which the quasi-parallel model cannot predict. For all choice pair combinations, it predicts choice probabilities larger than 0.5, i.e., in favor of the delayed reward option.

Figure 8.5 shows the accounts of six parallel model versions. The model labeling the parallel version is as before. S1: starting with System 1; S2: starting with System 2; S12: starting with any system with equal probability; 1r: identical distribution

Table 8.1 AIC for the 7 fitted model versions (serial (S) and 6 parallel (P))

Model start with system		$E(T)$	#Parameters	AIC
S	1	$1/r_1$	5	1663
P	1	$1/r_1 = 1/r_2$	5	2516
P	2	$1/r_1 = 1/r_2$	5	3303
P	Any	$1/r_1 = 1/r_2$	5	10567
P	1	$1/r_1 \neq 1/r_2$	6	1460
P	2	$1/r_1 \neq 1/r_2$	6	2188
P	Any	$1/r_1 \neq 1/r_2$	6	4656

For comparison of the parallel models, the time schedule was set and did not count as an additional parameter. A proper comparison with the serial model version, the number of parameters of the parallel versions increases by 1

with the same expected values for operating times in both systems; and 2r: identical distributions but with different expected values for each system. Obviously, the versions allowing two different expected values give a better account of the data than those with the same expected values in each system.

Table 8.1 shows the AIC of all models accounts. Because I fixed the order schedule here for illustration purposes, I did not count it as an additional parameter. As can be seen, the AIC for the serial version and the parallel version, starting with System 1 and allowing the mean operating time in both systems to be different are similar. This is easy to explain because the estimated $1/r_2$ can be adjusted (small or large) to mimic the serial model version for the choice probabilities.

For a possible differentiation, I predicted the mean choice response times for these two versions using the same parameter estimates obtained for predicting the choice probabilities and only estimated the mean residual non-decision times RT_{res} from 128 data points (64 mean response times for choosing the immediate reward, and 64 mean choice response times for choosing the delayed reward). The observed mean RT range from 0.968 to 1.471 sec for choosing R_D and from 0.731 to 1.502 sec for choosing R_I. The response time patterns are very heterogeneous but tend to decrease for choosing R_I for larger delayed rewards R_D.

Figure 8.6 shows the model accounts for the serial and quasi-parallel versions $PS1, 2r$. The model account of both model versions is relatively poor, particularly, for the immediate reward mean RT_I. Within a given time delay t, both versions predict decreasing mean RT_D as a function of immediate reward values and decreasing mean RT_D across the delay reward factors. Also, they both predict increasing mean RT_D as a function to the delay time t. Thus, the patterns are the same, but the magnitude is not. A further analysis of the RT is not pursued here.

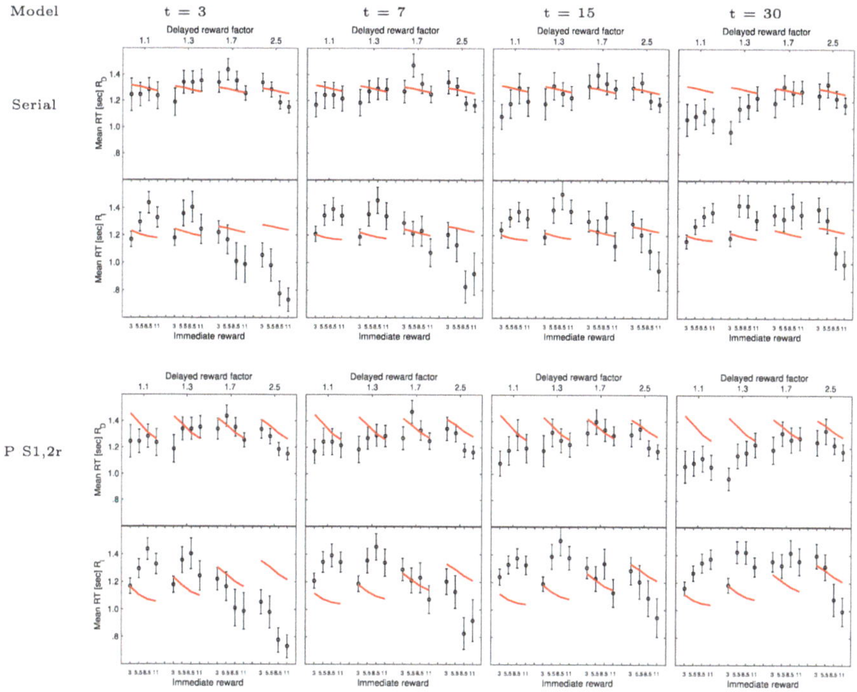

Fig. 8.6 Observed and predicted mean choice responses times for the two best fitting models. Note that only 1 parameter, RT_{res}, was estimated from 128 data points. AIC = 1241 for the serial version; AIC = 2311 for the quasi-parallel version

8.5 Discussion

Intertemporal choice situations are ubiquitous in daily life and have been studied by social scientists, particularly by economists and psychologists, for decades. The focus of most approaches has been on how the value of the delayed reward is discounted (e.g., Ballard et al. 2023; Ericson and Laibson 2019, for overviews). Most traditional models, regardless of the proposed discounting function, are deterministic (preferences are set 'in stone' and do not change when the DM is offered identical choice pairs at different trials/occasions) and static (deliberation time to choose an alternative is not part of the model) (Dai and Busemeyer 2014). This also holds true for a recently proposed dual-process model by Loewenstein et al. (2015).

Recently, a few models have been proposed that can account for the dynamic aspects of intertemporal choice. Dai and Busemeyer (2014) modified Decision Field Theory (Busemeyer and Townsend 1993, DFT) to account for intertemporal choice data. In a single process, the drift rate was determined by the difference in valence between the immediate option and the delayed option, similar to the approach proposed here. Rodriguez et al. (2014) used the Linear Ballistic Accumulator model (Brown and Heathcote 2005, LBA) to account for intertemporal choice data. The model assumes an independent race between the immediate and the delayed rewards, and the winner determines which option is chosen, as discussed before. Dai et al. (2018) tested 18 variations of DFT; 18 lexicographic semi-order models for predicting choice probabilities and using the LBA model for predicting choice response times; and 4 random-utility-with-discrimination-threshold models for predicting choice probabilities and ad hoc distributions for predicting choice response times. The model classes with various discounting functions were tested and evaluated based on their performance in accounting for several empirical regularities of intertemporal choice. Among these regularities are the *magnitude effect*, which describes the observation that larger payoff amounts have lower delay discounting rates compared to smaller ones, and the *common difference effect*, which refers to the phenomenon where increasing or decreasing the delay durations of two options by a common additive constant can lead to a preference reversal between the options (e.g., Loewenstein and Prelec 1992, for definitions).

Zhao et al. (2019) applied a single-stage Wiener process model to intertemporal choice and tested the existence of automatic biases by permitting stimuli-independent influences on the starting point and the drift rate.

Note that all these models assume a single process for intertemporal choice and do not test any dual process-level hypotheses. Because this paper aims at testing a mathematical framework that specifies the timing and interplay of the two systems explicitly rather than whether there is one, two, or more processes, I will not discuss single process models here any further (see, e.g., Augusto 2024; DeNeys 2021; Evans and Stanovich 2013; Keren and Schul 2009, for debates).

The model framework proposed here explicitly incorporates the two system as two entities, where in time information from one system is feeding into the other system. Mathematically, the approach is modeled as a non-time homogeneous stochastic processes (Diederich 1997; Diederich and Busemeyer 2003; Diederich and Oswald 2014). The preference accumulation process in each system is defined by drift rate rates indicating the direction and intensity of the preference towards selecting either option R_D or R_I. Note that the these drift rates are related to models on cognitive processes, particularly here to preferential choice models. As an example, the preferences in each system were modeled according to the dual-process model proposed by Loewenstein et al. (2015). In their static-deterministic model, System 1 refers to the affective process and System 2 to the deliberate. The subjective value of a given choice option is the sum of the two systems (see Eq. 8.8). Furthermore, they assumed exponential discounting functions for both systems with the affective system being more myopic than the deliberative system. The model predictions were concerned with the size of the myopic effect related to the motivational function $M(R, a)$ weighted by the willpower strength W and cognitive demands σ function $h(W, \sigma)$ rather than accounting for effects or anomalies (Loewenstein and Prelec 1992). The modeling ideas were utilized to map the preferences onto the respective drift rates, one for System 1 and one for System 2. Obviously, this procedure would allow to incorporate alternative discounting functions, for instance, hyperbolic, double exponential or hypoboloid (e.g. Ballard et al. 2023, for a list of possible functions) into preference building functions. Because the focus here was on the architecture of the dual process models, i.e., "default-interventionist" versus "parallel-competitive" view, no further examples are considered here.

The two-stage dynamic-stochastic dual model, a serial version within the framework meets the "default-interventionist" (Evans 2008) view for which System 1 starts first, and System 2 intervenes later. The parallel dynamic-stochastic dual process, a quasi-parallel versions of the framework, resembles the "parallel-competitive" view (Evans 2008) and allows rapid switching between the two systems.[1]

Comparing the predictions and model fits of the serial and the basic quasi-parallel model versions shows that the serial model gives a very good account of the choice proportions. The basic quasi-parallel model cannot account for the observed patters because it does not predict choice probabilities that cross the 0.5 line. The quasi-parallel model with System 1 leading and different expected values for the operating times in each system can mimic the serial version by adjusting the time in the second process. The predicted mean choice response times of both model versions did not agree with the data. However, the observed reaction times varied substantially within and across participants. Indeed, as Keidel et al. (2021, 2022) observed and investigated, temporal discounting is subject to considerable individual differences, associated, for instance, with socio-economic and health-related variables but also with personality and cognitive variables. This was not taken into account here and requires a different sample than those in Zhao et al. (2019). Furthermore, to account for individual differences, we may adopt a Bayesian approach as in (Zhao et al. 2019). However, the focus here was to show that formalizing verbal descriptions of the dual-process will solve some of the issues proposers and opponents of "default-interventionist" versus "parallel-competitive" have. Our approach is tested on published data providing more insight than a simple data fitting tool.

In subsequent investigations, it would be valuable to explore other discounting functions within the Loewenestein dual process approach and the potential distinctions between one-stage dynamic-stochastic models of intertemporal choice as outlined by Dai and Busemeyer (2014) and the framework presented here, aiming to capture behavioral patterns and effects/anomalies in intertemporal choice, as elucidated by Dai et al. (2018).

Acknowledgements The research was supported by German Research Foundation (DFG) grant DI 506/19-1.

[1] Note that an independent parallel model, i.e. a race between System 1 and System 2 and the winner determines the response is problematic. Given that System 1 is faster than System 2 (e.g., Kahneman 2011), System 1 would win the race in almost all cases. Then, as Diederich and Trueblood (2018) pointed out, decisions are almost entirely determined by preferences specified in System 1, making System 2 (and thus the notion of dual processes) useless (see also Alós-Ferrer 2018, for a critical evaluation of race models in this context). Furthermore, research assuming simultaneous operations in both systems (aka integration of two system) invokes almost always a (weighted) sum of the two processes (e.g., Baron and Gürcay 2017; Loewenstein et al. 2015; Mukherjee 2010). This procedure, however, reduces a dual-process model to a single-process model with two distinct components, again making the dual-process assumption useless (e.g., Keren and Schul 2009; Osman 2004).

Supplement: Parameter Estimation

The model was implemented in terms of the matrix approach (Diederich and Busemeyer 2003) and 5 $(\theta, \delta_1, \delta_2, h, 1/r_1)$ or 6 model parameters $(\theta, \delta_1, \delta_2, h, 1/r_1, 1/r_2)$. δ_1, δ_2, h were parameters for defining the drift rates in Eqs. 8.9 and 8.10. The operating time followed a geometric distribution with expected value $E(T) = 1/r$. It was weighted by the normalized difference in delayed and immediate rewards.

The parameters were estimated from 64 data points by minimizing the function (Smith and Vickers 1988)

$$\chi^2 = \sum \left(\frac{Pr_{obs} - Pr_{pred}}{SE_{Pr_{obs}}} \right)^2 \tag{8.11}$$

using the optimization routine fminsearchbnd of Matlab. The fminsearchbnd routine is similar to the standard fminsearch routine except that the range of the parameters of the parameters can be predetermined. The fminsearch uses the Nelder-Mead simplex search method of Lagarias et al. (1998). $SE_{Pr_{obs}}$ refer to the standard error of the observed choice proportions. The parameter θ (criterion) was fixed within one round of estimation iterations and varied across them. With the estimated parameters obtained here the means choice response times were predicted; only the non-decision time T_{res} was estimated from 128 data points by minimizing the function

$$\chi^2 = \sum \left(\frac{RT_{obs} - RT_{pred}}{SE_{RT_{obs}}} \right)^2. \tag{8.12}$$

References

Alós-Ferrer C (2018) A dual-process diffusion model. J Behav Decis Mak 31:203–218. https://doi.org/10.1002/bdm.1960

Augusto R (2024) Two kinds of process or two kinds of processing? Disambiguating dual process theories. Rev Philos Psychol 15:277–298. https://doi.org/10.1007/s13164-023-00673-1

Ballard T, Luckman A, Konstantinidis E (2023) A systematic investigation into the reliability of inter-temporal choice model parameters. Psychon Bull Rev 30:1294–1322. https://doi.org/10.3758/s13423-022-02241-7

Baron J, Gürcay B (2017) A meta-analysis of response-time tests of the sequential two-systems model of moral judgment. Memory Cogn 45(4):566–575. https://doi.org/10.3758/s1342

Brown S, Heathcote A (2005) A ballistic model of choice response time. Psychol Rev 112:117–128

Busemeyer JR, Townsend JT (1993) Decision field theory: a dynamic cognition approach to decision making. Psychol Rev 100:432–459

Dai J, Busemeyer JR (2014) A probabilistic, dynamic, and attribute-wise model of intertemporal choice. J Exp Psychol Gen 143(4):1489–1514

Dai J, Pleskac TJ, Pachur T (2018) Dynamic cognitive models of intertemporal choice. Cogn Psychol 104:29–56. https://doi.org/10.1016/j.cogpsych.2018.03.001

DeNeys W (2021) On dual- and single-process models of thinking. Perspect Psychol Sci 16(6):1413–1427
Diederich A (1997) Dynamic stochastic models for decision making with time constraints. J Math Psychol 41(3):260–274
Diederich A (2016) A multistage attention-switching model account for payoff effects on perceptual decision tasks with manipulated processing order. Decision 3(2):81
Diederich A (2024) A dynamic dual process model for binary choices: Serial versus parallel architecture. Comput Brain Behav 7:37–64. https://doi.org/10.1007/s42113-023-00186-1
Diederich A, Busemeyer J (2003) Simple matrix methods for analyzing diffusion models of choice probability, choice response time and simple response time. J Math Psychol 47(3):304–322
Diederich A, Oswald P (2014) Sequential sampling model for multiattribute choice alternatives with random attention time and processing order. Front Human Neurosci 8(697). https://doi.org/10.3389/fnhum.2014.00697
Diederich A, Trueblood JS (2018) A dynamic dual process model of risky decision making. Psychol Rev 125(2):270–292
Diederich A, Zhao W (2019) A dynamic dual process model of intertemporal choice. Spanish J Psychol 22:1–14
Ericson KM, Laibson D (2019) Intertemporal choice. In: Handbook of behavioral economics: applications and foundations 1, vol 2, North-Holland, pp 1–67. https://doi.org/10.1016/bs.hesbe.2018.12.001
Evans J (2006) The heuristic-analytic theory of reasoning: extension and evaluation. Psychonomic Bull Rev 13(3):378–395
Evans J (2008) Dual-processing accounts of reasoning, judgment, and social cognition. Ann Rev Pychol 59:255–278
Evans J, Stanovich K (2013) Dual-process theories of higher cognition: advancing the debate. Persepct Psychol Sci 8(3):223–241
Figner B, Knoch D, Johnson EJ, Krosch AR, Lisanby SH, Fehr E, Weber EU (2010) Lateral prefrontal cortex and self-control in intertemporal choice. Nat Neurosci 13(5):538–539
Fudenberg D, Levine D (2006) A dual-self model of impulse control. Am Econ Rev 96(5):1449–1476
Gawronski B, Creighton LA (2013) Dual process theories. In: Carlston D (ed) The Oxford handbook of social cognition. Oxford library of psychology. Oxford University Press, New York, pp 282–312
Grayot JD (2020) Dual process theories in behavioral economics and neuroeconomics: a critical review. Rev Philos Psychol 11(1):105–136. https://doi.org/10.1007/s13164-019-00446-9
Gürçay B, Baron J (2017) Challenges for the sequential two-system model of moral judgement. Think Reason 23(1):49–80. https://doi.org/10.1080/13546783.2016.1216011
Hammond K (1996) Human judgment and social policy. Oxford University Press
Hare TA, Camerer CF, Rangel A (2009) Self-control in decision-making involves modulation of the vmPFC valuation system. Science 324(5927):646–648
Kahneman D (2011) Thinking, fast and slow. Farrar, Straus and Giroux, New York
Kahneman D, Frederick S (2002) Representativeness revisited: Attribute substitution in intuitive judgment. In: Gilovich T, Griffin D, Kahneman D (eds) Heuristics and biases: the psychology of intuitive judgment. Cambridge University Press, New York, pp 49–81
Keidel K, Rramani Q, Weber B, Murawski C, Ettinger U (2021) Individual differences in intertemporal choice. Front Psychol 1–14. https://doi.org/10.3389/fpsyg.2021.643670
Keidel K, Ettinger U, Murawski C, Polner B (2022) The network structure of impulsive personality and temporal discounting. J Res Pers 96:1–15. https://doi.org/10.1016/j.jrp.2021.104166
Keren G, Schul Y (2009) Two is not always better than one: a critical evaluation of two-system theories. Perspect Psychol Sci 4(6):533–550
Klauer K, Beller S, Hütter M (2010) Conditional reasoning in context: a dual-source model of probabilistic inference. J Exp Psychol Learn Mem Cogn 36(2):298–323

Krajbich I, Bartling B, Hare T, Fehr E (2015) Rethinking fast and slow based on a critique of reaction-time reverse inference. Nat Commun 1–9. https://doi.org/10.1038/ncomms8455
Lagarias J, Reeds J, Wright M, Wright P (1998) Convergence properties of the Nelder-Mead simplex method in low dimensions. SIAM J Optim 9:112–147
Loewenstein G, Prelec D (1992) Anomalies in intertemporal choice: evidence and an interpretation. Q J Econ 107:573–597
Loewenstein G, O'Donoghue T, Bhatia S (2015) Modeling the interplay between affect and deliberation. Decision 2(2):55–81
Mallahi-Karai K, Diederich A (2019) Decision with multiple alternatives: geometric models in higher dimensions—the cube model. J Math Psychol 93(102):294. https://doi.org/10.1016/j.jmp.2019.102294
Mallahi-Karai K, Diederich A (2021) Decision with multiple alternatives: geometric models in higher dimensions—the disk model. J Math Psychol 100(102):493
Milli S, Lieder F, Griffiths TL (2021) A rational reinterpretation of dual-process theories. Cognition 217(104):881
Mukherjee K (2010) A dual system model of preferences under risk. Psychol Rev 117(1):243
Osman M (2004) An evaluation of dual-process theories of reasoning. Psychonomic Bull Rev 11(6):988–1010
Peters J, Büchel C (2011) The neural mechanisms of inter-temporal decision-making: understanding variability. Trends Cogn Sci 15(5):227–239
Rodriguez CA, Turner BM, McClure SM (2014) Intertemporal choice as discounted value accumulation. PLOS One 9(2):e90138. https://doi.org/10.1371/journal.pone.0090138
Schneider W, Shiffrin RM (1977) Controlled and automatic human information processing: I. detection, search, and attention. Psychol. Rev. 84(1):1–66
Sloman SA (1996) The empirical case for two systems of reasoning. Psychol Bull 119(1):3
Smith PL, Vickers D (1988) The accumulator model of two-choice discrimination. J Math Psychol 32:135–168
Stanovich K, West R (2000) Individual differences in reasoning: implications for the rationality debate? Behav Brain Sci 23(5):645–665
Zhao WJ, Diederich A, Trueblood JS, Bhatia S (2019) Automatic biases in intertemporal choice. Psychonomic Bull Rev 26:661–668

Chapter 9
The Effect of Transcranial Direct Stimulation of the Brain on Judgment and Decision-Making

Volker Thoma

Abstract Over two decades of research using non-invasive brain stimulation (NIBS) techniques such as transcranial direct stimulation (tDCS) resulted in insightful investigations of neural underpinnings of a range of cognitive processes. However, there is no clear picture of the effects of NIBS on higher cognitive functions such as judgment and decision-making (JDM). The current chapter summarises systematic reviews, meta-analyses and experiments that relate to JDM processes, such as working memory, inhibition processes, impulsivity, temporal discounting, judgment performance tests, choice tasks, and cognitive reflection. While people with cognitive impairments or behavioral addictions seem to benefit from NIBS interventions, for healthy participants there is still great variability regarding effects of tDCS on higher cognitive functions performance, partly owing to differences in methodology (montages, stimulation parameters, etc.), individual differences, and study designs. However, overall, there are clear indications for significant effects on JDM-related behavior after frontal (in particular dorsolateral) anodal stimulation in non-clinical participants. Future research will need to integrate developments in methodology and task design as well as assessing individual differences to better specify when and what exactly NIBS can contribute to establish causal links between brain function and higher cognitive processes.

Keywords Brain stimulation · Cognitive control · Decision making · Judgment · Executive functions

9.1 Introduction

Non-invasive brain stimulation (NIBS) has developed into a promising tool for research and for interventions to improve aspects of human cognition (for a broad overview, including clinical applications beyond cognition, see Mattioli et al. 2024).

V. Thoma (✉)
School of Psychology, University of East London, London, UK
e-mail: v.thoma@uel.ac.uk

U. Ettinger et al. (eds.), *Decision Making*, Studies in Neuroscience, Psychology and Behavioral Economics, https://doi.org/10.1007/978-3-032-00880-0_9

In particular, the potential of NIBS techniques to positively affect processes associated with executive functioning (e.g., task switching, inhibition, working memory updating) has caused excitement in clinical research (Herrera-Melendez et al. 2020; Jog et al. 2019; Kekic et al. 2016), largely because executive functions are generally linked to neuropsychiatric disorders or impairments (Brown 2008; McTeague et al. 2016; Scult et al. 2017; Snyder et al. 2015). For example, meta-analyses of repeated transcranial magnetic stimulation (rTMS; see Box 1) studies showed positive effects on global cognition (measuring executive function, attention, orientation, recall, calculation, and language, using e.g., Montreal Cognitive Assessment, MoCA, Nasreddine et al. 2005) in Alzheimer patients and patients with mild cognitive impairment (Teselink et al. 2021). Similar results were found for stroke patients (Yan et al. 2020) receiving transcranial direct current stimulation (tDCS; see Box 1).

Non-invasive brain stimulation aimed at improving cognitive performance test scores has also been studied in healthy adults (Antal et al. 2022; Coffman et al. 2014), including older (> 65 years) participants (Li et al. 2024). The effects of single or repeated sessions of tDCS on cognition have been examined across many domains, including face recognition (Fan et al. 2020; Yao et al. 2020; Civile et al. 2021; Barbieri et al. 2016; Gonzalez et al. 2019), attention (Coffman et al. 2014; Gladwin et al. 2012), working memory, (e.g., Hill et al., 2016), risk-taking (Khaleghi et al. 2020; León et al. 2020), probability judgment (Hecht et al. 2010) and planning (Dockery et al. 2009; Heinze et al. 2014), among others.

One particular area in which non-invasive brain stimulation has potential and has created interest as a research tool is human judgment and decision-making (JDM), to help address important questions about the underlying neural and cognitive mechanisms (Nitsche and Paulus 2001, Polania et al. 2018). The wider field of JDM research has had a stellar ‘career’ over the last two decades, in particular since the Nobel Prizes in economics for Daniel Kahneman (in 2002) and Richard Thaler (in 2017), improving our understanding of human thinking and deciding. However, there are no comprehensive reviews or meta-analyses of NIBS effects focussing on improvement of JDM in healthy populations (Farhat et al. 2022). The current chapter aims to give an overview of the literature, focussing on the NIBS technique most commonly used and most accessible for research in higher cognitive processes, transcranial direct current stimulation (tDCS).

Judgment research is generally concerned with the ways in which human beings use various cues (external or internal) to draw inferences about situations, people, objects, and events. In contrast, decision making typically refers to choosing among various options (Hardman 2009) such as choosing between two (or more) gambles described by expected values and risk (i.e., probability of winning/losing an amount at stake), or between two or more consumer items (e.g., Thoma et al. 2021).

Human judgment and decision-making engages a network of neural regions, including dorsolateral prefrontal cortex (Hare et al. 2009), orbitofrontal cortex (Knoch et al. 2006), and dorsal anterior cingulate (Modirrousta and Fellows 2008), as well as the amygdala (Bechara et al. 1999; De Martino et al. 2010), as evident from lesion data (Knoch et al. 2006; Sutterer et al. 2015) and neuroimaging studies (Cohen et al. 2005). In order to directly modulate decision-making processes, tDCS has been applied to regions of these neural networks, in particular the orbitofrontal and dorsolateral prefrontal cortices.

This chapter will first review research that focuses on cognitive functions typically implicated in JDM processes, notably attention and executive functions. Then, the focus will turn to research using tDCS applied in tasks more closely associated with judgment and decision-making.

Box 1 Non-invasive Brain Stimulation Techniques

Transcranial magnetic stimulation (TMS) manipulates brain activity by producing a magnetic field that induces a current in the brain tissue (Hallett 2000), which is achieved by passing an electric current through a coil above the scalp. The most common TMS protocol is repetitive TMS (rTMS), which involves applying TMS pulses in trains, either at high frequency (HF) to enhance neural activation or low frequency (LF) to inhibit neural activation (Rossi et al. 2009). One rTMS variation is theta burst stimulation (TBS) with three high-frequency (50 Hz) bursts (pulses at e.g., 5 Hz), which by using intermittent TBS (iTBS)—rather than a continuous stimulation pattern—increases the cortical excitability (Huang et al. 2005; Maeda et al. 2000).

Transcranial electrical stimulation (tES) is an umbrella term for various types of brain stimulation using small currents (1–2 mA) created via a battery-driven stimulator between two or multiple electrodes attached to the scalp (Reed and Cohen Kadosh 2018). Transcranial direct current stimulation (tDCS) is a technique that delivers a constant low current through electrodes (typically sponges soaked in NaCl solution) placed on the scalp (Nitsche et al. 2008; Nitsche and Paulus 2001). tDCS transiently alters neuronal excitability by depolarizing or hyperpolarizing a neuron's resting membrane potentials (for a critical primer, see Bestmann and Walsh 2017). Anodal stimulation typically depolarizes local neurons, making them require less dendritic input to fire, whereas cathodal stimulation hyperpolarizes neurons (Boros et al. 2008), needing more dendritic input to achieve the same effect (Nitsche and Paulus 2001) lasting for a short period after stimulation. Additionally, tDCS has been found to involve more complex mechanisms such as long-term potentiation (LTP) and long-term depression (LTD) at the synaptic level, affecting neural activity via the anode and cathode, respectively. These changes in LTP and LTD are mediated by various synaptic mechanisms, including NMDA receptors, GABAergic activity, glutamatergic activity, intracellular calcium concentration, among others (Liebetanz 2002; Nitsche et al. 2004; Stagg et al. 2009; Stagg and Nitsche 2011). Non-synaptic mechanisms may also play a role in the long-term effects of tDCS (Ardolino et al. 2005).

Unlike tDCS, transcranial alternating current stimulation (tACS; Antal and Paulus 2013) uses alternating currents to entrain endogenous neural rhythms

(Reato et al. 2010; Woods et al. 2016), potentially enhancing cognitive functions (Al Qasem et al. 2022; Antal et al. 2022; Helfrich et al. 2014) and alleviating neuropsychiatric symptoms (Daughters et al. 2020). Finally, transcranial random noise stimulation (tRNS) is delivered as a weak current, similar to white noise (Terney et al. 2008), with frequencies between 0.1 and 640 Hz (full-range) to increase neuronal excitability (see Brancucci et al. 2023).

9.2 NIBS and "Hot" and "Cold" Cognition

Executive functions are conceived as higher-order mental processes that enable the planning, initiation, control and monitoring of complex, goal-directed behavior and thoughts (EFs, Diamond 2013; Friedman and Robbins 2022). On a more basic level, EFs include processes such as attentional control, inhibition, working memory, and cognitive flexibility. Higher-order executive functions are thought to draw on the use of multiple basic executive functions (Baddeley 2012; Oberauer 2019) and include problem-solving, reasoning, and planning, and are sometimes used synonymously with the term "fluid intelligence" (Au et al. 2015; Jaeggi et al. 2008).

Executive functions related to decision-making are often divided into cognitive ("cold") processes that are independent of emotional involvement and affective/reward-related ("hot") processes that are biased by emotional (or reward-related) contexts (Friedman and Robbins 2022; Ongur 2000; Stuss 2011). Cold cognition is typically associated with processes including working memory updating, inhibition, and task shifting, as identified by Miyake et al. (2000) in their Unity-Diversity model (Miyake and Friedman 2012). Emotional or motivational executive functions are primarily associated with the orbitofrontal cortex (OFC) and other medial PFC regions (Fuster 2001). The OFC is presumed to play a role in integrating affective and cognitive information as well mediating motivated goal-directed behaviors (Salehinejad et al. 2021; Stuss 2011) and to be involved in value-based and economic decision-making (see Padoa-Schioppa and Conen 2017, for a review,). 'Cold' processes often involve monitoring conflict situations and inhibiting automatic responses, and are associated with lateral PFC (Nejati et al. 2018; Salehinejad et al. 2021).

9.3 NIBS and Executive Function (EF)

9.3.1 NIBS and Cortical Stimulation

If decision-making can be broken down into sub-processes linked to attention, executive functions and planning (e.g., Evans and Stanovich 2013; Stanovich 2018) then one could expect that related test scores can be modulated by tDCS in the same direction as JDM performance as a whole. Regarding working memory, for example, tDCS is thought to affect cortical balance between excitability in prefrontal areas (Senkowski et al. 2022).

In a recent and very comprehensive meta-analysis of 63 studies (using either rTMS and tDCS) including a variety of tasks at different cognitive levels, de Boer et al. (2021) showed that excitatory stimulation (tDCS) had a small positive effect on performance in some inhibition tasks (Stop Signal Task and Go/No-Go Task), while inhibitory stimulation had a small negative effect on Flanker Task performance. However, non-invasive brain stimulation did not alter performance levels in working memory (e.g., updating) and switching tasks, which broadly concurs with findings of two previous meta-analyses (Horvath et al. 2015; Mancuso et al. 2016). The effects on tasks involving planning processes (e.g., Tower of Hanoi) were inconclusive. In summary, deBoer et al. report that (single session or repetitive) tDCS significantly improved performance in some working memory processes associated with judgment and decision-making (in particular, tasks involving inhibition processes).

Other meta-analyses such as Hill et al. (2016) concentrated on the effects of anodal tDCS on working memory only. Results showed a significant but small reduction in reaction time for the n-back, Sternberg and digit span tasks following anodal tDCS stimulation, which was applied prior to these working memory tasks (i.e. offline stimulation). There was a trend for improved error rates. Online stimulation showed no effects. In the included studies, tDCS had been applied to DLPFC in 14 studies and to OFC in four studies. Considering the application of tDCS to DLPFC specifically, Dedoncker et al.'s (2016) meta-analysis reported an improvement in reaction time for cognitive tasks with a single session of anodal tDCS in healthy participants, but again there were no changes in accuracy. The analysis found an improvement in working memory (WM) accuracy in neuropsychiatric cohorts, but not for response times. Similar results for DLPFC single-session stimulation were found in a meta-analysis by Imburgio and Orr (2018), although they did find an improvement for updating tasks only (not for inhibition or switching tasks). These results broadly fit with an earlier meta-analysis by Brunoni and Vanderhasselt (2014) summarising results from studies on the effects of either rTMS or a-tDCS targeting the DLPFC in neuropsychiatric and healthy cohorts, ultimately finding an overall improvement in both cohorts, but mainly driven by the rTMS studies. Again, for the WM tasks in question, results showed that tDCS only improved response times, not accuracy. A more focussed recent meta-analysis on 10 studies (Müller et al. 2022) including only high-density tDCS montages showed no effect on reaction time (RT), and only a trend for better WM accuracy. Slightly more positive results were reported for tACS studies on WM (Nissim et al. 2023). Neuronal synchrony (e.g., in the theta band) between frontal and parietal cortices has previously been linked to WM performance (Polanía et al. 2012). Based on a meta-analysis of 10 studies, tACS showed a small-to medium effect size (Nissim et al. 2023), with larger effect sizes for various parameters (e.g., parietal and frontal areas over fronto-parietal and fronto-temporal; and for increasing number of sessions).

Overall, there appears to be only weak evidence for tES effects improving executive functions, specifically for working memory performance in healthy samples. Interestingly, however, a small meta-analysis by di Rosa et al. (2024) found greater tDCS effects on EF in individuals with greater depressive symptoms (negative mood). Senkowski et al. (2022) found in their review of tDCS and tACS that single-session tDCS stimulation studies show no reliable effects, while there is evidence that in multisession experiments tDCS has moderate beneficial effects on WM training (see also Mancuso et al. 2016). However, tACS can indeed result in improvements of WM performance even after a single session (or result in negative effects, depending on the stimulation parameters).

9.3.2 tDCS and Cerebrellar Stimulation

Recently, non-cortical areas such as the cerebellum have also been implicated in higher order cognitive functions. Pezetta et al. (2024) performed a meta-analysis that showed no overall effects of NIBS on cognitive processes, while anodal tDCS (to the medial cerebellum) improved performance in cognitive tasks. This result is broadly consistent with a previous meta-analysis (Oldrati and Schutter 2018) that showed that tDCS to the cerebellum can enhance cognitive functioning.

9.3.3 tDCS and Dopamine in Executive Function

Few studies also examined tES effects in combination with neurotransmitters. Leow et al. (2023) tested the effect of tDCS on the speed-accuracy trade-off in a dot-motion task, which requires strategic cognitive processes. Cathodal tDCS was applied over the left PFC and the superior medial frontal cortex in healthy participants. Dopamine in combination with tDCS reduced decision accuracy, but also negatively affected discriminability. This result confirms the role of pharmacological manipulations of dopamine, which has been found previously to alter effects of tDCS on plasticity in the motor cortex (Nitsche et al. 2006).

9.4 tDCS Effects on Cognitive Reflection Tasks

One of the most prolific areas in psychology and behavioral economics in recent years is based on (or aims to test) dual process theories of JDM. Although the theories differ (see e.g., De Neys and Osman 2014; Evans and Stanovich 2013), most dual-process theories propose qualitatively different cognitive processes in decision-making. Type 1 processes—often assumed to be akin to mental shortcuts or heuristics (Tversky and Kahneman 1974)—are associated with fast, intuitive decisions based on rapid and effortless associations with stored representations, while Type 2 processes are linked to slow, reflective consideration of the judgment or decision situation (Evans and Stanovich 2013). The former are employed when cognitive resources are scarce, and often lead to judgment (or decision) biases—in contrast to Type 2 processes, which are associated with analytic mental processes that lead to more appropriate responses. A test often employed to test reflective versus heuristic processing is the Cognitive Reflection Test (CRT) (Frederick 2005; Toplak et al. 2014), which contains short math-like problems designed to elicit an incorrect intuitive response which can be easily corrected with time and inclination to reflect (e.g., the now famous bat-and-ball question, Frederick 2003). The CRT is especially interesting, for at least two reasons. First, the CRT not only returns the correct score, but also a score of how many of committed errors were due to systematically following the lure (the implied answer)—indicating the degree of cognitive inhibition. Second, performance on the CRT correlates with a range of other judgment and decision-making measures, such as temporal discounting, framing, and expected-value gambles (see Toplak et al. 2014). For example, high CRT scorers have been found to be less susceptible to subscribing to paranormal beliefs (Pennycook et al. 2012), less ‘unusual experiences’ (i.e., ‘jumping to conclusions’; Broyd et al. 2019) and are better at discriminating fake news from real ones (Bronstein et al. 2019; Pennycook and Rand 2019).

In one of the few stimulation studies using the CRT as the dependent variable, Oldrati et al. (2016) found that cathodal tDCS to the left DLPFC (decreasing cortical excitability) reduces performance, presumably by decreasing inhibitory control. Edgcumbe et al. (2019) also used tDCS on an extended CRT (e.g., Toplak et al. 2014) and observed higher CRT scores and improved performance when solving incongruent base-rate vignettes after a single session of offline anodal tDCS over the right DLPFC, relative to sham or left DLPFC stimulation. The effects could not be explained by potential confounding factors, such as IQ, thinking dispositions (e.g., attitudinal tendency for reflective thinking), and other demographic variables (e.g., religiosity). However, the processes underlying this effect were not clear, and could be due to tDCS reducing a person’s general impulsivity (Teti Mayer et al. 2020), improving their working memory (Müller et al. 2022), or their inhibition performance (de Boer et al. 2021). In a follow-up study, Edgcumbe et al. (2024) controlled for impulsiveness and added WM (updating) tests. This study largely replicated the CRT effects for right DLPFC stimulation, and further showed increased JDM effects after repeated stimulation sessions. However, there was no effect of stimulation on working memory performance.

In summary, there are replicable effects of brain stimulation on widely used judgment tasks such as the CRT and base-rate tests. Edgcumbe et al.'s and Oldrati et al.'s work indicates an involvement of the right DLPFC in classic JDM tasks, and would fit with this brain area's presumed role in conflict monitoring and cognitive inhibition (Friedman and Robbins 2022), while WM performance appears to be associated with the left DLPFC (Müller et al. 2022). Interestingly, improvements in cognitive reflection cannot be easily accounted for by improvements in executive functions (see Toplak et al. 2011), also reflected by the inconclusive meta-analyses (see above) on executive functions (in particular, working memory) performance after tDCS stimulation. However, effect sizes are small to medium, and more research—using a more diverse set of tasks—is needed.

9.5 tDCS Effects on Decision-Making Tasks

As described above, "hot cognition" tasks are those EF processes that are goal-directed in contexts that involve emotion, motivation, and conflicts between immediate gratification and long-term rewards (Salehinejad et al. 2021). While it is suggested that the DLPFC is crucial for brain networks regulating executive functions and employing cognitive control by way of top-down input for task-relevant behaviours, 'hot cognition' has been linked—among other non-cortical areas—to the OFC and ventrolateral PFC (Ongur 2000; Stuss 2011).

9.5.1 tDCS Effects on Impulsivity

Impulsivity is a heterogenous psychological construct describing the experienced urge and heightened behavioural tendency to initiate actions (with or without external stimulation) that are usually seen as premature, without much deliberation or appropriate consideration of their consequences (Keidel et al. 2022; Swann et al. 2002). Impulsivity is often linked to delay aversion, higher risk-taking, reduced planning ability and poor selection of relevant or reduced inhibition of irrelevant stimuli (Aichert et al. 2012; Brevet-Aeby et al. 2016; Keidel et al. 2022). A related concept is self-regulation to allow goal-directed behavior (Baumeister 2014; Baumeister et al. 2018) such as when a person is resisting to check their smart phone for messages or adheres to a dietary plan. Self-regulation concerns processes that enable humans to guide their thoughts, feelings, and behaviors in a goal-directed way. A systematic review by Teti Mayer et al. (2020) investigated tDCS effects on impulsivity in healthy and clinical populations, as did Kelley et al. (2019) for self-regulation, and both authors found overall positive effects of tDCS, albeit with great variation in study parameters, in particular dosage and consideration of individual differences. An earlier review by Brevet-Aeby et al. (2016) associates the DLPFC with a range of functions related to impulsivity, while the inferior frontal gyrus (IFG) was associated more specifically with control of inhibition. The following subsections summarise tDCS studies that link impulsivity to specific decision-making tasks.

9.5.2 tDCS Effects on Temporal Discounting

In a temporal discounting task, participants are given a choice between a smaller immediate reward (e.g., "£100 now") or a greater later reward (e.g., "£120 in a week"). The degree of choosing lower immediate rewards more frequently over larger later rewards is the degree of temporal discounting, i.e., a lower subjective value of delayed rewards. In the review and meta-analysis by Teti Mayer et al. (2020), two out of six tDCS studies with healthy participants reported a reduction in temporal discounting after unilateral anodal stimulation over the left DLPFC both online (Shen et al. 2016; HD-tDCS) and offline (He et al. 2016). The meta-analysis by de Boer et al. (2021) mentioned above, only cited two studies: One using TMS (Cho et al. 2010) and one using tDCS (Nejati et al. 2018). There was no significant effect overall, though the latter study found decreased discount rates after stimulation of the anodal DLPFC/cathodal OFC (both left and right). A meta-analysis by Lin and Feng (2024) with 9 studies (and 35 effect sizes) showed that excitatory NIBS targeting the left DLPFC (5 of the 9 were tDCS) reduced delay discounting rates. It seems, therefore, that tDCS stimulation of the (in particular left) DLPFC can reduce temporal discounting (see also Fusco et al. 2024, for mixed results on impulsivity using tACS). Recently, a study using tDCS stimulation over OFC—thought to be involved in hot cognition and the evaluation of external stimuli—has also been found to reduce temporal discounting (Moro et al. 2023).

9.5.3 tDCS Effects on Gambling tasks

Another domain relevant to decision-making that is linked to impulsivity concerns behavioral addictions (see a review comparing NIBS techniques for behavioral addictions, Gomis-Vicent et al. 2019). One particular area of interest here is gambling, and research has aimed to compare gambling-related tasks for healthy participants with patients suffering from gambling disorder (GD). In a systematic review of tDCS in healthy subjects and GD patients, Pettorusso et al. (2021) showed that while anodal stimulation (activation) of the right (and cathodal/inhibition of the left) DLPFC led to an improvement (increase) in risk aversion during trials that incurred a monetary loss (Ye et al. 2016), inhibition of the left DLPFC caused the opposite pattern (Ye et al. 2016) in healthy participants. Soyata et al. (2019) found that a similar montage (i.e., similar parameters relating to electrode placement) improved scores (Iowa Gambling Task, IGT; Wisconsin Card Sorting Test, WCST) for 20 GD patients.

In a systematic review of 40 publications on mostly healthy participants, but also including six studies focusing on GD and Internet Gaming Disorder (IGD) patients, as well as other addictions, Stankovic et al. (2023) reported mostly beneficial effects after tDCS. The most frequently targeted area was the DLPFC, and measures used were computerized cognitive tasks on risk taking and decision making, e.g., the balloon analogue risk task, IGT, Cambridge gambling task, etc. The systematic review indicated that 70% of the (tES) studies showed neuromodulatory effects. However, there was considerable variability in the results, particularly for healthy participants, including the direction of the effects, with similar montages and stimulation parameters showing opposite outcomes (of reducing or increasing impulsive risk-taking).

9.5.4 tDCS Effecs in Decision-Making Under Risk

In a seminal study, Fecteau et al. (2007) observed substantially reduced risk-taking in the Balloon-Analogue-Risk-taking task (BART; i.e., the number of pumps of the balloons that did risk bursting before toy-money could be banked) after bilateral tDCS of the DLPFC compared to unilateral tDCS. In a meta-analysis of 48 studies Bell and DeWall (2018) investigated whether tDCS applied to the prefrontal cortex would affect a range of social behaviors (such as aggression, overeating, impulsivity, bias, dishonesty, and risk-taking). A small but reliable effect was found for risk-taking (13 studies) and overeating (7 studies) only. Moderators such as tDCS duration and dose (milliamps) did not show any effects. Bell and deWall (2018) further claimed that their meta-analysis showed that while risk-taking was significantly affected, impulsivity measures were not. A systematic review on tDCS effects by Teti Mayer et al. (2020) reported positive effects in eight out of 16 trials for risk-taking tasks, whereas only one reported negative results. Khaleghi et al. (2020) also conducted a meta-analysis of 16 tES effects in risk-taking studies, indicating a small overall effect for tDCS over the DLPFC. In general, it appears that right (or right and left) anodal DLPFC stimulation reduces risk-taking behavior.

9.5.5 tDCS Effects on Economic Decision-Making and Cooperation

So far, this chapter has summarised meta-analyses and systematic reviews that focus on higher order cognitive tasks that are typically related to decision-making, such as risky gambles. However, economic decision-making often happens in a social context and relies on (actual or perceived) cooperation and norms. Maréchal et al. (2017) found that honesty in a die-throwing task (that allowed cheating by misreporting results) can be increased with tDCS over the right dorsolateral prefrontal cortex. Similar to the conclusions by Edgcumbe et al. (2024) on cognitive reflection tasks, these results may indicate that stimulation helps to resolve conflicting processes (here honesty and self-interest).

Furthermore, a study on trust games (Nihonsugi et al. 2015) showed a role for the right DLPFC in economic games, when anodal tDCS showed increased engagement in economic cooperation based on feelings of guilt (intention-based decisions), in contrast to activity (measured with fMRI) in ventral areas, which was associated with reward (outcome-based) decision-making. Lin and Feng's (2024) extensive meta-analysis on impulsivity-related effects found similar results for inhibitory NIBS (including TMS) over the right DLPFC, increasing self-interested choice of unfair offersin an economic game. Previously, Ruff et al. (2013) had shown that right lateral prefrontal cortex tDCS stimulation increased social norm compliance in a version of an economic trust game. These results agree with the review by Kelley et al. (2019) that reported tDCS effects in self-regulation tests, such as in aggression (with right DLPFC stimulation reducing aggressive tendencies, increasing fearful ones), as well as a reduction of food cravings (or intake), in particular for unhealthy foods, again mostly after right anodal DLPFC stimulation. Somewhat related to economic games is moral reasoning (arguably replacing consideration of utilities with personal interests and moral values), which has shown left anodal stimulation reducing appropriateness judgments of utilitarian solutions to moral dilemmas (Kuehne et al. 2015). However, Zheng et al. (2018) found this effect by inhibiting the right DLPFC, and some research points to the involvement of the right temporo-parietal area (rTPJ) in judgments regarding moral reasoning (Leloup et al. 2016).

Another area that may play a role in economic decision-making is the posterior parietal cortex (PPC), which has neural connections with the DLPFC (Woo et al. 2022). In a multi-option decision experiment involving previously valued snacks on a grid, Woo et al. (2022) found that HD-tDCS to the DLPFC affects filtering choice-irrelevant information, while HD-tDCS over PPC indicates a causal role in value-based (spatial) selection. Previously, stimulation of the PPC has been associated with episodic memory retrieval (Pergolizzi and Chua 2015), which may potentially be linked to judgment processes.

9.6 Summary

Overall, the systematic reviews and meta-analyses as well as the individual JDM studies (targeting specific processes) reported in this chapter show (mostly positive) effects of NIBS on decision-making related tasks in human participants, although effect sizes tend to be small (or small to medium). Many studies reviewed above employed a bipolar (pre-)frontal tDCS montage (putting two electrodes of opposite charge on the head to create a unidirectional current flow). Traditionally, the left DLPFC is considered significant for regulating negative emotions and controlling impulsive behavior while the right DLPFC is thought to be more associated with reward processing—including emotions and motivations related to rewards.

However, the results are often variable (see e.g., Salehinejad et al. 2021), and stimulation of the right DLPFC has also been sometimes shown to inhibit cognitive processes (e.g., Edgcumbe et al, 2019).

There are several challenges to obtaining a clearer picture of the effectiveness of tDCS, such as the use of small samples in many studies, unclear randomisation procedures, insufficient blinding, and other factors as discussed by summary articles (Di Rosa et al. 2024; Stanković et al. 2023; Teti Mayer et al. 2020). Also, it seems that often there is no clear rationale for the use of specific montages, stimulation parameters, and behavioral measures, as well as confusion with regards to what constitutes a decision-making test (for a comprehensive review of these and other factors, see Kelley et al. 2019).

Some patterns, however, are clearly distinguishable. While tDCS results for healthy adults are reported with small effect sizes and sometimes contradicting neural origins, the positive effects of tDCS on processes relevant to decision-making are more reliable in older adults, patients with neurological impairments (Yan et al. 2020), or maladaptations such as behavioural addictions (e.g., GD; Pettorusso et al. 2021), although the duration of these effects is still elusive. Second, anodal stimulation seems to have positive effects on gambling-related risky decision-making (Bell 2018; Pettorusso et al. 2021), and while the causal link to impulsivity is less clear (Bell and deWall 2018), there are indications that tDCS to frontal areas reduces impulsive behaviour (Brevet-Aeby et al. 2016; Teti Mayer et al. 2020). In terms of cognitive functions, both right and left PFC appear causally connected to performance in selective, sustained and divided attention tests. However, the evidence favours the right DLPFC as linked to inhibitory processes for higher cognitive functions, such as judgment and reflective thinking (Kelley et al. 2019; Edgcumbe et al. 2019) and economic decision-making (e.g., Marechal et al. 2017; Ruff et al. 2013). Currently, there is little evidence from this literature linking tDCS (or tES in general) effects with improving JDM performance via simply increasing WM performance, which is interesting for theories of JDM processes. There is less data available on the effectiveness of stimulation of the OFC and VMPFC (with connections to ventral striatum) coding for external or internal values, which are potentially modulated by intention-based processes associated with the DLPFC (Nihonsogu et al. 2015).

To conclude, research and scholarship in human judgment and decision-making spans many disciplines, including cognitive neuroscience, which offers exciting avenues to better understand cognitive functions and related behaviours that are not only important for individuals or patient groups, but determine the wellbeing of our societies and habitats. Studies using NIBS offer an opportunity to investigate the causal links between brain processes and behaviour and allow us to better understand the underlying mechanisms of JDM related processes, on topics such as cognitive control, reflective judgment, economic decisions, and risk-taking. Furthermore, relatively cheap and accessible NIBS techniques such as tDCS offer the opportunity to improve cognitive performance, not only in clinical patients, but also sub-clinical and healthy populations (for a discussion regarding ethical aspects, see e.g. Voarino et al. 2017). Hence, there is a need for further high-quality experiments as well as systematic reviews and meta-analyses specific to JDM. One emerging focus and area of research is the combination of neurostimulation with cognitive experiments, brain imaging, psychopharmacology, and computational modelling.

References

Aichert DS, Wöstmann NM, Costa A, Macare C, Wenig JR, Möller H-J, Rubia K, Ettinger U (2012) Associations between trait impulsivity and prepotent response inhibition. J Clin Exp Neuropsychol 34(10):1016–1032

Al Qasem W, Abubaker M, Kvašňák E (2022) Working memory and transcranial-alternating current stimulation—state of the art: findings, missing, and challenges. Front Psychol 13:822545. https://doi.org/10.3389/fpsyg.2022.822545

Antal A, Luber B, Brem A-K, Bikson M, Brunoni AR, Cohen Kadosh R, Dubljević V, Fecteau S, Ferreri F, Flöel A, Hallett M, Hamilton RH, Herrmann CS, Lavidor M, Loo C, Lustenberger C, Machado S, Miniussi C, Moliadze V, Paulus W (2022) Non-invasive brain stimulation and neuroenhancement. Clin Neurophysiol Pract 7:146–165. https://doi.org/10.1016/j.cnp.2022.05.002

Antal A, Paulus W (2013) Transcranial alternating current stimulation (tACS). Front Hum Neurosci 7.https://doi.org/10.3389/fnhum.2013.00317

Ardolino G, Bossi B, Barbieri S, Priori A (2005) Non-synaptic mechanisms underlie the after-effects of cathodal transcutaneous direct current stimulation of the human brain. J Physiol 568(2):653–663. https://doi.org/10.1113/jphysiol.2005.088310

Au J, Sheehan E, Tsai N, Duncan GJ, Buschkuehl M, Jaeggi SM (2015) Improving fluid intelligence with training on working memory: a meta-analysis. Psychon Bull Rev 22(2):366–377. https://doi.org/10.3758/s13423-014-0699-x

Baddeley A (2012) Working memory: theories, models, and controversies. Annu Rev Psychol 63(1):1–29. https://doi.org/10.1146/annurev-psych-120710-100422

Barbieri M, Negrini M, Nitsche MA, Rivolta D (2016) Anodal-tDCS over the human right occipital cortex enhances the perception and memory of both faces and objects. Neuropsychologia 81:238–244

Baumeister RF (2014) Self-regulation, ego depletion, and inhibition. Neuropsychologia 65:313–319

Baumeister RF, Bratslavsky E, Muraven M, Tice DM (2018) Ego depletion: is the active self a limited resource? In: Self-regulation and self-control. Routledge, pp 16–44

Bechara A, Damasio H, Damasio AR, Lee GP (1999) Different contributions of the human amygdala and ventromedial prefrontal cortex to decision-making. J Neurosci 19(13):5473–5481

Bell SB, DeWall N (2018) Does transcranial direct current stimulation to the prefrontal cortex affect social behavior? A meta-analysis. Social Cognitive and Affective Neuroscience 13(9):899–906. https://doi.org/10.1093/scan/nsy069

Bestmann S, Walsh V (2017) Transcranial electrical stimulation. Curr Biol 27(23):R1258–R1262. https://doi.org/10.1016/j.cub.2017.11.001

Boros K, Poreisz C, Münchau A, Paulus W, Nitsche MA (2008) Premotor transcranial direct current stimulation (tDCS) affects primary motor excitability in humans. Eur J Neurosci 27(5):1292–1300. https://doi.org/10.1111/j.1460-9568.2008.06090.x

Brancucci A, Rivolta D, Nitsche MA, Manippa V (2023) The effects of transcranial random noise stimulation on motor function: a comprehensive review of the literature. Physiol Behav 261:114073. https://doi.org/10.1016/j.physbeh.2023.114073

Brevet-Aeby C, Brunelin J, Iceta S, Padovan C, Poulet E (2016) Prefrontal cortex and impulsivity: interest of noninvasive brain stimulation. Neurosci Biobehav Rev 71:112–134

Bronstein MV, Pennycook G, Bear A, Rand DG, Cannon TD (2019) Belief in fake news is associated with delusionality, dogmatism, religious fundamentalism, and reduced analytic thinking. J Appl Res Mem Cogn 8(1):108–117

Brown TE (2008) ADD/ADHD and impaired executive function in clinical practice. Curr Psychiatry Rep 10(5):407–411. https://doi.org/10.1007/s11920-008-0065-7
Broyd A, Ettinger U, Thoma V (2019) Thinking dispositions and cognitive reflection performance in schizotypy. Judgm Decis Mak 14(1):80–90
Brunoni AR, Vanderhasselt M-A (2014) Working memory improvement with non-invasive brain stimulation of the dorsolateral prefrontal cortex: a systematic review and meta-analysis. Brain Cogn 86:1–9. https://doi.org/10.1016/j.bandc.2014.01.008
Cho SS, Ko JH, Pellecchia G, Van Eimeren T, Cilia R, Strafella AP (2010) Continuous theta burst stimulation of right dorsolateral prefrontal cortex induces changes in impulsivity level. Brain Stimul 3(3):170–176. https://doi.org/10.1016/j.brs.2009.10.002
Civile C, McLaren R, Milton F, McLaren IPL (2021) The effects of transcranial direct current stimulation on perceptual learning for upright faces and its role in the composite face effect. J Exp Psychol: Anim Learn Cogn 47(1):74
Coffman BA, Clark VP, Parasuraman R (2014) Battery powered thought: Enhancement of attention, learning, and memory in healthy adults using transcranial direct current stimulation. Neuroimage 85:895–908. https://doi.org/10.1016/j.neuroimage.2013.07.083
Cohen MX, Heller AS, Ranganath C (2005) Functional connectivity with anterior cingulate and orbitofrontal cortices during decision-making. Cogn Brain Res 23(1):61–70
Daughters SB, Yi JY, Phillips RD, Carelli RM, Fröhlich F (2020) Alpha-tACS effect on inhibitory control and feasibility of administration in community outpatient substance use treatment. Drug Alcohol Depend 213:108132. https://doi.org/10.1016/j.drugalcdep.2020.108132
De Boer NS, Schluter RS, Daams JG, Van Der Werf YD, Goudriaan AE, Van Holst RJ (2021) The effect of non-invasive brain stimulation on executive functioning in healthy controls: a systematic review and meta-analysis. Neurosci Biobehav Rev 125:122–147. https://doi.org/10.1016/j.neubiorev.2021.01.013
De Martino B, Camerer CF, Adolphs R (2010) Amygdala damage eliminates monetary loss aversion. Proc Natl Acad Sci 107(8):3788–3792
De Neys W, Osman M (eds) (2014) New approaches in reasoning research. Psychology Press
Dedoncker J, Brunoni AR, Baeken C, Vanderhasselt M-A (2016) A systematic review and meta-analysis of the effects of transcranial direct current stimulation (tDCS) over the dorsolateral prefrontal cortex in healthy and neuropsychiatric samples: Influence of stimulation parameters. Brain Stimul: Basic Transl Clin Res Neuromodulation 9(4):501–517
Di Rosa E, Masina F, Pastorino A, Galletti E, Gambarota F, Altoè G, Edelstyn N, Mapelli D (2024) Mood moderates the effects of prefrontal tDCS on executive functions: a meta-analysis testing the affective state-dependency hypothesis. J Affect Disord 351:920–930. https://doi.org/10.1016/j.jad.2024.02.009
Diamond A (2013) Executive functions. Annu Rev Psychol 64(1):135–168. https://doi.org/10.1146/annurev-psych-113011-143750
Dockery CA, Hueckel-Weng R, Birbaumer N, Plewnia C (2009) Enhancement of planning ability by transcranial direct current stimulation. J Neurosci 29(22):7271–7277. https://doi.org/10.1523/JNEUROSCI.0065-09.2009
Edgcumbe DR, Rivolta D, Nitsche MA, Thoma V (2024) Single session and repeated anodal transcranial direct current stimulation over the right dorsolateral prefrontal cortex increases reflective thinking but not working memory updating performance. Heliyon e36078. https://doi.org/10.1016/j.heliyon.2024.e36078
Edgcumbe DR, Thoma V, Rivolta D, Nitsche MA, Fu CHY (2019) Anodal transcranial direct current stimulation over the right dorsolateral prefrontal cortex enhances reflective judgment and decision-making. Brain Stimulhttps://doi.org/10.1016/j.brs.2018.12.003
Evans JSB, Stanovich KE (2013) Dual-process theories of higher cognition: advancing the debate. Perspect Psychol Sci 8(3):223–241
Fan B, Mao W, Jin J, Ma Q (2020) Modulating activity in the dorsolateral prefrontal cortex alter corruption behavior: a transcranial direct current stimulation study. Behav Brain Res 382:112479

Farhat LC, Carvalho AF, Solmi M, Brunoni AR (2022) Evidence-based umbrella review of cognitive effects of prefrontal tDCS. Soc Cogn Affect Neurosci 17(1):43–60. https://doi.org/10.1093/scan/nsaa084

Fecteau S, Knoch D, Fregni F, Sultani N, Boggio P, Pascual-Leone A (2007) Diminishing risk-taking behavior by modulating activity in the prefrontal cortex: a direct current stimulation study. J Neurosci 27(46):12500–12505. https://doi.org/10.1523/JNEUROSCI.3283-07.2007

Frederick S (2005) Cognitive reflection and decision making. J Econ Perspect 19(4):25–42. https://www.jstor.org/stable/4134953

Friedman NP, Robbins TW (2022) The role of prefrontal cortex in cognitive control and executive function. Neuropsychopharmacology 47(1):72–89. https://doi.org/10.1038/s41386-021-01132-0

Fusco G, Scandola M, Lin H, Inzlicht M, Aglioti SM (2024) Modulating preferences during intertemporal choices through exogenous midfrontal transcranial alternating current stimulation: a registered report. Cortex 171:435–464. https://doi.org/10.1016/j.cortex.2023.09.019

Fuster JM (2001) The prefrontal cortex—an update. Neuron 30(2):319–333. https://doi.org/10.1016/S0896-6273(01)00285-9

Gladwin TE, Den Uyl TE, Fregni FF, Wiers RW (2012) Enhancement of selective attention by tDCS: interaction with interference in a Sternberg task. Neurosci Lett 512(1):33–37. https://doi.org/10.1016/j.neulet.2012.01.056

Gomis-Vicent E, Thoma V, Turner JJD, Hill KP, Pascual-Leone A (2019) Review: non-invasive brain stimulation in behavioral addictions: insights from direct comparisons with substance use disorders. Am J Addict 28(6):431–454. https://doi.org/10.1111/ajad.12945

Gonzalez-Perez M, Wakui E, Thoma V, Nitsche MA, Rivolta, D (2019) Transcranial alternating current stimulation (tACS) at 40 Hz enhances face and object perception. Neuropsychologia 135:107237. https://doi.org/10.1016/j.neuropsychologia.2019.107237

Hallett M (2000) Transcranial magnetic stimulation and the human brain. Nature 406(6792):147–150. https://doi.org/10.1038/35018000

Hardman D (2009) Judgment and decision making: psychological perspectives. BPS Blackwell

Hare TA, Camerer CF, Rangel A (2009) Self-control in decision-making involves modulation of the vmPFC valuation system. Science 324(5927):646–648

He Q, Chen M, Chen C, Xue G, Feng T, Bechara A (2016) Anodal stimulation of the left DLPFC increases IGT scores and decreases delay discounting rate in healthy males. Front Psychol 7.https://doi.org/10.3389/fpsyg.2016.01421

Hecht D, Walsh V, Lavidor M (2010) Transcranial direct current stimulation facilitates decision making in a probabilistic guessing task. J Neurosci 30(12):4241–4245. https://doi.org/10.1523/JNEUROSCI.2924-09.2010

Heinze K, Ruh N, Nitschke K, Reis J, Fritsch B, Unterrainer JM, Rahm B, Weiller C, Kaller CP (2014) Transcranial direct current stimulation over left and right DLPFC: lateralized effects on planning performance and related eye movements. Biol Psychol 102:130–140. https://doi.org/10.1016/j.biopsycho.2014.07.019

Helfrich RF, Knepper H, Nolte G, Strüber D, Rach S, Herrmann CS, Schneider TR, Engel AK (2014) Selective modulation of interhemispheric functional connectivity by HD-tACS shapes perception. PLoS Biol 12(12):e1002031. https://doi.org/10.1371/journal.pbio.1002031

Herrera-Melendez A-L, Bajbouj M, Aust S (2020) Application of transcranial direct current stimulation in psychiatry. Neuropsychobiology 79(6):372–383. https://doi.org/10.1159/000501227

Hill AT, Fitzgerald PB, Hoy KE (2016) Effects of anodal transcranial direct current stimulation on working memory: a systematic review and meta-analysis of findings from healthy and neuropsychiatric populations. Brain Stimul 9(2):197–208. https://doi.org/10.1016/j.brs.2015.10.006

Horvath JC, Forte JD, Carter O (2015) Quantitative review finds no evidence of cognitive effects in healthy populations from single-session transcranial direct current stimulation (tDCS). Brain Stimul 8(3):535–550. https://doi.org/10.1016/j.brs.2015.01.400

Huang Y-Z, Edwards MJ, Rounis E, Bhatia KP, Rothwell JC (2005) Theta burst stimulation of the human motor cortex. Neuron 45(2):201–206. https://doi.org/10.1016/j.neuron.2004.12.033

Imburgio MJ, Orr JM (2018) Effects of prefrontal tDCS on executive function: Methodological considerations revealed by meta-analysis. Neuropsychologia 117:156–166. https://doi.org/10.1016/j.neuropsychologia.2018.04.022

Jaeggi SM, Buschkuehl M, Jonides J, Perrig WJ (2008) Improving fluid intelligence with training on working memory. Proc Natl Acad Sci 105(19):6829–6833. https://doi.org/10.1073/pnas.0801268105

Jog MV, Wang DJJ, Narr KL (2019) A review of transcranial direct current stimulation (tDCS) for the individualized treatment of depressive symptoms. Pers Med Psychiatry 17–18:17–22. https://doi.org/10.1016/j.pmip.2019.03.001

Keidel K, Ettinger U, Murawski C, Polner B (2022) The network structure of impulsive personality and temporal discounting. J Res Pers 96:104166

Kekic M, Boysen E, Campbell IC, Schmidt U (2016) A systematic review of the clinical efficacy of transcranial direct current stimulation (tDCS) in psychiatric disorders. J Psychiatr Res 74:70–86. https://doi.org/10.1016/j.jpsychires.2015.12.018

Kelley NJ, Gallucci A, Riva P, Romero Lauro LJ, Schmeichel BJ (2019) Stimulating self-regulation: a review of non-invasive brain stimulation studies of goal-directed behavior. Front Behav Neurosci 12:337. https://doi.org/10.3389/fnbeh.2018.00337

Khaleghi A, Pirzad Jahromi G, Zarafshan H, Mostafavi S-A, Mohammadi MR (2020) Effects of transcranial direct current stimulation of prefrontal cortex on risk-taking behavior. Psychiatry Clin Neurosci 74(9):455–465

Knoch D, Gianotti LR, Pascual-Leone A, Treyer V, Regard M, Hohmann M, Brugger P (2006) Disruption of right prefrontal cortex by low-frequency repetitive transcranial magnetic stimulation induces risk-taking behavior. J Neurosci 26(24):6469–6472

Kuehne M, Heimrath K, Heinze H-J, Zaehle T (2015) Transcranial direct current stimulation of the left dorsolateral prefrontal cortex shifts preference of moral judgments. PLoS ONE 10(5):e0127061. https://doi.org/10.1371/journal.pone.0127061

Leloup L, Miletich DD, Andriet G, Vandermeeren Y, Samson D (2016) Cathodal transcranial direct current stimulation on the right temporo-parietal junction modulates the use of mitigating circumstances during moral judgments. Front Hum Neurosci 10.https://doi.org/10.3389/fnhum.2016.00355

León JJ, Sánchez-Kuhn A, Fernández-Martín P, Páez-Pérez MA, Thomas C, Datta A, Sánchez-Santed F, Flores P (2020) Transcranial direct current stimulation improves risky decision making in women but not in men: a sham-controlled study. Behav Brain Res 382:112485

Leow L-A, Marcos A, Nielsen E, Sewell D, Ballard T, Dux PE, Filmer HL (2023) Dopamine alters the effect of brain stimulation on decision-making. J Neurosci 43(41):6909–6919. https://doi.org/10.1523/JNEUROSCI.1140-23.2023

Li S, Tang Y, Zhou Y, Ni Y (2024) Effects of transcranial direct current stimulation on cognitive function in older adults with and without mild cognitive impairment: a systematic review and meta-analysis of randomized controlled trials. Gerontology 70(5):544–560. https://doi.org/10.1159/000537848

Liebetanz D (2002) Pharmacological approach to the mechanisms of transcranial DC-stimulation-induced after-effects of human motor cortex excitability. Brain 125(10):2238–2247. https://doi.org/10.1093/brain/awf238

Lin Y, Feng T (2024) Lateralization of self-control over the dorsolateral prefrontal cortex in decision-making: a systematic review and meta-analytic evidence from noninvasive brain stimulation. Cogn Affect Behav Neurosci 24(1):19–41. https://doi.org/10.3758/s13415-023-01148-7

Maeda F, Keenan JP, Tormos JM, Topka H, Pascual-Leone A (2000) Modulation of corticospinal excitability by repetitive transcranial magnetic stimulation. Clin Neurophysiol 111(5):800–805. https://doi.org/10.1016/S1388-2457(99)00323-5

Mancuso LE, Ilieva IP, Hamilton RH, Farah MJ (2016) Does transcranial direct current stimulation improve healthy working memory?: a meta-analytic review. J Cogn Neurosci 28(8):1063–1089. https://doi.org/10.1162/jocn_a_00956

Maréchal MA, Cohn A, Ugazio G, Ruff CC (2017) Increasing honesty in humans with noninvasive brain stimulation. Proc Natl Acad Sci 114(17):4360–4364. https://doi.org/10.1073/pnas.1614912114

Mattioli F, Maglianella V, D'Antonio S, Trimarco E, Caligiore D (2024) Non-invasive brain stimulation for patients and healthy subjects: current challenges and future perspectives. J Neurol Sci 456:122825. https://doi.org/10.1016/j.jns.2023.122825

Mayer JT, Chopard G, Nicolier M, Gabriel D, Masse C, Giustiniani J, Vandel P, Haffen E, Bennabi D (2020) Can transcranial direct current stimulation (tDCS) improve impulsivity in healthy and psychiatric adult populations? A systematic review. Prog Neuropsychopharmacol Biol Psychiatry 98:109814

McTeague LM, Goodkind MS, Etkin A (2016) Transdiagnostic impairment of cognitive control in mental illness. J Psychiatr Res 83:37–46. https://doi.org/10.1016/j.jpsychires.2016.08.001

Miyake A, Friedman NP (2012) The nature and organization of individual differences in executive functions: four general conclusions. Curr Dir Psychol Sci 21(1):8–14. https://doi.org/10.1177/0963721411429458

Miyake A, Friedman NP, Emerson MJ, Witzki AH, Howerter A, Wager TD (2000) The unity and diversity of executive functions and their contributions to complex "frontal lobe" tasks: a latent variable analysis. Cogn Psychol 41(1):49–100. https://doi.org/10.1006/cogp.1999.0734

Modirrousta M, Fellows LK (2008) Dorsal medial prefrontal cortex plays a necessary role in rapid error prediction in humans. J Neurosci 28(51):14000–14005

Moro AS, Saccenti D, Vergallito A, Scaini S, Malgaroli A, Ferro M, Lamanna J (2023) Transcranial direct current stimulation (tDCS) over the orbitofrontal cortex reduces delay discounting. Front Behav Neurosci 17:1239463. https://doi.org/10.3389/fnbeh.2023.1239463

Müller D, Habel U, Brodkin ES, Weidler C (2022) High-definition transcranial direct current stimulation (HD-tDCS) for the enhancement of working memory—a systematic review and meta-analysis of healthy adults. Brain Stimul 15(6):1475–1485. https://doi.org/10.1016/j.brs.2022.11.001

Nasreddine ZS, Phillips NA, Bédirian V, Charbonneau S, Whitehead V, Collin I, Cummings JL, Chertkow H (2005) The montreal cognitive assessment, MoCA: a brief screening tool for mild cognitive impairment. J Am Geriatr Soc 53(4):695–699. https://doi.org/10.1111/j.1532-5415.2005.53221.x

Nejati V, Salehinejad MA, Nitsche MA (2018) Interaction of the left dorsolateral prefrontal cortex (l-DLPFC) and right orbitofrontal cortex (OFC) in hot and cold executive functions: evidence from transcranial direct current stimulation (tDCS). Neuroscience 369:109–123. https://doi.org/10.1016/j.neuroscience.2017.10.042

Nihonsugi T, Ihara A, Haruno M (2015) Selective increase of intention-based economic decisions by noninvasive brain stimulation to the dorsolateral prefrontal cortex. J Neurosci 35(8):3412–3419. https://doi.org/10.1523/JNEUROSCI.3885-14.2015

Nissim NR, McAfee DC, Edwards S, Prato A, Lin JX, Lu Z, Coslett HB, Hamilton RH (2023) Efficacy of transcranial alternating current stimulation in the enhancement of working memory performance in healthy adults: a systematic meta-analysis. Neuromodulation: Technol Neural Interface 26(4):728–737. https://doi.org/10.1016/j.neurom.2022.12.014

Nitsche MA, Cohen LG, Wassermann EM, Priori A, Lang N, Antal A, Paulus W, Hummel F, Boggio PS, Fregni F, Pascual-Leone A (2008) Transcranial direct current stimulation: State of the art 2008. Brain Stimul 1(3):206–223. https://doi.org/10.1016/j.brs.2008.06.004

Nitsche MA, Lampe C, Antal A, Liebetanz D, Lang N, Tergau F, Paulus W (2006) Dopaminergic modulation of long-lasting direct current-induced cortical excitability changes in the human motor cortex. Eur J Neurosci 23(6):1651–1657. https://doi.org/10.1111/j.1460-9568.2006.04676.x

Nitsche MA, Liebetanz D, Schlitterlau A, Henschke U, Fricke K, Frommann K, Lang N, Henning S, Paulus W, Tergau F (2004) GABAergic modulation of DC stimulation-induced motor cortex excitability shifts in humans. Eur J Neurosci 19(10):2720–2726. https://doi.org/10.1111/j.0953-816X.2004.03398.x

Nitsche MA, Paulus W (2001) Sustained excitability elevations induced by transcranial DC motor cortex stimulation in humans. Neurology 57(10):1899–1901

Oberauer K (2019) Working memory and attention—a conceptual analysis and review. J Cogn 2(1):36. https://doi.org/10.5334/joc.58

Oldrati V, Patricelli J, Colombo B, Antonietti A (2016) The role of dorsolateral prefrontal cortex in inhibition mechanism: a study on cognitive reflection test and similar tasks through neuromodulation. Neuropsychologia 91:499–508. https://doi.org/10.1016/j.neuropsychologia.2016.09.010

Oldrati V, Schutter DJLG (2018) Targeting the Human Cerebellum with Transcranial Direct Current Stimulation to Modulate Behavior: A Meta-Analysis. The Cerebellum 17(2):228–236. https://doi.org/10.1007/s12311-017-0877-2

Ongur D (2000) The organization of networks within the orbital and medial prefrontal cortex of rats, Monkeys and Humans. Cerebral Cortex 10(3):206–219. https://doi.org/10.1093/cercor/10.3.206

Padoa-Schioppa C, Conen KE (2017) Orbitofrontal cortex: a neural circuit for economic decisions. Neuron 96(4):736–754. https://doi.org/10.1016/j.neuron.2017.09.031

Pennycook G, Cheyne JA, Seli P, Koehler DJ, Fugelsang JA (2012) Analytic cognitive style predicts religious and paranormal belief. Cognition 123(3):335–346. https://doi.org/10.1016/j.cognition.2012.03.003

Pennycook G, Rand DG (2019) Lazy, not biased: Susceptibility to partisan fake news is better explained by lack of reasoning than by motivated reasoning. Cognition 188:39–50

Pergolizzi D, Chua EF (2015) Transcranial direct current stimulation (tDCS) of the parietal cortex leads to increased false recognition. Neuropsychologia 66:88–98. https://doi.org/10.1016/j.neuropsychologia.2014.11.012

Pettorruso M, Miuli A, Di Natale C, Montemitro C, Zoratto F, De Risio L, d'Andrea G, Dannon PN, Martinotti G, Di Giannantonio M (2021) Non-invasive brain stimulation targets and approaches to modulate gambling-related decisions: a systematic review. Addict Behav 112:106657. https://doi.org/10.1016/j.addbeh.2020.106657

Pezzetta R, Gambarota F, Tarantino V, Devita M, Cattaneo Z, Arcara G, Mapelli D, Masina F (2024) A meta-analysis of non-invasive brain stimulation (NIBS) effects on cerebellar-associated cognitive processes. Neurosci Biobehav Rev 157:105509. https://doi.org/10.1016/j.neubiorev.2023.105509

Polanía R, Nitsche MA, Korman C, Batsikadze G, Paulus W (2012) The importance of timing in segregated theta phase-coupling for cognitive performance. Curr Biol 22(14):1314–1318. https://doi.org/10.1016/j.cub.2012.05.021

Polanía R, Nitsche MA, Ruff CC (2018) Studying and modifying brain function with non-invasive brain stimulation. Nat Neurosci 21(2):174–187. https://doi.org/10.1038/s41593-017-0054-4

Reato D, Rahman A, Bikson M, Parra LC (2010) Low-intensity electrical stimulation affects network dynamics by modulating population rate and spike timing. J Neurosci 30(45):15067–15079. https://doi.org/10.1523/JNEUROSCI.2059-10.2010

Reed T, Cohen Kadosh R (2018) Transcranial electrical stimulation (tES) mechanisms and its effects on cortical excitability and connectivity. J Inherit Metab Dis 41(6):1123–1130. https://doi.org/10.1007/s10545-018-0181-4

Rossi S, Hallett M, Rossini PM, Pascual-Leone A (2009) Safety, ethical considerations, and application guidelines for the use of transcranial magnetic stimulation in clinical practice and research. Clin Neurophysiol 120(12):2008–2039. https://doi.org/10.1016/j.clinph.2009.08.016

Ruff CC, Ugazio G, Fehr E (2013) Changing social norm compliance with noninvasive brain stimulation. Science 342(6157):482–484. https://doi.org/10.1126/science.1241399

Salehinejad MA, Ghanavati E, Rashid MHA, Nitsche MA (2021) Hot and cold executive functions in the brain: a prefrontal-cingular network. Brain Neurosci Adv 5:239821282110077. https://doi.org/10.1177/23982128211007769

Scult MA, Paulli AR, Mazure ES, Moffitt TE, Hariri AR, Strauman TJ (2017) The association between cognitive function and subsequent depression: a systematic review and meta-analysis. Psychol Med 47(1):1–17. https://doi.org/10.1017/S0033291716002075

Senkowski D, Sobirey R, Haslacher D, Soekadar SR (2022) Boosting working memory: uncovering the differential effects of tDCS and tACS. Cereb Cortex Commun 3(2):tgac018. https://doi.org/10.1093/texcom/tgac018

Shen B, Yin Y, Wang J, Zhou X, McClure SM, Li J (2016) High-definition tDCS alters impulsivity in a baseline-dependent manner. Neuroimage 143:343–352. https://doi.org/10.1016/j.neuroimage.2016.09.006

Snyder HR, Miyake A, Hankin BL (2015) Advancing understanding of executive function impairments and psychopathology: bridging the gap between clinical and cognitive approaches. Front Psychol 6.https://doi.org/10.3389/fpsyg.2015.00328

Soyata AZ, Aksu S, Woods AJ, İşçen P, Saçar KT, Karamürsel S (2019) Effect of transcranial direct current stimulation on decision making and cognitive flexibility in gambling disorder. Eur Arch Psychiatry Clin Neurosci 269(3):275–284. https://doi.org/10.1007/s00406-018-0948-5

Stagg CJ, Best JG, Stephenson MC, O'Shea J, Wylezinska M, Kincses ZT, Morris PG, Matthews PM, Johansen-Berg H (2009) Polarity-sensitive modulation of cortical neurotransmitters by transcranial stimulation. J Neurosci 29(16):5202–5206. https://doi.org/10.1523/JNEUROSCI.4432-08.2009

Stagg CJ, Nitsche MA (2011) Physiological basis of transcranial direct current stimulation. Neuroscientist 17(1):37–53. https://doi.org/10.1177/1073858410386614

Stanković M, Bjekić J, Filipović SR (2023) Effects of transcranial electrical stimulation on gambling and gaming: a systematic review of studies on healthy controls, participants with gambling/gaming disorder, and substance use disorder. J Clin Med 12(10):3407. https://doi.org/10.3390/jcm12103407

Stanovich KE (2018) Miserliness in human cognition: The interaction of detection, override and mindware. Think Reason 24(4):423–444. https://doi.org/10.1080/13546783.2018.1459314

Stuss DT (2011) Functions of the frontal lobes: relation to executive functions. J Int Neuropsychol Soc 17(05):759–765. https://doi.org/10.1017/S1355617711000695

Sutterer MJ, Koscik TR, Tranel D (2015) Sex-related functional asymmetry of the ventromedial prefrontal cortex in regard to decision-making under risk and ambiguity. Neuropsychologia 75:265–273

Swann AC, Bjork JM, Moeller FG, Dougherty DM (2002) Two models of impulsivity: relationship to personality traits and psychopathology. Biol Psychiat 51(12):988–994. https://doi.org/10.1016/s0006-3223(01)01357-9

Terney D, Chaieb L, Moliadze V, Antal A, Paulus W (2008) Increasing human brain excitability by transcranial high-frequency random noise stimulation. J Neurosci 28(52):14147–14155. https://doi.org/10.1523/JNEUROSCI.4248-08.2008

Teselink J, Bawa KK, Koo GK, Sankhe K, Liu CS, Rapoport M, Oh P, Marzolini S, Gallagher D, Swardfager W, Herrmann N, Lanctôt KL (2021) Efficacy of non-invasive brain stimulation on global cognition and neuropsychiatric symptoms in Alzheimer's disease and mild cognitive impairment: A meta-analysis and systematic review. Ageing Res Rev 72:101499. https://doi.org/10.1016/j.arr.2021.101499

Teti Mayer J, Chopard G, Nicolier M, Gabriel D, Masse C, Giustiniani J, Vandel P, Haffen E, Bennabi D (2020) Can transcranial direct current stimulation (tDCS) improve impulsivity in healthy and psychiatric adult populations? A systematic review. Prog Neuropsychopharmacol Biol Psychiatry 98:109814. https://doi.org/10.1016/j.pnpbp.2019.109814

Thoma V, Rodway P, Tamlyn G (2021) Gut thinking and eye tracking: Evidence for a central preference heuristic. J Cogn Psychol 33(8):919–930. https://doi.org/10.1080/20445911.2021.1969942

Toplak ME, West RF, Stanovich KE (2011) The cognitive reflection test as a predictor of performance on heuristics-and-biases tasks. Mem Cognit 39(7):1275–1289. https://doi.org/10.3758/s13421-011-0104-1

Toplak ME, West RF, Stanovich KE (2014) Assessing miserly information processing: an expansion of the cognitive reflection test. Think Reason 20(2):147–168

Tversky A, Kahneman D (1974) Judgment under uncertainty: heuristics and biases. Science (New York, N.Y.) 185(4157):1124–1131. https://doi.org/10.1126/science.185.4157.1124

Voarino N, Dubljević V, Racine E (2017) tDCS for memory enhancement: analysis of the speculative aspects of ethical issues. Front Hum Neurosci 10.https://doi.org/10.3389/fnhum.2016.00678

Woo T-F, Law C-K, Ting K-H, Chan CCH, Kolling N, Watanabe K, Chau BKH (2022) Distinct causal influences of dorsolateral prefrontal cortex and posterior parietal cortex in multiple-option decision making. Cereb Cortex 32(7):1390–1404. https://doi.org/10.1093/cercor/bhab278

Woods AJ, Antal A, Bikson M, Boggio PS, Brunoni AR, Celnik P, Cohen LG, Fregni F, Herrmann CS, Kappenman ES, Knotkova H, Liebetanz D, Miniussi C, Miranda PC, Paulus W, Priori A, Reato D, Stagg C, Wenderoth N, Nitsche MA (2016) A technical guide to tDCS, and related non-invasive brain stimulation tools. Clin Neurophysiol 127(2):1031–1048. https://doi.org/10.1016/j.clinph.2015.11.012

Yan R, Zhang X, Li Y, Hou J, Chen H, Liu H (2020) Effect of transcranial direct-current stimulation on cognitive function in stroke patients: a systematic review and meta-analysis. PLoS ONE 15(6):e0233903. https://doi.org/10.1371/journal.pone.0233903

Yao Y, Jia X, Luo J, Chen F, Liang P (2020) Involvement of the right dorsolateral prefrontal cortex in numerical rule induction: a transcranial direct current stimulation study. Front Hum Neurosci 14

Ye H, Huang D, Wang S, Zheng H, Luo J, Chen S (2016) Activation of the prefrontal cortex by unilateral transcranial direct current stimulation leads to an asymmetrical effect on risk preference in frames of gain and loss. Brain Res 1648:325–332. https://doi.org/10.1016/j.brainres.2016.08.007

Zheng H, Lu X, Huang D (2018) tDCS over DLPFC leads to less utilitarian response in moral-personal judgment. Front Neurosci 12:193. https://doi.org/10.3389/fnins.2018.00193

Chapter 10
Value-Based Decision-Making in the Wild: Opportunities and Challenges

Anne Kühnel, Lilly Thurn, Trevor Steward, and Nils B. Kroemer

Abstract Value-based decisions are an integral part of our life. Many disorders are characterized by aberrant choices with, at times, detrimental consequences for the physical and mental health of an individual. In the past decades, the field has progressed by optimizing tasks and computational models to better describe canonical decision-making processes. However, concerning clinical applications, important gaps remain that relate to various limitations of conventional laboratory-based assessments. In this chapter, we will outline opportunities and challenges of leaving the lab to attain more naturalistic applications that may ultimately help improve the prediction of behavior and the classification of disorder-related states. To this end, we will discuss psychometric principles, provide examples for suitable designs, and briefly review hierarchical modeling approaches to exploit the information contained in large, naturalistic data sets. Consequently, these emerging lines of research offer great promise for computational psychiatry in monitoring treatment effects over time and tailoring therapy-related decisions to changes in latent decision-making processes.

Keywords Ecological momentary assessment · Bayesian Modeling · Value-based decision-making · Gamified tasks

A. Kühnel · L. Thurn · N. B. Kroemer (✉)
Section of Medical Psychology, Department of Psychiatry and Psychotherapy, Faculty of Medicine, University of Bonn, Bonn, Germany
e-mail: nkroemer@uni-bonn.de

N. B. Kroemer
Department of Psychiatry and Psychotherapy, Tübingen Center for Mental Health, University of Tübingen, Tübingen, Germany

German Center for Mental Health (DZPG), Partner Site Tübingen, Tübingen, Germany

T. Steward
Melbourne School of Psychological Sciences, Faculty of Medicine, Dentistry, and Health Sciences, University of Melbourne, Parkville, VIC, Australia

U. Ettinger et al. (eds.), *Decision Making*, Studies in Neuroscience, Psychology and Behavioral Economics, https://doi.org/10.1007/978-3-032-00880-0_10

10.1 Life is About Choices

Every day, we make numerous decisions based on the subjective values assigned to competing options. Should we have a healthy apple or a delicious piece of cake for dessert? Do we walk for 20 min to pick up groceries or take the car for a smooth 4-min drive? Is it better to invest our monthly surplus for a gain in one year or do we use it for a shopping spree now? Such choices may span across various domains and might be resolved via estimated subjective values of available options (Grabenhorst and Rolls 2011; Roy et al. 2012). While their effects on our life may seem trivial, even a series of short-term oriented decisions may lead to undesired long-term consequences on health and well-being. Important choices in life that single-handedly shape our prospects are thought to recruit the same basic decision-making circuitry in the brain (Kronke et al. 2020; Scherbaum et al. 2022). To evaluate the consequences of our decisions, we track the discrepancies between expected and actual outcomes using prediction errors. These processes form the backbone of our motivational system in the brain that is crucially involved in decision making (Peters and Büchel 2011). Since aberrant value-based choices and learning are prevalent across many different disorders, understanding their causes and consequences is pivotal to advance treatments that limit negative health effects.

In this chapter, we will briefly review the recent progress in model-based approaches toward value-based decision-making and learning. Then, we will discuss limitations of conventional assessments that lead to translational gaps, which currently hamper the use of rigorous decision-making models for clinical applications in routine care settings. To further illustrate these opportunities and challenges, we will introduce examples of suitable designs for deep behavioral phenotyping of value-based decision-making and learning that are sufficiently scalable for use outside the lab. In closing, we will argue that gamified online assessments may effectively bridge the gap between well-controlled lab experiments and extensive sampling of individual choice trajectories in real-world settings that are necessary for clinical applications. Since such envisioned applications also call for refined statistical and computational models that harness the full potential of deep individual phenotyping, we will outline the advantages of hierarchical models to parse state- and trait-related aspects of behavior. Collectively, these steps may help unlock the enormous potential of understanding decision-making for psychiatry and clinical psychology, as well as health-related behavior more broadly.

10.2 Lab-Based Assessments of Decision-Making

A wide range of assessment tools have been developed to investigate facets of decision-making in animals and humans. Behavioral tasks that are traditionally performed in a lab setting can either produce decision-related measures (e.g., choices but also reaction times) or estimate cognitive measures through computational modeling (e.g., learning rates) (Zorowitz and Niv 2023). Common decision-making paradigms include reinforcement learning tasks, discounting tasks, mixed gambles, and auction tasks. Reinforcement learning tasks ask an individual to decide between two or more options, which differ in terms of the distributions of outcomes, for example, in win probability or rewards. By repeatedly performing the tasks, people form increasingly precise expectations of the distribution of outcomes and adapt choices accordingly (Browning et al. 2023; Sutton and Barto 2018). Bandit tasks are a classic example of a reinforcement learning task, in which participants are presented with several bandits with an unknown win probability. Only through initial exploration, participants can learn which bandit has the highest win probability and exploit it to maximize wins (Steyvers et al. 2009).

In contrast, people's tendency to discount the value of future rewards, a phenomenon called delay discounting, can be captured by intertemporal choice tasks, in which participants choose between a lower immediate reward or a higher delayed reward (Du et al. 2002; Keidel et al. 2021). Loss aversion is captured by the mixed gambles task, where participants are asked to accept or decline a gamble with a 50% chance of winning a reward and a 50% chance of losing a stake (Tom et al. 2007). The Becker-DeGroot-Marschak (BDM) auction task is a decision-making task resembling a traditional auction process to determine an individual's subjective value via their willingness to pay (De Martino et al. 2013; Hare et al. 2008). Participants must submit a bid for a presented item after which a random price is generated for this item. The task encourages participants to bid their "true" subjective value because they only get the item if their bid exceeds the random price (Newton-Fenner et al. 2023).

Behavior on these decision-making tasks is used to investigate key parameters of decision making such as learning rates (i.e., how quickly values are updated), reward sensitivity (i.e., how strongly an individual adjusts their behavior in response to external rewards), loss aversion (i.e., the tendency to avoid losses over acquiring gains), risk taking (i.e., the willingness to face uncertainty in the pursuit of potential rewards), or discount rates (i.e., the rate at which future rewards lose their value compared to immediate rewards). While these tasks have been useful in investigating individual differences in value-based decision-making, they are often tied to lab-based assessments, providing cross-sectional measures in a setting that may not be representative of behavior "in the wild".

10.3 Studying Decision-Making Outside the Lab

Since virtually everybody in Western countries possesses a smartphone, data acquisition in naturalistic environments is much more feasible today. Yet, this comes with different challenges. Whereas online tools and smartphone applications have become increasingly popular to surpass time and space constraints, many conventional lab-based assessments are not suitable for less-controlled environments. This also requires user-oriented optimization of the task design. However, once appropriately adapted, the advantages of online assessments are that they can be easily scaled to reach larger and more diverse populations in terms of age, socioeconomic background, and symptoms of disorders (Gillan et al. 2016; Spook et al. 2013). Moreover, tasks can be conducted repeatedly across various environments and states (Gillan and Rutledge 2021). By collecting data with higher variability in settings that are representative of a person's environment, the resulting choices are potentially more generalizable, leading to better predictions of real-life outcomes (Gillan and Rutledge 2021).

To make repeated assessments intrinsically more appealing, researchers can learn from successful game developers. Popular smartphone apps often have similar mechanics as decision-making tasks, and principles of gamification, such as storytelling and feedback on progression, facilitate replay (Neuser et al. 2023). Consequently, gamified versions of traditional tasks have been successfully used to assess individuals' choice behavior over extended periods (Gillan and Daw 2016; Gillan et al. 2016; Neuser et al. 2023; Rutledge et al. 2014) and cognitive measures closely mimic the outcomes from lab-based studies. Such behavioral readouts can also be combined with passive sensor data (e.g., movement data; Saragosa-Harris et al. 2022) or complement neuroimaging data collected in the lab (Gillan and Rutledge 2021; Rutledge et al. 2017). Overall, using smartphones to collect decisions from tasks provides an easily accessible way to sample large and diverse groups of individuals, as well as choices over extended periods and across realistic settings.

Computational psychiatry involves the description, classification, prediction, and mechanistic understanding of cognitive processes in mental disorders (Bennett et al. 2019; Wang and Krystal 2014). To this end, tasks that require, for example, value-based decisions and learning are often used to uncover individual differences in cognition. Assessing cognitive measures, such as learning rates, reward sensitivities, or risk aversion, can unveil aberrant processes of mental disorders. Understanding and treating dysfunctional processes, rather than focusing on the diagnostic category of an individual, is one goal of computational psychiatry (Nair et al. 2020). Aberrant cognitive processing is observed in various psychiatric conditions. For example, the processing of rewards (i.e., reward sensitivity) is up- or down-regulated in bipolar and unipolar depression, respectively (Alloy et al. 2016). Moreover, abnormalities in reinforcement learning of patients with depression mainly arise from reduced reward sensitivity, not altered learning rates (Browning et al. 2023). Such mechanistic understanding can help guide treatment approaches that are tailored toward the individual's symptom profile of the disorder. However, lab assessments of patients are akin to snapshots, failing to capture the development or fluctuation of symptoms over time. Consequently, it has been proposed that dense sampling of mental states and monitoring of symptom fluctuations through ecological momentary assessments (EMA) can facilitate the prediction of symptom development over time (Panaite et al. 2020; Smyth et al. 2009). Therefore, longitudinal assessments may monitor dysfunctional cognition at extended time scales to track individual symptom trajectories. Such insights into dynamics could help adapt treatment regimens.

Although promising, conventional lab-based assessments of decision-making have so far shown only limited potential to classify individuals or predict real-life outcomes (Nair et al. 2020; Scholl and Klein-Flügge 2018). Due to the lack of studies monitoring decision-related processes over extended periods, it is still unclear whether changes in behavior similarly track changes in clinical symptoms with sufficient precision. Although decision-making measures cannot replace clinical diagnoses by a trained clinician and may only supplement the therapeutic process (Gillan and Rutledge 2021), reliable measures of intra- and inter-individual differences may allow researchers and clinicians to better understand a patient's condition, ultimately aiding in treatment selection and the prediction of disorder progression.

10.4 Psychometric and Practical Arguments for Deep Individual Phenotyping

Clinicians make use of diverse measures to better describe, compare, and understand individuals and their symptoms of disorders. However, task measures must have three essential psychometric qualities to be useful as clinical decision aids in practice. First, a measure must have discriminatory power, reflecting variability of a variable of interest (Zorowitz and Niv 2023). If a task produces the same outcome for every participant, it is of little use to assess individual differences. Second, a measure must be valid, truly reflecting the concept it intends to measure (Zorowitz and Niv 2023). For example, a working memory task should measure working memory capacity and not an individual's prior knowledge or long-term memory (i.e., show both construct and discriminant validity). Furthermore, to guide real-life decisions, the measure should capture behavior representative of real-life decisions (i.e., show ecological validity). For example, by giving participants the chance to win the outcome of an auction task, the choice behavior should closely resemble real-life behavior. Third, task measures have to be reliable, that is, they must produce comparable outcomes if a test is repeated (i.e., show test–retest reliability). Thus, the reliability of a task measure indicates how much of the outcome variance is due to robust inter-individual differences relative to (irreproducible) measurement noise (Hedge et al. 2018; Zorowitz and Niv 2023). For diagnostic classification, suitable measures should have minimal measurement noise requiring high internal consistency (as changes in rank order over time may also reflect the differential progression of a disorder). To monitor development over time, it is further essential that differences in measures are not overshadowed by measurement noise but reflect actual changes in the construct. For example, if a clinician were to use a smartphone-based assessment tool to measure their patient's reward sensitivity alongside therapeutic sessions, changes in the outcome measure should primarily track treatment-induced changes in reward sensitivity, not noise due to sampling from a limited number of observations. Ideally, clinicians should be able to monitor both intended and unintended changes in latent decision-making processes to adapt the treatment accordingly. Taken together, for measures to classify and predict mental states and function, they must show suitable psychometric properties: they must be able to discriminate individual differences, they must be valid measures of what they intend to measure, and they must reliably produce comparable outcomes if the test is repeated.

Although decision-making is often conceptualized as a trait-like concept, common tasks have not always been shown to be highly reliable. For example, delay discounting tasks often yield measures with high reliability, whereas reward learning tasks typically yield measures with low or medium reliability (for a detailed discussion, see Karvelis et al. 2023). Likewise, the reliability of estimates from computational models are often low to moderate across various studies. Illustratively, Moutoussis and colleagues assessed parameters of learning and motivation through lab-based decision-making tasks and reported low test–retest correlations among parameter estimates ranging from 0.07 [noise parameter] to 0.37 [model fit] (Moutoussis et al. 2018). Another lab-based study using a bandit task reported small to moderate intra-class correlations, ranging from 0.23 to 0.54 (Smith et al. 2022). Two lab-based studies with reinforcement learning tasks found very low to fair test–retest reliabilities of learning parameters (0.22–0.4 (Vrizzi et al. 2023); 0.33 $\pm$ 0.25 (Pratt et al. 2021)). Other cognitive processes assessed with online tools tended to yield measures with higher test–retest reliability compared to cross-sectional lab-based studies. For example, Donegan and colleagues developed a gamified smartphone application to assess model-based estimates of planning behavior and reported a test–retest reliability of 0.63 (Donegan et al. 2023). Another study using an online version of a traditional visuo-perceptual decision-making task to study metacognitive biases even reported a test–retest reliability of 0.86 (Fox et al. 2023). Taken together, studies using value-based decision-making and reward learning tasks typically show low to moderate test–retest reliability for most parameters. Although limited psychometric data are available to make direct comparisons with gamified online decision-making tasks, they seem to provide similar or better psychometric characteristics, suggesting potential for the clinical monitoring of individual symptom trajectories.

One example of an online application deliberately designed to assess value-based decision-making and learning over time is the open-source cross-platform task, Influenca (Neuser et al. 2023). Influenca has been designed as a gamified version of a classic reinforcement learning paradigm (Behrens et al. 2007) with reward probabilities that fluctuate over time. Alongside customizable EMA items that measure potential state variables (e.g., mood), the user can play up to 31 runs of the game. Based on extensive replay data, the reliability of behavioral parameters can be evaluated, including meta-learning over runs (Neuser et al. 2023). Using a large validation sample, Neuser et al. showed an improvement in reliability over the course of several runs, although inter-individual variability of the parameters tended to decrease over time. These results suggest that multiple assessment runs can also improve the stability of the estimates compared to single, cross-sectional assessments, likely because participants may require more extensive training on a task to unmask trait-like behavioral differences (Neuser et al. 2023). The design of the application allows researchers to disentangle inter-individual and intra-individual differences in value-based decision-making, which is advantageous for clinical applications (Hedge et al. 2018). Therefore, repeated online assessments of reward learning represent a promising and cost-effective avenue for research on individual differences in value-based decision-making, holding the potential to track crucial motivational processes over extended periods.

To summarize, appropriate psychometric qualities are the cornerstone for any assessment to provide tangible benefits for applications in clinical and other professional settings. For most decision-making and reward learning tasks, studies have reported low to moderate reliability estimates for many parameters of interest. Taking behavioral tasks from the lab to naturalistic environments might not only enable us to reach larger and more diverse samples, but also facilitate more extensive and distributed data collection. Such deep phenotyping studies could permit users to extract reliable inter- and intra-individual parameter estimates without substantial increases in cost or burden imposed on the participants. However, acquiring more data within participants brings new challenges in analyzing the resulting nested data. To decompose trait components of behavior, state components (e.g., treatment response, the onset of a disorder episode, or state-related fluctuations), and mere measurement noise, optimized estimation methods are necessary (see section: *Disentangling trait- and state-like aspects of decision-making with hierarchical models*). Hierarchical Bayesian models enable increasingly efficient estimation and computation among large, complex data sets to enhance task reliability. In conclusion, innovative smartphone-based assessments might provide an important new avenue to translate values-based decisions into clinical applications.

10.5 A Roadmap Toward Precision Psychiatry Using Decision-Making Assessments

Computational psychiatry has rapidly grown as a research field in the past decade. Initially, computational models were used to elucidate aberrant mechanistic processes in patients with mental disorders, operating mostly on the group level. For example, alterations in delay discounting have been reliably related to addiction (Amlung et al. 2017) and altered reward processing has been related to major depressive disorder (Huys et al. 2013). However, to translate those group-level differences into routine clinical applications, model parameters have to provide diagnostic value at the individual level as well. A promising example of reliable decision-making assessments are delay discounting tasks that yield stable rank orders even across multiple years during adolescence (Odum et al. 2020). On the one hand, such high reliability makes delay discounting a prime candidate for cross-sectional assessments to classify and predict health-related behavior such as substance use and addiction (Alessi and Petry 2003; Amlung et al. 2017; Barlow et al. 2016; Keidel et al. 2024, 2021; Story et al.

2015, 2014). On the other hand, if delay discounting is largely stable even across transformational periods, such as adolescence, it raises the question of whether it may also capture state-dependent fluctuations, potentially reflecting a characteristic episode of a disorder. Illustrating this trait-like character, it has been shown that inter-individual differences in delay discounting between smokers and non-smokers were largely unaffected by a cognitive behavioral therapy as intervention to stop smoking while choice consistency was a better predictor of relapse (Grosskopf et al. 2021).

How reliable (or stable) decision-related parameters should be for clinical use depends on the exact goal. Whereas capturing trait-like differences might be optimal for identifying at-risk groups, capturing fluctuations in response to disorder-related states, such as a current depressive episode, is desirable to tailor treatment strategies to individual trajectories over time (Dubois and Adolphs 2016). Although the reliability of a potential diagnostic measure is a requirement for its successful use, the reporting has only recently accelerated (Karvelis et al. 2023; Zorowitz and Niv 2023) and is often limited to lab-based settings. Moreover, developmental studies may lead to an underestimation of reliability due to expected differences in individual trajectories (Zorowitz and Niv 2023). Crucially, online and smartphone-based assessments now allow us to repeatedly measure value-based decision-making in large samples, providing ideal conditions for the optimization of psychometric characteristics across many repetitions in less-controlled environments.

After initial optimism regarding the quick translation of decision-making parameters derived from computational models to clinical use, the necessity of optimizing experimental designs and analyses to improve psychometric properties has been increasingly recognized. In response, promising examples have been proposed of how reliability and clinical utility can be improved (Zorowitz and Niv 2023). They largely fall into two categories: (i) acquiring more data to improve the reliability of estimates (Hedge et al. 2020) or (ii) improving task designs or analysis approaches to improve reliability with the same amount of data (Kwon et al. 2023).

Since collecting more data arguably improves precision on average by reducing sampling noise, it is often preferable to simply collect more data. To improve group-level inferences, either the number of trials that are included in a task (Shahar et al. 2019), the number of repeated assessments for each participant (Zech et al. 2023), or the number of participants in a study have to be increased (Hedge et al. 2020). However, there are multiple constraints in practice and we will therefore focus instead on ways to increase effective data collection according to the goal of the study (e.g., comparison of groups, individual-level prediction, or individual-level monitoring) given a comparable time window for data collection. Mobile apps have successfully capitalized on the widespread availability of smartphones to measure different value-based decision-making tasks in large and diverse samples that would have been hardly possible to recruit through conventional means. For example, the Great Brain Experiment (Rutledge et al. 2014) started about 10 years ago and by now, more than 45,000 participants have completed at least one game (Rutledge 2023). Crucially, since participants can repeatedly complete the games, the data can be used to investigate within-person changes (e.g., circadian effects; Bedder et al. 2023) and inter-individual differences (Blain and Rutledge 2020; Mcnab et al. 2015; Rutledge et al. 2017, 2016). Other examples are the Neureka app (Donegan et al. 2023; Fox et al. 2023), which provides well-established learning and decision-making tasks in gamified settings to track changes in behavior over time and the Brain Explorer, which combines multiple decision-making tasks with mental health questionnaires (Skvortsova and Hauser 2022). In addition to reaching large, diverse populations, more frequent repetitions may help disentangle trait-like differences from state-like changes such as increased stress. Accordingly, an online experiment where 90 participants completed the same decision-making tasks 12 times has shown that both training and mood affect performance (Schurr et al. 2024). Extending this approach to clinical populations might provide crucial insights into how behavioral adaptations in natural environments are linked to psychopathology.

Although acquiring more data has become relatively inexpensive with online methods, participants still have to invest their time—with likely larger competition for their attention. Therefore, the second target for improving reliability is optimizing the design of the tasks themselves. First, low levels of reliability can occur when tasks are too easy or too hard causing floor or ceiling effects in performance. Therefore, difficulty levels of decision-making tasks should be calibrated for the population at hand (e.g., adolescents or older populations vs. university students). Second, trials should be informative for parameter estimation and choice options can be selected to maximize information gain over trials. For example, a classical version of a valenced go/no-go task consists of 240 trials (30 for each cue with two cues per condition) with one fixed contingency for each condition. Hence, it only shows moderate levels of reliability at best (Moutoussis et al. 2018; Zorowitz et al. 2023). In a modified version of the task, Zorowitz et al. (2023) included 6 distinct cues for the same conditions but decreased the number of repetitions per cue. In the optimized version, participants had to learn the action associated with a cue more often while the number of less informative trials during asymptotic performance was reduced. These modifications significantly improved reliability (Zorowitz et al. 2023). A similar idea is underlying Bayesian adaptive task designs (Kwon et al. 2023; Pooseh et al. 2018). In adaptive designs, parameters are estimated after each trial, and the next trial (e.g., the options in a delay discounting task) is chosen to maximize the gain in information for parameter estimation (Pooseh et al. 2018). Third, many learning tasks show substantial training effects when the task is completed multiple times. Such meta-learning can affect psychometric characteristics if training effects differ between participants (Zorowitz and Niv 2023). Training effects are also an issue when learning tasks are measured multiple times to assess the effects of interventions (Zorowitz and Niv 2023) and design choices can reduce potential confounds (e.g., by using random Gaussian walks instead of fixed transitions every *n* trials). Clear instructions with quizzes and sufficient practice trials (Zorowitz and Niv 2023) can ensure correct understanding and help to reach stable levels of performance during data acquisition (Neuser et al. 2023), which is likely more important in online settings. Fourth, participants should remain engaged and focused on the task. In addition to paying performance-based rewards, gamification might increase engagement and long-term motivation. Gamification of cognitive tasks has become more popular and smartphone-based task batteries often rely on gamified scenarios (Donegan et al. 2023; Fox et al. 2023) that embed the task in a storyline. The idea of progression increases the motivation to replay the game multiple times in longitudinal studies, improving retention of sufficiently committed participants (Bernecker and Ninaus 2021; Hu et al. 2022; Wouters et al. 2013).

To summarize, value-based decision-making tasks are a promising tool for understanding altered choices and preferences in mental disorders. Smartphone-based assessments hold great potential to transform computational psychiatry by presenting gamified tasks that have been optimized for both psychometric properties and enjoyment. Gathering large and diverse data sets over extended periods will allow us to disentangle trait- and state components of value-based decision-making and learning. Such tools may help to evaluate the long-term trajectories of behavioral parameters in patients, for example, when predicting the onset or progression of disorder-related episodes (e.g., a depressive phase or binge eating episode) or when monitoring the success of treatment modules. To fully capitalize on the potential of densely sampled data, innovative analysis approaches must be developed and applied. Computational models are traditionally fitted using maximum likelihood estimation (MLE), which, however, can yield very noisy estimates. One solution is using hierarchical Bayesian models that regularize estimates by including prior information on likely observations. These models can handle complex data structures that are common in extensive longitudinal assessments, but they are computationally expensive. To conclude, leveraging smartphone-based assessments together with improved data analysis methods may provide crucial new opportunities to assess value-based decision-making in real-life settings and monitor trajectories of behavior across many conditions and states.

10.6 Disentangling Trait- and State-Like Aspects of Decision-Making with Hierarchical Models

Repeatedly tracking decision-making over extended periods may offer invaluable insights into the progression of mental disorders. For example, individual trajectories could be used as outcomes in monitoring responses to psychotherapeutic or pharmacological treatment (Gillan and Rutledge 2021; Nair et al. 2020; Norbury et al. 2024) or to predict the onset of relapses or depressive episodes (Fried et al. 2023). However, densely sampled data also bring challenges for data analysis: How can we differentiate stable (trait-like) inter-individual differences in decision-making from meaningful (state-like) fluctuations in a person (e.g., due to mood changes)? How can we model mere noise?

To reliably differentiate trait- and state components of decision-making behavior, enough data over time (i.e., repeated runs) and in each state (i.e., trials within a run) must be acquired. As a worked example of how much data (trials and repetitions) is needed, we re-analyzed data from the reinforcement learning game Influenca (see section: *Psychometric and practical arguments for deep individual phenotyping*) that participants ($N = 207$) completed 31 times over several weeks (Neuser et al. 2023). First, we simulated choices for single runs using the other 30 runs of each participant to determine their average parameter estimates as best approximation for the held-out (i.e., not included and thus capturing a prediction to unseen data) run. To determine how well choices can be captured by the model if we know the underlying parameters (i.e., the upper bound given the data, Fig. 10.1a, yellow), we then calculated the loglikelihood of simulated choices per trial of the held-out run. This reflects the optimal fit that could be reached if the estimate fully captures both trait and state components of behavior in this run. Second, we compared this with model fits of observed behavior in held-out runs based on the average parameters of the other 30 runs (Fig. 10.1a: blue). As expected, they fit the observed choices worse compared to the upper bound after simulating choices indicating that a substantial proportion of the variance in behavior can likely be explained by state-dependent fluctuations. Third, we fitted maximum-likelihood parameters to the held-out run based on 10–120 trials of the run and evaluated how well they explained choices in the last 30 trials of the run (Fig. 10.1a, pink). When the fit is based on only a few trials, we observe severe overfitting which leads to worse model fit for held-out trials in the same run compared to average parameters from independent runs. Crucially, the generalizability to held-out trials improves beyond average parameters when parameters are fit on more than 70 trials. Therefore, at least 70 trials of this task are necessary to capture reliable state parameter information beyond trait-like behavior and noise.

To determine how many runs are necessary to derive stable average parameters per participant, we also varied the number of runs the average parameters were based on. Then, we determined how well decisions of a held-out run were explained by the averages (Fig. 10.1b). As expected, averaging across more runs led to better model fits in held-out data, although gains largely diminished after 10 runs potentially reflecting a ceiling effect of additional data for generalization. To conclude, in designing studies that capture behavioral fluctuations across states, one should include sufficient repetitions across states to reliably determine trait-like average parameters as well as sufficient trials to potentially capture state-like effects better than the typical behavior. Although we provide a worked example, the exact numbers depend on the task and the model (e.g., the number of free parameters) that are used, so simulations can help substantially improve the power of a design.

To fully capitalize on the rich information of densely sampled repeated-measures data, their nested structure has to be modeled as well. Latent State Trait (LST) models are a well-established approach to disentangle trait- and state-like components of repeated measures data across different fields in psychology (Steyer et al. 2015). They use the covariance matrix of aggregated outcome measures to estimate both components (Meyhöfer et al. 2016). For example, such models have shown that inhibitory control across tasks was mostly explained by trait components although not by one single dimension of inhibition (Fassbender et al. 2023). Another promising approach to tackle such structures is hierarchical Bayesian modeling that relies on Bayesian principles to model the complete generative process underlying the choices in each trial (Haines et al. 2020; Annis and Palmeri 2018; Lee 2011; Lee and Wagenmakers

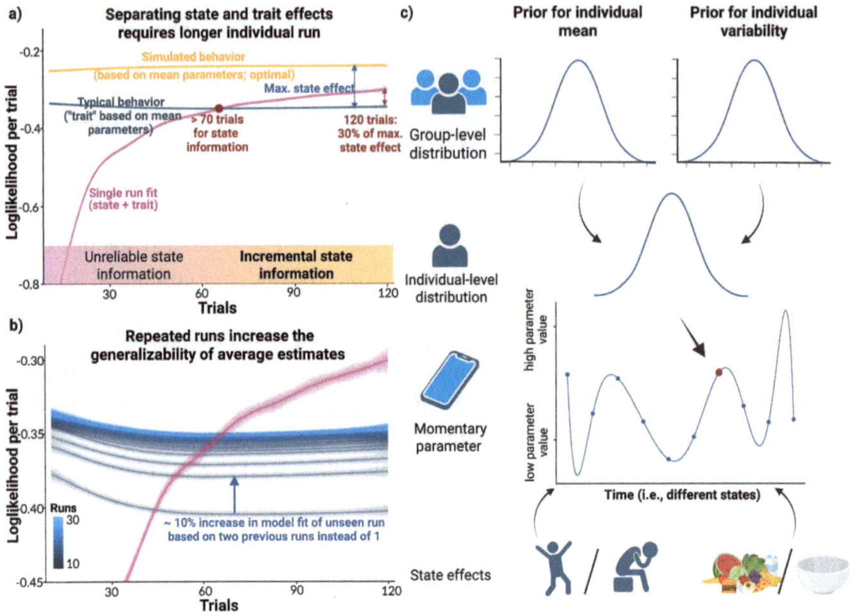

Fig. 10.1 Leveraging repeated assessments to disentangle state and trait components of behavior. a Average parameter estimates from 30 runs explain held-out data better compared to parameter estimates from short runs (< 70 trials) of the same session. Compared to an upper bound (i.e., derived from simulated choices of an agent with known parameters), average parameter estimates explain data less well. This is likely due to state-dependent fluctuations that are not explained by the average parameters or sampling noise. If individual runs are sufficiently long (i.e., > 70 trials), they can capture state-dependent variance that generalizes to held-out data in a run. Crucially, more trials increase the extent of state information that can be extracted. **b** With each additional run included in average parameter estimates, generalization to held-out runs improves. Gains are initially large, but the model fit for held-out data only improves slightly after 10 runs. **c** Hierarchical Bayesian models can be used to model the entire data structure of densely sampled longitudinal assessments by including three levels: First, group-level distributions (i.e., across participants) are given as priors for individual-level distributions that are described by the mean and variability of the parameters for each individual separately. Second, the mean and variability of an individual-level distribution are 'drawn' from the corresponding group-level distributions. Last, the momentary parameters for a single run are then each drawn from the individual-level distribution of the corresponding participant. Importantly, the systematic effects of momentary states or inter-individual differences can also be incorporated into the model and affect the momentary parameters. Figure created with biorender.

2014). In contrast to MLE, hierarchical Bayesian models do not independently estimate the parameters of each run. Instead, all data is modeled at once, and each parameter estimate is regularized by higher-level distributions (e.g., across the group), so that extreme observations are shrunk towards their mean. The regularization of point estimates can improve the reliability of the estimates (Brown et al. 2020; Haines et al. 2020; Zech et al. 2023) and more faithfully recover underlying data (Boehm et al. 2018; Katahira 2016). Crucially, these models are highly flexible and can be extended to include multiple hierarchy levels, for example, if participants repeatedly perform a task in a longitudinal study. They can be used to simultaneously estimate individual trait-like components of behavior (i.e., individual means), the state-like fluctuations over time (i.e., individual variability), and systematic effects of states explaining behavioral variability due to practice or mood (Neuser et al. 2023; Schurr et al. 2024). For example, if value-based decision-making is measured repeatedly over weeks, the model can include distributions reflecting individual mean and variance components in addition to group-level priors (Fig. 10.1c). Consequently, it provides point estimates for each measurement and estimates of the average (trait-like) and variability (state-like) of parameters for each participant.

Estimating inter-individual differences in means and variability may provide important insights into mental disorders as they might not be well characterized by differences in average behavior. For example, binges may not be explained by a general decrease in reward sensitivity but more variable behavior over time or a greater susceptibility to mood (Neuser et al. 2020). Hierarchical Bayesian models can incorporate state variables that explain fluctuations in behavior (Schurr et al. 2024). Such analyses may provide more nuanced insights into the motivational and affective dynamics of value-based decision-making in mental disorders. To conclude, hierarchical Bayesian models provide a powerful tool to (i) derive robust, regularized parameter estimates and (ii) model various characteristics in behavior, such as overall variability and responsivity to specific states, within one model by leveraging large, nested data. In the long run, identifying states that affect value-based decision-making behavior on an individual level might pave the way for clinical applications.

Since hierarchical Bayesian models use shrinkage or regularization on parameter estimates, they also improve their reliability and generalizability (Haines et al. 2020; Katahira 2016). This increase in reliability has been strikingly shown in Brown et al (2020) or Waltman et al. (2022), where most reward learning parameters had excellent test–retest correlations compared to other studies using MLE. Although it is tempting to see hierarchical Bayesian modeling as an easy solution to the reliability problem, reliability derived from the models should not be directly compared to conventional intra-class correlations because shrinkage alters their interpretation. Put simply, regularization pulls parameter estimates toward the average. If there is limited data available, this is the best guess to approximate typical behavior. In turn, if shrinkage increases similarity across sessions, it will also increase reliability. To illustrate this idea, imagine an extreme case of shrinkage, where all values of a parameter are simply replaced with the individual mean (e.g., when prior variability is assumed to be very small or there is too little data to move the prior mean). In that case, reliability would be perfect since the posterior estimates of all time points are consistent for each participant. Although this scenario violates our intuition of improved reliability, shrinkage is stronger if limited data is available or if observations are extreme. Consequently, if data is noisy, such as model parameters fit to a small number of observations, the effect is even more pronounced (Harrison et al. 2018), potentially leading to an inflation of derivative measures including test–retest reliability. Depending on the priors, this might lead to scenarios where reliability is driven by the precision of a few participants, whereas data from other, less precise participants is shrunk more (Katahira et al. 2024, for in-depth discussion). Extensive shrinkage to mean parameters might even reduce potentially meaningful variability related to extreme mood or physiological states (Zech et al. 2022). Although it is important to discern measurement noise from meaningful fluctuations, this requires many repeated measurements (see Fig. 10.1). Since decision-making tasks are envisioned to work as diagnostic tools in clinical settings with just one or two assessments and without the possibility of acquiring enough data for a full model, such assessments should still be optimized to fulfill the classical psychometric criteria to determine their potential for robust diagnostics. To conclude, when evaluating and reporting psychometric qualities, it is crucial to provide various reliability estimates with and without estimation priors, so that researchers and practitioners evaluate suitable use cases depending on their objectives (Katahira et al. 2024).

To summarize, densely sampled longitudinal assessments of value-based decision-making are a crucial next step to closing the translational gap and bringing scientific advances into the clinic. Longitudinal data can provide more reliable estimates of stable behavioral tendencies while also capturing transient state-dependent fluctuations in decision-making (Lee and Wagenmakers 2014). Both state and trait components of behavior are relevant in evaluating clinical trajectories of patients over time. Leveraging state-of-the-art data analyses, such as hierarchical Bayesian models, allows users to differentiate trait and state components as well as the influence of specific states or events on behavior. Ultimately, disentangling the effects of trait-like behavioral tendencies from state-dependent behavioral adaptions to mood or life events might enable the use of value-based decision-making tasks to monitor progression across disorder-related states, predict the progression of episodes, or evaluate the treatment response of specific symptoms.

10.7 Overarching Summary, Recommendations, and Conclusions

Making good value-based choices is a pivotal aspect of our life and this may require some flexibility across diverse situations that we encounter. For example, we know that we should not shop on an empty stomach because it may lead to impulsive decisions and if we are elated, we sometimes ignore the potential negative consequences of a decision such as having another drink at a party. In this chapter, we have reviewed the recent progress toward deep behavioral phenotyping of value-based decision-making for various applications in mental health research. To unlock the potential of computational psychiatry to transform clinical diagnoses and predictions, we have outlined the benefits as well as necessary safeguards of moving into the wild to capture richer insights into the dynamics of individual choices that contribute to our long-term health and mental well-being.

Based on well-established principles of psychometric research and recent advances in computational modeling, we have provided recommendations for the design of tasks that assess value-based decision-making for clinical applications. Crucially, there is no "one-size-fits-all" solution, and the best approach is contingent on the research objectives as the optimal design for robust classification may differ from the optimal design to track subtle changes over time. To illustrate this tradeoff, we have used a worked example with model-based simulations that can guide data-driven decision-making whether one should rather increase the planned sample size or include more repeated measures within a participant. In general, task designs can and should be optimized to improve reliability by either increasing the trial number or making trials more informative as in adaptive testing (Fröhner et al. 2019; Pooseh et al. 2018). To improve the commitment and adherence of participants, gamification of tasks is a successful strategy, and playing for real-world incentives increases compliance. Crucially, if participants perform the tasks as committedly as in a lab-based setting, this improves data quality and reliability of the estimates. To obtain high-quality data, participants must fully understand the task since learning of game mechanics during early runs may alter individual estimates. This might be harder to ensure in online only studies as interactions with the participants and inquiries about the understanding of instructions are often not possible. Therefore, games typically use tutorials, and the knowledge of the instructions can be verified using quizzes before the start of the proper experiment. As initial variance in parameter estimates might reflect learning on the task (Neuser et al. 2023; Schurr et al. 2024), it is advisable to include sufficient "burn-in" trials or runs before the data of interest is acquired (Neuser et al. 2023). Repeated assessment of behavior across different mental or metabolic states (Neuser et al. 2023; Schurr et al. 2024) may also provide a more generalizable estimation of trait-like differences and the inclusion of systematic state-related effects may improve our understanding of (mal)adaptive choice behavior in everyday life.

To conclude, online assessments offer the opportunity to translate evidence-based knowledge of value-based decision-making into promising tools for clinical applications. By providing reliable and validated tasks to track the effectiveness of interventions over time in individuals, it might become feasible to deliver tailored treatment modules that target specifically value-based decision-making to facilitate the long-term quality of life through better choices.

Acknowledgements The work was supported by the German Research Foundation (DFG), grants KR 4555/7-1, KR 4555/9-1, and KR 4555/10-1.

Author Contributions AK, LT, & NBK wrote the first draft of the manuscript and TS provided crucial feedback on all sections. All authors contributed to the interpretation of findings, provided critical revision of the manuscript for important intellectual content, and approved the final version for publication.

Financial Disclosure The authors declare no competing financial interests.

References

Alessi SM, Petry NM (2003) Pathological gambling severity is associated with impulsivity in a delay discounting procedure. Behav Proc 64(3):345–354. https://doi.org/10.1016/S0376-6357(03)00150-5

Alloy LB, Olino T, Freed RD, Nusslock R (2016) Role of reward sensitivity and processing in major depressive and bipolar spectrum disorders. Behav Therapy 47(5):600–621. https://doi.org/10.1016/j.beth.2016.02.014

Amlung M, Vedelago L, Acker J, Balodis I, MacKillop J (2017) Steep delay discounting and addictive behavior: a meta-analysis of continuous associations. Addiction 112(1):51–62. https://doi.org/10.1111/add.13535

Annis, J., & Palmeri, T. J. (2018). Bayesian statistical approaches to evaluating cognitive models. Wiley Interdiscip Rev-Cogn Sci 9(2). ARTN e1458. 1002/wcs.1458

Barlow P, Reeves A, McKee M, Galea G, Stuckler D (2016) Unhealthy diets, obesity and time discounting: a systematic literature review and network analysis. Obes Rev 17(9):810–819. https://doi.org/10.1111/obr.12431

Bedder RL, Vaghi MM, Dolan RJ, Rutledge RB (2023) Risk taking for potential losses but not gains increases with time of day. Sci Rep 13(1). ARTN 5534. https://doi.org/1038/s41598-023-31738-x

Behrens TEJ, Woolrich MW, Walton ME, Rushworth MFS (2007) Learning the value of information in an uncertain world. Nat Neurosci 10(9):1214–1221. https://doi.org/10.1038/nn1954

Bennett D, Silverstein SM, Niv Y (2019) The two cultures of computational psychiatry. JAMA Psychiat 76(6):563–564. https://doi.org/10.1001/jamapsychiatry.2019.0231

Bernecker K, Ninaus M (2021) No pain, no gain? Investigating motivational mechanisms of game elements in cognitive tasks. Comput Hum Behavior 114:106542. https://doi.org/10.1016/j.chb.2020.106542

Blain B, Rutledge RB (2020) Momentary subjective well-being depends on learning and not reward. Elife 9.https://doi.org/10.7554/eLife.57977

Boehm U, Marsman M, Matzke D, Wagenmakers E-J (2018) On the importance of avoiding shortcuts in applying cognitive models to hierarchical data. Behav Res Methods 50(4):1614–1631

Brown VM, Chen J, Gillan CM, Price RB (2020) Improving the reliability of computational analyses: model-based planning and its relationship with compulsivity. Biol Psychiatry Cogn Neurosci Neuroimaging 5(6):601–609. https://doi.org/10.1016/j.bpsc.2019.12.019

Browning M, Paulus M, Huys QJM (2023) What is computational psychiatry good for? Biol Psychiatry 93(8):658–660. https://doi.org/10.1016/j.biopsych.2022.08.030

De Martino B, Fleming SM, Garrett N, Dolan RJ (2013) Confidence in value-based choice. Nat Neurosci 16(1):105–110. https://doi.org/10.1038/nn.3279

Donegan KR, Brown VM, Price RB, Gallagher E, Pringle A, Hanlon AK, Gillan CM (2023) Using smartphones to optimise and scale-up the assessment of model-based planning. Commun Psychol 1(1):31. https://doi.org/10.1038/s44271-023-00031-y

Du W, Green L, Myerson J (2002) Cross-Cultural comparisons of discounting delayed and probabilistic rewards. The Psychological Record 52(4):479–492. https://doi.org/10.1007/BF03395199

Dubois J, Adolphs R (2016) Building a science of individual differences from fMRI. Trends Cogn Sci 20(6):425–443

Fassbender K, Meyhöfer I, Ettinger U (2023) Latent state-trait and latent growth curve modeling of inhibitory control. J Exp Psychol-Gen 152(5):1396–1419. https://doi.org/10.1037/xge0001344

Fox CA, McDonogh A, Donegan KR, Hanlon A, Gallagher E, Rouault M, Gillan C (2023) Reliable, rapid, and remote measurement of metacognitive bias

Fried EI, Proppert RKK, Rieble CL (2023) Building an early warning system for depression: rationale, objectives, and methods of the WARN-D study. Clin Psychol Eur 5(3):e10075. https://doi.org/10.32872/cpe.10075

Fröhner JH, Teckentrup V, Smolka MN, Kroemer NB (2019) Addressing the reliability fallacy in fMRI: similar group effects may arise from unreliable individual effects. Neuroimage 195:174–189. https://doi.org/10.1016/j.neuroimage.2019.03.053

Gillan CM, Daw ND (2016) Taking psychiatry research online. Neuron 91(1):19–23. https://doi.org/10.1016/j.neuron.2016.06.002

Gillan CM, Kosinski M, Whelan R, Phelps EA, Daw ND (2016) Characterizing a psychiatric symptom dimension related to deficits in goal-directed control. Elife 5https://doi.org/10.7554/eLife.11305

Gillan CM, Rutledge RB (2021) Smartphones and the neuroscience of mental health. Annu Rev Neurosci 44:129–151. https://doi.org/10.1146/annurev-neuro-101220-014053

Grabenhorst F, Rolls ET (2011) Value, pleasure and choice in the ventral prefrontal cortex. Trends Cogn Sci 15(2):56–67. https://doi.org/10.1016/j.tics.2010.12.004

Grosskopf CM, Kroemer NB, Pooseh S, Böhme F, Smolka MN (2021) Temporal discounting and smoking cessation: choice consistency predicts nicotine abstinence in treatment-seeking smokers. Psychopharmacology 238(2):399–410. https://doi.org/10.1007/s00213-020-05688-5

Haines N, Kvam PD, Irving LH, Smith C, Beauchaine TP, Pitt MA, Ahn W-Y, Turner BM (2020) Theoretically informed generative models can advance the psychological and brain sciences: lessons from the reliability paradox. psyarxiv

Hare TA, O'Doherty J, Camerer CF, Schultz W, Rangel A (2008) Dissociating the role of the orbitofrontal cortex and the striatum in the computation of goal values and prediction errors. J Neurosci 28(22):5623–5630. https://doi.org/10.1523/JNEUROSCI.1309-08.2008

Harrison XA, Donaldson L, Correa-Cano ME, Evans J, Fisher DN, Goodwin CE, Robinson BS, Hodgson DJ, Inger R (2018) A brief introduction to mixed effects modelling and multi-model inference in ecology. PeerJ 6:e4794

Hedge C, Bompas A, Sumner P (2020) Task reliability considerations in computational psychiatry. Biol Psychiatry-Cogn Neurosci Neuroimaging 5(9):837–839. https://doi.org/10.1016/j.bpsc.2020.05.004

Hedge C, Powell G, Sumner P (2018) The reliability paradox: Why robust cognitive tasks do not produce reliable individual differences. Behav Res Methods 50(3):1166–1186. https://doi.org/10.3758/s13428-017-0935-1

Hu YZ, Urakami J, Wei HT, Vomberg LH, Chignell M (2022) Longitudinal analysis of sustained performance on gamified cognitive assessment tasks. Appl Neuropsychol Adult 1–25.https://doi.org/10.1080/23279095.2022.2039931

Huys QJ, Pizzagalli DA, Bogdan R, Dayan P (2013) Mapping anhedonia onto reinforcement learning: a behavioural meta-analysis [journal article]. Biol Mood Anxiety Disord 3(1):12. https://doi.org/10.1186/2045-5380-3-12

Karvelis P, Paulus MP, Diaconescu AO (2023) Individual differences in computational psychiatry: a review of current challenges. Neurosci Biobehav Rev 148:105137. https://doi.org/10.1016/j.neubiorev.2023.105137

Katahira K (2016) How hierarchical models improve point estimates of model parameters at the individual level. J Math Psychol 73:37–58. https://doi.org/10.1016/j.jmp.2016.03.007

Katahira K, Oba T, Toyama A (2024) Does reliability of computational models truly improve with hierarchical modeling? Some recommendations and considerations for the assessment of model parameter reliability

Keidel K, Lu X, Suzuki S, Murawski C, Ettinger U (2024) Association of temporal discounting with transdiagnostic symptom dimensions. Npj Ment Health Res 3(1):13. https://doi.org/10.1038/s44184-024-00060-3

Keidel K, Rramani Q, Weber B, Murawski C, Ettinger U (2021) Individual differences in intertemporal choice [review]. Front Psychol 12.https://doi.org/10.3389/fpsyg.2021.643670

Kronke KM, Wolff M, Mohr H, Kraplin A, Smolka MN, Buhringer G, Goschke T (2020) Predicting real-life self-control from brain activity encoding the value of anticipated future outcomes. Psychol Sci 956797619896357.https://doi.org/10.1177/0956797619896357

Kwon M, Lee SH, Ahn WY (2023) Adaptive design optimization as a promising tool for reliable and efficient computational fingerprinting. Biol Psychiatry-Cogn Neurosci Neuroimaging 8(8):798–804. https://doi.org/10.1016/j.bpsc.2022.12.003

Lee MD (2011) How cognitive modeling can benefit from hierarchical Bayesian models. J Math Psychol 55(1):1–7. https://doi.org/10.1016/j.jmp.2010.08.013

Lee MD, Wagenmakers E-J (2014) Bayesian cognitive modeling: a practical course. Cambridge University Press

Mcnab F, Zeidman P, Rutledge RB, Smittenaar P, Brown HR, Adams RA, Dolan RJ (2015) Age-related changes in working memory and the ability to ignore distraction. Proc Natl Acad Sci USA 112(20):6515–6518. https://doi.org/10.1073/pnas.1504162112

Meyhöfer I, Bertsch K, Esser M, Ettinger U (2016) Variance in saccadic eye movements reflects stable traits. Psychophysiology 53(4):566–578. https://doi.org/10.1111/psyp.12592

Moutoussis M, Bullmore ET, Goodyer IM, Fonagy P, Jones PB, Dolan RJ, Dayan P (2018) Change, stability, and instability in the Pavlovian guidance of behaviour from adolescence to young adulthood. PLoS Comput Biol 14(12):e1006679. https://doi.org/10.1371/journal.pcbi.1006679

Nair A, Rutledge RB, Mason L (2020) Under the hood: using computational psychiatry to make psychological therapies more mechanism-focused. Front Psychiatry 11:140. https://doi.org/10.3389/fpsyt.2020.00140

Neuser MP, Kühnel A, Kräutlein F, Teckentrup V, Svaldi J, Kroemer NB (2023) Reliability of gamified reinforcement learning in densely sampled longitudinal assessments. PLOS Digital Health 2(9):e0000330. https://doi.org/10.1371/journal.pdig.0000330

Neuser MP, Kühnel A, Svaldi J, Kroemer NB (2020) Beyond the average: The role of variable reward sensitivity in eating disorders. Physiol Behav 112971

Newton-Fenner A, Hewitt D, Henderson J, Roberts H, Mari T, Gu Y, Gorelkina O, Giesbrecht T, Fallon N, Roberts C, Stancak A (2023) Economic value in the Brain: a meta-analysis of willingness-to-pay using the Becker-DeGroot-Marschak auction. PLoS ONE 18(7):e0286969. https://doi.org/10.1371/journal.pone.0286969

Norbury A, Hauser TU, Fleming SM, Dolan RJ, Huys QJM (2024) Different components of cognitive-behavioral therapy affect specific cognitive mechanisms. Sci Adv 10(13):eadk3222. https://doi.org/10.1126/sciadv.adk3222

Odum AL, Becker RJ, Haynes JM, Galizio A, Frye CCJ, Downey H, Friedel JE, Perez DM (2020) Delay discounting of different outcomes: review and theory. J Exp Anal Behav 113(3):657–679. https://doi.org/10.1002/jeab.589

Panaite V, Rottenberg J, Bylsma LM (2020) Daily affective dynamics predict depression symptom trajectories among adults with major and minor depression. Affect Sci 1(3):186–198

Peters J, Büchel C (2011) The neural mechanisms of inter-temporal decision-making: understanding variability. Trends Cogn Sci 15(5):227–239. https://doi.org/10.1016/j.tics.2011.03.002

Pooseh S, Bernhardt N, Guevara A, Huys QJM, Smolka MN (2018) Value-based decision-making battery: a Bayesian adaptive approach to assess impulsive and risky behavior. Behav Res Methods 50(1):236–249. https://doi.org/10.3758/s13428-017-0866-x

Pratt DN, Barch DM, Carter CS, Gold JM, Ragland JD, Silverstein SM, MacDonald AW (2021) Reliability and replicability of implicit and explicit reinforcement learning paradigms in people with psychotic disorders. Schizophr Bull 47(3):731–739. https://doi.org/10.1093/schbul/sbaa165

Roy M, Shohamy D, Wager TD (2012) Ventromedial prefrontal-subcortical systems and the generation of affective meaning. Trends Cogn Sci 16(3):147–156. https://doi.org/10.1016/j.tics.2012.01.005

Rutledge RB (2023) Risky decision and happiness task: the great brain experiment smartphone app. https://doi.org/10.5061/dryad.prr4xgxkk

Rutledge RB, Moutoussis M, Smittenaar P, Zeidman P, Taylor T, Hrynkiewicz L, Lam J, Skandali N, Siegel JZ, Ousdal OT, Prabhu G, Dayan P, Fonagy P, Dolan RJ (2017) Association of neural and emotional impacts of reward prediction errors with major depression. JAMA Psychiat 74(8):790–797. https://doi.org/10.1001/jamapsychiatry.2017.1713

Rutledge RB, Skandali N, Dayan P, Dolan RJ (2014) A computational and neural model of momentary subjective well-being. Proc Natl Acad Sci U S A 111(33):12252–12257. https://doi.org/10.1073/pnas.1407535111
Rutledge RB, Smittenaar P, Zeidman P, Brown HR, Adams RA, Lindenberger U, Dayan P, Dolan RJ (2016) Risk taking for potential reward decreases across the lifespan. Curr Biol 26(12):1634–1639. https://doi.org/10.1016/j.cub.2016.05.017
Saragosa-Harris NM, Cohen AO, Reneau TR, Villano WJ, Heller AS, Hartley CA (2022) Real-World exploration increases across adolescence and relates to affect, risk taking, and social connectivity. Psychol Sci 33(10):1664–1679. https://doi.org/10.1177/09567976221102070
Scherbaum S, Lade SJ, Siegmund S, Goschke T, Dshemuchadse M (2022) From single decisions to sequential choice patterns: extending the dynamics of value-based decision-making. PLoS ONE 17(4):e0267249. https://doi.org/10.1371/journal.pone.0267249
Scholl J, Klein-Flügge M (2018) Understanding psychiatric disorder by capturing ecologically relevant features of learning and decision-making. Behav Brain Res 355:56–75. https://doi.org/10.1016/j.bbr.2017.09.050
Schurr R, Reznik D, Hillman H, Bhui R, Gershman SJ (2024) Dynamic computational phenotyping of human cognition. Nat Hum Behav.https://doi.org/10.1038/s41562-024-01814-x
Shahar N, Hauser TU, Moutoussis M, Moran R, Keramati M, Consortium N, Dolan RJ (2019) Improving the reliability of model-based decision-making estimates in the two-stage decision task with reaction-times and drift-diffusion modeling. PLoS Comput Biol 15(2):e1006803. https://doi.org/10.1371/journal.pcbi.1006803
Skvortsova V, Hauser T (2022) Distinct computational mechanisms underlying cognitive flexibility deficits in impulsivity and compulsivity.
Smith R, Taylor S, Stewart JL, Guinjoan SM, Ironside M, Kirlic N, Ekhtiari H, White EJ, Zheng H, Kuplicki R, Investigators T, Paulus MP (2022) Slower learning rates from negative outcomes in substance use disorder over a 1-year period and their potential predictive utility. Comput Psychiatry. https://doi.org/10.5334/cpsy.85
Smyth JM, Wonderlich SA, Sliwinski MJ, Crosby RD, Engel SG, Mitchell JE, Calogero RM (2009) Ecological momentary assessment of affect, stress, and binge-purge behaviors: day of week and time of day effects in the natural environment. Int J Eat Disord 42(5):429–436. https://doi.org/10.1002/eat.20623
Spook JE, Paulussen T, Kok G, Van Empelen P (2013) Monitoring dietary intake and physical activity electronically: feasibility, usability, and ecological validity of a mobile-based ecological momentary assessment tool. J Med Internet Res 15(9):e214. https://doi.org/10.2196/jmir.2617
Steyer R, Mayer A, Geiser C, Cole DA (2015) A theory of states and traits–revised. Annu Rev Clin Psychol 11:71–98. https://doi.org/10.1146/annurev-clinpsy-032813-153719
Steyvers M, Lee MD, Wagenmakers E-J (2009) A Bayesian analysis of human decision making on bandit problems. J Math Psychol 53(3):168–179. https://doi.org/10.1016/j.jmp.2008.11.002
Story GW, Moutoussis M, Dolan RJ (2015) A computational analysis of aberrant delay discounting in psychiatric disorders. Front Psychol 6:1948. https://doi.org/10.3389/fpsyg.2015.01948
Story GW, Vlaev I, Seymour B, Darzi A, Dolan RJ (2014) Does temporal discounting explain unhealthy behavior? A systematic review and reinforcement learning perspective. Front Behav Neurosci 8:76. https://doi.org/10.3389/fnbeh.2014.00076
Sutton RS, Barto AG (2018) Reinforcement learning: an introduction, 2nd edn. The MIT Press
Tom SM, Fox CR, Trepel C, Poldrack RA (2007) The neural basis of loss aversion in decision-making under risk. Science 315(5811):515–518. https://doi.org/10.1126/science.1134239
Vrizzi S, Najar A, Lemogne C, Palminteri S, Lebreton M (2025) Comparing the test-retest reliability of behavioral, computational and self-reported individual measures of reward and punishment sensitivity in relation to mental health symptoms. Nat Mental Health 1–13. https://doi.org/10.1038/s44220-025-00427-1
Waltmann M, Schlagenhauf F, Deserno L (2022) Sufficient reliability of the behavioral and computational readouts of a probabilistic reversal learning task. Behav Res Methods. https://doi.org/10.3758/s13428-021-01739-7

Wang XJ, Krystal JH (2014) Computational psychiatry. Neuron 84(3):638–654. https://doi.org/10.1016/j.neuron.2014.10.018

Wouters P, Van Nimwegen C, Van Oostendorp H, Van Der Spek ED (2013) A meta-analysis of the cognitive and motivational effects of serious games. J Educ Psychol 105(2):249

Zech H, Waltmann M, Lee Y, Reichert M, Bedder RL, Rutledge RB, Deeken F, Wenzel J, Wedemeyer F, Aguilera A, Aslan A, Bach P, Bahr NS, Ebrahimi C, Fischbach PC, Ganz M, Garbusow M, Grosskopf CM, Heigert M, ReCoDe-Consortium (2023) Measuring self-regulation in everyday life: Reliability and validity of smartphone-based experiments in alcohol use disorder. Behav Res Methods 55(8):4329–4342.https://doi.org/10.3758/s13428-022-02019-8

Zech HG, Reichert M, Ebner-Priemer UW, Tost H, Rapp MA, Heinz A, Dolan RJ, Smolka MN, Deserno L (2022) Mobile data collection of cognitive-behavioral tasks in substance use disorders: where are we now? Neuropsychobiology 1–13.https://doi.org/10.1159/000523697

Zorowitz S, Karni G, Paredes N, Daw N, Niv Y (2023) Improving the reliability of the pavlovian Go/no-go task

Zorowitz S, Niv Y (2023) Improving the reliability of cognitive task measures: a narrative review. Biol Psychiatry Cogn Neurosci Neuroimaging 8(8):789–797. https://doi.org/10.1016/j.bpsc.2023.02.004

Chapter 11
Mechanisms and Correlates of the Date/Delay Effect in Intertemporal Choice

Kristof Keidel, Carsten Murawski, and Ulrich Ettinger

Abstract Intertemporal choice refers to decisions involving a trade-off in reward size and delay to receiving the reward ('a smaller reward now versus a larger reward later'). When making such decisions, people typically devalue future rewards, a phenomenon known as temporal discounting. Individual differences in temporal discounting have been shown to be relatively stable across time but choice behaviour is also malleable, as revealed by various framing effects. Here, we provide a narrative review of the date/delay effect, the phenomenon of reduced temporal discounting when time until receipt of rewards is given as dates (e.g., on 24th November 2024) rather than delay units (e.g., in 89 days). The date/delay effect has been replicated by many studies, with medium effect size. It is robust across most task designs and participant samples, but some of its determinants (e.g., task parameters such as delay length) remain to be systematically tested. Furthermore, only few studies have examined the effect's cognitive and neural mechanisms. Experimental evidence to date suggests that the date framing condition leads to shorter estimations of delays compared to the delay framing condition. Additionally, individual differences studies have shown that people comparing time attributes and computing intervals more in the date than in the delay framing condition might show a lower date/delay effect, whereas people focusing more on rewards in the decision process might show a higher date/delay effect. Such individual differences in the date/delay effect are not only important to further elucidate its mechanisms but also to examine relations with psychopathology, as general population studies have suggested the effect to be particularly strong in vulnerable individuals. Ultimately, further studies in non-clinical and clinical populations are needed to examine individual determinants of the date/delay effect as well as its applicability in clinical settings.

Keywords Attention · Choice strategy · Date/delay effect · Intertemporal choice · Temporal discounting · Time perception

K. Keidel (✉) · U. Ettinger
Department of Psychology, University of Bonn, Bonn, Germany
e-mail: kristof.keidel@uni-bonn.de

K. Keidel · C. Murawski
Centre for Brain, Mind and Markets, Faculty of Business and Economics, The University of Melbourne, Carlton, Victoria, Australia

U. Ettinger et al. (eds.), *Decision Making*, Studies in Neuroscience, Psychology and Behavioral Economics, https://doi.org/10.1007/978-3-032-00880-0_11

11.1 Introduction

Intertemporal choice refers to decisions involving a trade-off between the reward size and the delay until the reward is received (Frederick et al. 2002). In other words, it relates to decision-making scenarios in which the consequences of different actions materialise at different points in time. Such decisions are ubiquitous in everyday life and important in areas such as education, finance, politics and health care. Consider, for example, a university student facing the decision between going out to a bar with their friends or studying for an exam scheduled for the following day. While the decision to have a night out may be tempting as it has positive short-term consequences, getting a good grade in the exam may have a higher long-term payoff. Another example is the decision between spending money on an expensive meal in a restaurant or saving it towards a long-term goal, such as an overseas holiday. In a broader context, societal and political decisions about whether to implement measures to mitigate climate change or not also involve intertemporal choices.

In its most canonical form, intertemporal choice manifests in a decision between a smaller, sooner and a larger, later reward. Most decision-makers will choose a larger reward over a smaller reward and a sooner reward over a later reward, ceteris paribus. Modelling the trade-off between the size of a reward and delay until its receipt has been studied for a long time (for a review, see Frederick et al. 2002) and a large number of computational models have been proposed (Doyle 2013; also see the chapter by Diederich in this volume). Typically, intertemporal choice tasks are used to infer an individual's degree of *temporal discounting*, that is, the degree to which future rewards are devalued. In such tasks, decision-makers are presented with a range of choices between smaller, sooner and larger, later rewards, typically using monetary payoffs (e.g., "Would you prefer €49 today or €60 in 89 days?"; Kirby et al. 1999). Participants' decisions are used to estimate the degree of discounting, typically referred to as *discount rate*.

It is often assumed that, when faced with an intertemporal choice, a decision-maker will combine the different attributes (i.e., reward size and time) of an option into a common unit (i.e., a subjective value) and choose the option with the higher value on that unit (*discounted utility model*; Samuelson 1937). That is, the subjective value of an option is computed by discounting the objective reward as a function of time using a so-called discount function. Discount functions can have different functional forms (e.g., exponential, hyperbolic, quasi-hyperbolic) and typically take as arguments the delay as well as a discount rate. A higher discount rate reflects higher devaluation of a future reward per unit of time, which is typically interpreted as more impulsive decision-making (Hamilton et al. 2015; Peters and Büchel 2011). Figure 11.1a illustrates how a reward may be devalued based on the hyperbolic discount function given by $SV = R/(1 + kt)$ (Mazur 1987), where the current subjective value SV of an objective future reward R is computed by discounting the reward R by a discount fraction depending on the discount rate k and the delay until receipt t. Using this discount function in the decision example given above, the larger, later reward of €60 to be received in 89 days has a current subjective value of €55.10

for decision-maker A with $k = 0.001$ (i.e., it is higher than the smaller, sooner reward of 49€), whereas it has a current subjective value of €47.36 for decision-maker B with $k = 0.003$ and of €36.96 for decision-maker C with $k = 0.007$ (i.e., it is lower than the smaller, sooner reward 49€). Accordingly, decision-maker A would be expected to choose the larger, later reward, whereas decision-makers B and C would be expected to choose the smaller, sooner reward.

It should be noted that more recent approaches have begun to challenge the assumption that reward size and delay to receipt of reward of a decision option are integrated before options are compared with each other. That is, in contrast to *option-based models* such as the discounted utility model (Samuelson 1937), which assumes a computation of subjective values, *attribute-based* models propose that rewards and times are compared separately and decisions result from the trade-off of these comparisons (Amasino et al. 2019; Dai and Busemeyer 2014; Marzilli Ericson et al. 2015; Read et al. 2013; Scholten and Read 2010). So far, the debate on which group of models better explains actual choice behaviour is ongoing.

Impulsive decision-making, captured by the discount rate, varies substantially between individuals, and these differences tend to be relatively stable across time

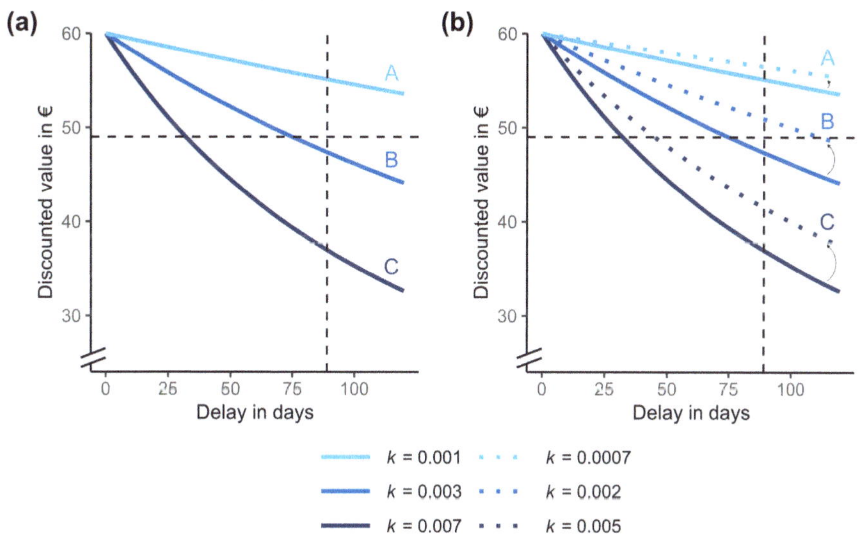

Fig. 11.1 Illustration of temporal discounting. The panels depict hyperbolic temporal discounting of three decision-makers faced with the decision "Would you prefer €49 today or €60 in 89 days?" (Kirby et al. 1999). Darker lines represent steeper temporal discounting (i.e., more impulsive decision-making). The dashed horizontal line indicates the value of the smaller, sooner reward (€49), the dashed vertical line indicates the delay of the larger, later reward (89 days). In panel (**a**), only decision-maker A would be expected to choose the larger-later reward, whereas decision-makers B and C would be expected to choose the smaller, sooner reward. Panel (**b**) illustrates how a within-subject manipulation may reduce temporal discounting (dotted lines), so that decision-maker B would now also be expected to choose the larger, later reward, whereas decision-maker C would still be expected to choose smaller, sooner reward. For details, see text

(Kirby 2009; Strickland et al. 2023). Interestingly, individual differences in temporal discounting appear to translate into everyday life decisions. This becomes apparent when considering numerous health-related and non-health-related correlates of individual differences in temporal discounting (see Keidel et al. 2021, for an overview). For instance, people with higher discount rates (i) engage less in preventive health behaviour such as regularly attending dental visits or getting vaccinated (Axon et al. 2009; Bradford 2010), (ii) show more risky health behaviour, such as smoking, drinking or problematic eating (Audrain-McGovern et al. 2009; Fröhner et al. 2022; Kekic et al. 2020), and (iii) are generally less successful in their academic and work lives (de Ruijter et al. 2023; Freeney and O'Connell 2010; Munasinghe and Sicherman 2005). Moreover, temporal discounting is elevated in a most psychiatric disorders (Amlung et al. 2019; Jackson and MacKillop 2016; Weinsztok et al. 2021) and has been discussed as a potential transdiagnostic marker (Amlung et al. 2019; Bailey et al. 2021; Bickel et al. 2019; Keidel et al. 2024a; Lempert et al. 2019; Levitt et al. 2023), underlining its clinical relevance.

Thus, given the high importance of intertemporal decision-making for numerous life outcomes, the question whether temporal discounting behaviour can be altered despite its relatively high temporal stability appears to be vital. One important determinant of decision-making in general is the *framing* of choices, that is, the way in which choice scenarios are presented (e.g., *Prospect Theory;* Kahneman and Tversky 1979; Tversky and Kahneman 1981). Specifically, extensive research has revealed that different framing manipulations of decision options influence choice behaviour in financial (Roszkowski and Snelbecker 1990), medical (Edwards et al. 2001; Gallagher and Updegraff 2012) or legal contexts (McCaffery et al. 1995). Importantly, framing also plays an important role in intertemporal decision-making, as outlined by several review articles (Lempert and Phelps 2016; Rung and Madden 2018; Scholten et al. 2019). These reviews provide overviews of different experimental manipulations and behavioural trainings that alter intertemporal choice behaviour, thereby revealing its context-dependency and malleability. With regards to framing, manipulations of the way reward or time attributes are presented may influence intertemporal decisions (e.g., reduce temporal discounting) by modifying attention allocation (Magen et al. 2008 2014; Reeck et al. 2017), reference-dependent valuation (Kahneman et al. 1991; Loewenstein 1988) and processing of information about time (LeBoeuf 2006; Peters and Büchel 2010; Read et al. 2005).

Interestingly, Lempert and Phelps (2016), Rung and Madden (2018) and Scholten et al. (2019) all come to the conclusion that experimental manipulations targeting temporal cognition (i.e., the processing of time information) may be particularly promising targets for altering temporal discounting. One experimental effect related to intertemporal decision-making involving time information is the so-called *date/delay effect* (LeBoeuf 2006; Read et al. 2005). It refers to the phenomenon that temporal discounting is reduced when time until receipt of rewards is presented in terms of dates (e.g., on 24th November 2024) rather than delay units (e.g., in 89 days). Figure 11.1b illustrates how such a manipulation may change an individual's discount rate. Specifically, here, it reduces each individual's discount rate by approximately one third. In the given example (delay framing: €49 today vs. €60 in 89 days | date framing: €49 on 27 August 2024 [=today] vs. €60 on 24 November 2024), the subjective values of the future reward in the date framing condition would therefore correspond to €56.48 for decision-maker A, €50.93 for decision-maker B and €41.52 for decision-maker B. Accordingly, not only decision-maker A but also decision-maker B would be expected to choose the larger, later reward in the date framing condition, since the manipulation leads to an increased current subjective value of the future reward, which now exceeds €49. Thus, this scenario exemplifies a framing-induced preference reversal in decision-maker B. On the other hand, decision-maker C would still be expected to choose the smaller, sooner reward, since the current subjective value of the future reward remains below €49.

In the remainder of this chapter, we will provide a comprehensive review of the literature on the date/delay effect. First, past research will be reviewed, with the aim to examine the robustness and potential determinants of the effect. Second, studies of the cognitive mechanisms and individual difference correlates of the effect will be discussed. The chapter will conclude with recommendations for future research on the date/delay effect and intertemporal choice more generally.

11.2 Robustness of the Date/Delay Effect

The date/delay effect was originally described by Read et al. (2005) and LeBoeuf (2006). In both articles, several experiments were reported testing whether framing time intervals as dates reduces temporal discounting compared to a delay unit framing. Read et al. (2005) conducted five between-subject experiments, all confirming the date/delay effect. They showed that the date/delay effect occurred when using either hypothetical choice (experiment 1) or matching (experiment 2) tasks. The effect was also found when participants had the chance of winning real rewards (experiment 3). When combining delay and date framing conditions, temporal discounting was similar to the delay framing condition alone (experiment 4). Finally, Read et al. (2005) showed that temporal discounting might become more exponential (i.e., less hyperbolic) in the date framing condition (experiment 5; however, see contradicting evidence below). At around the same time as Read et al. (2005), LeBoeuf (2006) conducted six between-subject experiments on the date/delay effect. Similar to experiments 1 and 2 in Read et al. (2005), he observed the date/delay effect in both choice (experiment 3) and matching tasks (experiments 1A, 1B and 2) in the contexts of prizes and investments. Moreover, the effect also replicated in a version of the task involving losses (experiment 4). Focusing on potential mechanisms, LeBoeuf (2006) revealed that temporal discounting was lower when time was framed as dates, both compared to conditions in which time was framed as days or months (in which temporal discounting did not differ significantly; experiment 5). He thereby showed that the date/delay effect is unlikely to be due to subadditivity only (i.e., reduced temporal discounting when the delay is separated into smaller intervals). Finally, LeBoeuf (2006) found that time intervals in the date framing condition are estimated to be shorter than in the delay framing condition (experiment 6). Later experiments built upon the original findings by Read et al. (2005) and LeBoeuf (2006) to replicate the date/delay effect, examine potential influential factors and explore its cognitive and neural mechanisms. While the latter will be subject of the next section of this chapter, here, we will summarise key findings on the effect itself as well as its potential determinants.

To our knowledge, 15 further studies have investigated the date/delay effect in intertemporal choice in total. Of these, 13 studies successfully replicated the date/delay effect in at least one experiment (DeHart and Odum 2015; Dshemuchadse et al. 2013; Jiang and Dai 2021; Keidel et al. 2024a, b, 2025a, b; Klapproth 2012; Naudé et al. 2018; Reyes-Huerta et al. 2023; Scherbaum et al. 2012; Schoemann et al. 2019; Wang et al. 2015), whereas two did not find significant differences in temporal discounting between conditions (Lempert et al. 2016; Sharma and Khan 2022). Overall, the effect size of the date/delay effect appears to be in the medium range (Hedge's $g = 0.45$; Rung and Madden 2018).

The successful replication studies of the date/delay effect employed different tasks and designs. The effect was revealed in studies using both between- (e.g., Jiang and Dai 2021; Klapproth 2012) and within-subject designs (e.g., Dshemuchadse et al. 2013; Keidel et al. 2025b). Moreover, it was replicated both when using preset tasks such as Kirby et al.'s (1999) Monetary Choice Questionnaire (e.g., Keidel et al. 2025b; Naudé et al. 2018) or titration procedures (e.g., DeHart and Odum 2015; Jiang and Dai 2021), as well as in designs involving hypothetical (e.g., Dshemuchadse et al. 2013; Keidel et al. 2025b; Naudé et al. 2018) or potentially real outcomes (e.g., Jiang and Dai 2021; Keidel et al. 2025a). Additionally, while most studies examining the date/delay effect used a date format in which the month was written out (e.g., "24 November 2024"; Keidel et al. 2024a; LeBoeuf 2006; Read et al. 2005), the effect was also replicated in studies where months were abbreviated (e.g., "24 Nov 2024"; Jiang and Dai 2021; although see Lempert et al. 2016; Sharma and Khan 2022) or where months were given as integers (e.g., "24.11.2024"; Dshemuchadse et al. 2013; Scherbaum et al. 2012). Regarding study populations, the date/delay effect was found in relatively small clinical samples (e.g., individuals with substance use disorders; Klapproth 2012) as well as in large general population samples of up to 1,000 participants (e.g., Keidel et al. 2025b). The studies comprising larger samples examined participants from Germany (Keidel et al. 2025b), the United States (Keidel et al. 2024a; Naudé et al. 2018) and China (Jiang and Dai 2021), providing at least some evidence for cultural generalisability. In total, research so far suggests that the date/delay effect is relatively robust across task designs and samples.

It should be noted, however, that some of the studies examining the date/delay effect suggest that the effect is more likely to occur in certain contexts. For instance, Lempert et al. (2016) proposed that their null result regarding the date/delay effect could be due to several task-related factors. Specifically, they speculated that the relatively short delays in their task, the used date notation (i.e., the absence of a year; e.g., "24 Nov") or the intermixed presentation of date and delay framing trials (i.e., no separate blocks) may have prevented the date/delay effect. To date, no study has systematically tested whether these factors are essential for the occurrence of the date/delay effect.

Furthermore, studies have begun to investigate how the date/delay effect interacts with other well-known effects related to intertemporal choice, such as the magnitude effect, the hidden-zero effect, the sign effect and the delay effect (Keidel et al. 2025b; Naudé et al. 2018; Reyes-Huerta et al. 2023). Such interactions may also shed light on the determinants of the date/delay effect.

First, the *magnitude effect* is well-established and shows that higher monetary amounts are discounted less steeply than lower monetary amounts (Green et al. 1997). Several studies have examined whether the date/delay effect interacts with this effect (Keidel et al. 2025b; Naudé et al. 2018; Reyes-Huerta et al. 2023). While two studies using larger, later amounts between $25 and $85 (or €) did not find a significant interaction (Keidel et al. 2025b; Naudé et al. 2018), one study found that the date/delay effect was significant for a larger, later amount of $2,000 only and not for a delayed amount of $16,000 (Reyes-Huerta et al. 2023). Thus, the date/delay effect may only be found when using monetary magnitudes up to several thousand dollars/euros.

Second, Naudé et al. (2018) examined the interaction between the date/delay effect and the hidden-zero effect. The *hidden-zero effect* describes that, compared to a standard condition (e.g., €49 today vs. €60 in 89 days), temporal discounting is reduced when additionally presenting the consequence (i.e., the explicit zero as opportunity cost) of the other option (e.g., €49 today and €0 in 89 days vs. €0 today and €60 in 89 days) (Magen et al. 2008 2014). In each of their two experiments, Naudé et al. (2018) replicated the date/delay effect but did not find the hidden-zero effect. There was no interaction between the two effects. While further replication is needed, especially since the hidden-zero effect is usually robust (Koffarnus and Bickel 2014; Radu et al. 2011; Wu and He 2012), these results suggest that the date/delay effect is robust across hidden-zero and explicit-zero conditions.

Third, Reyes-Huerta et al. (2023) examined interactions between the date/delay effect, the magnitude effect, the sign effect and the delay effect. The *sign effect* describes the phenomenon that losses are discounted less than gains (Estle et al. 2006; Myerson et al. 2017), and the *delay effect* describes the observation that temporal discounting is reduced when common delays are added to both option outcomes, that is, when both lie in the future (corresponding to hyperbolic discounting; Green et al. 1994; Kirby and Herrnstein 1995). Intriguingly, results showed that the date/delay effect for both gains and losses was only significant when the delayed amount was small (i.e., $2,000; see above) and when the smaller, sooner reward was immediate and not delayed by two years, thereby revealing a critical interaction with the delay effect. Since the hyperbolic discount function generally described the data best, this interaction might not be explained by more exponential discounting in the date condition (as proposed by Read et al. 2005) but by the idea that the saliency of a common delay is substantially higher in the delay framing condition than in the date framing condition. As a consequence, the delay effect reduces temporal discounting only in the delay framing condition, rendering discount rates in the two framing conditions comparable in size (Reyes-Huerta et al. 2023). Interestingly, another study, which used delays of 0 and 7 days, did not report significant effects of timing (Dshemuchadse et al. 2013). Thus, the length of the (common) delay might play a role not only in the occurrence of the date/delay effect but also in its interaction with the delay effect.

In sum, the date/delay effect in intertemporal choice has been consistently replicated since it was first described (LeBoeuf 2006; Read et al. 2005), with only few exceptions. It appears to be robust across most task designs and across populations from different countries, although specific factors such as the delay length, the date notation or the block-wise versus intermixed trial presentation should be systematically examined in future. Moreover, more studies examining interaction effects with other well-established phenomena in intertemporal choice may not only inform the circumstances under which the effect occurs but also shed light on the mechanisms of the date/delay effect and individual variation in its magnitude. The latter two will be examined in the next section.

11.3 Mechanisms and Correlates of the Date/Delay Effect

Despite the large number of studies confirming the date/delay effect, its cognitive mechanisms are not fully understood yet. Read et al. (2005) and LeBoeuf (2006) have introduced several hypotheses, and recent studies have begun to investigate them.

First, both Read et al. (2005) and LeBoeuf (2006) suggested that the date/delay effect may be due to relative changes in attention allocation on and weighting of option attributes in the date compared to the delay framing condition (*attention-focusing hypothesis*). That is, while the delay framing condition highlights the waiting time until a larger, later reward is received, the date framing condition draws the decision-maker's attention to the time of receipt of the reward, its value or post-delay transactions. As a consequence, the larger, later reward may be attended to more, become more accessible and may therefore be chosen more often (LeBoeuf 2006; Read et al. 2005).

Second, the more complex date notation could initially lead to a higher cognitive load and may result in changes of the computational strategy while making intertemporal decisions (*choice strategy hypothesis*; Read et al. 2005). While decision-makers may rely on an option-wise computation in the delay framing condition, integrating the continuous numeric values of outcome and time into a single subjective value, this strategy might be impeded by dates. That is, unlike delay units, which represent integers that can be integrated to obtain some kind of exchange rate of options, date notations require either a transformation of dates to delay units (to follow a similar option-wise strategy) or a change of strategy to adapt to the different notation. This could be, for instance, a change to a more attribute-wise strategy, in which time comparisons and reward comparisons are traded off against each other. If this strategy change is accompanied by a higher focus on or weighting of reward values, decision-makers may avoid an increased cognitive load and ultimately decide more often based on the outcome comparison, that is, in favour of the larger, later reward.

Third, subjective time perception could differ between delay and date framing conditions (*differential time estimation hypothesis*; LeBoeuf 2006; Read et al. 2005). Specifically, intervals given in delay units are thought to be perceived as subjectively longer than date intervals. This may, for example, be a consequence of the different foci on waiting time versus reward value (LeBoeuf 2006) or of the perceived similarity of time notations, which is arguably higher for dates than for delay units (Read et al. 2005). Ultimately, a subjectively shorter time interval in the date condition renders the larger, later reward relatively more attractive.

Finally, further hypotheses propose that the date condition may result in more positive valence (LeBoeuf 2006; Lempert et al. 2016), a more deliberate processing style (Scherbaum et al. 2012; but see Dshemuchadse et al. 2013) or higher confidence that the reward will actually be delivered (Jiang and Dai 2021; Read et al. 2005).

In view of the potential interactions of the different hypotheses, it is not unlikely that a combination of hypotheses is applicable. And indeed, the current state of research points into this direction.

The *differential time estimation hypothesis* has so far obtained the strongest empirical support. Several studies have confirmed that participants, when asked to indicate how long time intervals feel to them on a scale from "very short" to "very long", perceive intervals given in delay units as significantly longer than those given as date intervals (Jiang and Dai 2021; LeBoeuf 2006; Zauberman et al. 2009). In addition, the date/delay effect disappears when accounting for subjective time perception, as indicated by a study using a single-shot intertemporal matching task (Wang et al. 2015, Experiment 3). Furthermore, the date condition is associated with higher neural activity in brain regions related to episodic thinking (posterior cingulate cortex, precuneus and angular gyrus; Keidel et al. 2024b), which could reflect greater retrieval of past reward experiences or more vivid imagery of future rewards, which may ultimately result in a more contracted estimation of time and a lower psychological distance (Trope and Liberman 2003). While this finding suggests that differential time estimation between conditions is due to memory- or prospection-related processes, which result from the focus on the receipt of the reward rather than on waiting time in the date framing condition and ultimately make the larger, later reward appear more vivid, it cannot be ruled out that the similarity of notations (Read et al. 2005) may play a role as well. Future studies could assess evaluations of similarity in addition to time estimation tasks. Moreover, it would be interesting to investigate whether higher between-subject differences in time estimations are predictive of a higher date/delay effect, which would provide further evidence for the *differential time estimation hypothesis*.

By potentially highlighting past and future reward experiences, the neural findings reported by Keidel et al. (2024b) also support the hypothesis that (larger, later) rewards may be considered more strongly in the date than in the delay framing condition. However, this does not appear to be evident in measures of overt attention, e.g., fixation times on attributes of decision options, as participants fixate time attributes longer and make more saccades between them in the date than in the delay condition, whereas no difference is evident regarding reward attributes (Keidel et al. 2024b, 2025a; Sharma and Khan 2022). Potential confounds regarding differences in reading requirements in the two conditions aside, this finding contradicts the hypothesis that overt attention is more strongly drawn towards rewards in the date framing condition. However, although not visible in overt attention, rewards could still be weighted more strongly than time intervals in the decision process (*reward-weighting hypothesis*; Keidel et al. 2025a). Indeed, both a mouse-tracking and an eye-tracking study found that reward differences may be more predictive of choice behaviour in the date than in the delay framing condition (Dshemuchadse et al. 2013; Keidel et al. 2025a; but see Schoemann et al. 2019, who did not replicate the result from Dshemuchadse et al. 2013, when using a different mouse-tracking procedure).

Regarding the *choice strategy hypothesis*, it appears that the main difference between conditions is that decision-makers compare time attributes more often in the date than in the delay framing condition (Keidel et al. 2024b, 2025a). However, as mentioned before, the number of saccades between reward attributes does not differ between conditions (Keidel et al. 2024b, 2025a; Sharma and Khan 2022). Therefore, the computational strategy in the date framing condition may be best reflected by a hybrid model such as the *interval discounting model* (Scholten and Read 2006), according to which decision-makers evaluate options separately but discount values not only by the time from the present but also by the interval between options. Since the proportion of comparative (versus integrative) saccades is not only associated with lower temporal discounting in general (Amasino et al. 2019) but also with a lower date/delay effect (Keidel et al. 2024b, 2025a), and people with a lower date/delay effect appear to show higher activity in frontal brain areas during date compared to delay framing trials (Keidel et al. 2024b), it is conceivable that some decision-makers actually make use of their strategy to *prevent* the date/delay effect. That is, in line with the findings that reaction times are higher (Keidel et al. 2025a) and accumulation of time information may be slower (Scherbaum et al. 2012) in the date than in the delay framing condition, only some decision-makers may make a high effort to process time information in the date framing condition and compute time intervals using more comparative saccades between time attributes. These decision-makers would most likely decide similarly as in the delay framing condition. On the other hand, decision-makers who do not apply such an arithmetic strategy may rely on simpler heuristics such as reward focussing, which ultimately lead to the date/delay effect. Thus, a deeper look at individual differences and dynamic processes, such as intra- and inter-individual reaction time distributions as well as eye movements, may be a particularly promising pathway to further uncover the mechanisms of the date/delay effect.

Indeed, the study of individual differences in the date/delay effect may be of particular interest because so far, there are only few studies on the correlates of this effect. While individual difference correlates of temporal discounting have been thoroughly reviewed elsewhere (Keidel et al. 2021), here, we will briefly summarise the literature on correlates of the date/delay effect to round off this section.

Several studies have shown that the date/delay effect is particularly pronounced in persons whose temporal discounting in the standard condition is relatively high (Keidel et al. 2024a, 2025a; Klapproth 2012). While this may seem trivial, as decision-makers who show relatively shallow temporal discounting can only reduce their discount rates to a limited extent, it is nonetheless important to note that particularly those decision-makers with steeper discount rates show high reductions in their discount rates, especially in view of the many health-related and non-health-related correlates of more impulsive decision-making (see above; Keidel et al. 2021).

Moreover, in line with studies finding altered intertemporal decision-making in a broad range of mental disorders (Amlung et al. 2019; Jackson and MacKillop 2016; Weinsztok et al. 2021) and the debate on whether temporal discounting can be considered a transdiagnostic marker (Amlung et al. 2019; Bailey et al. 2021; Bickel et al. 2019; Lempert et al. 2019; Levitt et al. 2023), the date/delay effect has been examined in relation to psychopathology. Specifically, Klapproth (2012) examined participants with substance use disorders and controls and found that the date manipulation reduced temporal discounting particularly strongly in individuals with substance use disorders. Notably, it attenuated temporal discounting so much that temporal discounting only differed between the two groups in the delay but not in the date framing condition. More recently, two studies used dimensional approaches to examine how the date/delay effect covaries with psychopathological symptom patterns in the general population (Keidel et al. 2024a, 2025b). One study, focusing on schizotypy, a construct reflecting traits resembling schizophrenia symptoms on a subclinical level (Ettinger et al. 2014; Nelson et al. 2013), did not find any significant associations between schizotypy traits and temporal discounting, but revealed that the date/delay effect was more pronounced in participants scoring higher on a positive schizotypy dimension whilst controlling for other dimensions (Keidel et al. 2025b). Another study assessed a broad range of clinical symptoms in a large general population sample to examine associations of different clinical symptom patterns and factor-analytically derived transdiagnostic dimensions with temporal discounting and the date/delay effect (Keidel et al. 2024a). In that study, temporal discounting correlated significantly with scores on a broad range of symptom patterns loading on an anxious/depression factor and an inattention/impulsivity/overactivity factor. Importantly, the date/delay effect was particularly strong in people scoring higher on the inattention/impulsivity/overactivity factor (comprising uncontrolled eating, emotional eating, attention-deficit/hyperactivity, positive urgency, negative urgency, lack of premeditation) (Keidel et al. 2024a). While these studies suggest that targeting time perception and episodic processing may be particularly promising to alleviate elevated temporal discounting in people with positive schizotypy and inattention/impulsivity/overactivity symptoms, further studies in clinical populations are needed to deduce theefficacy of a potential clinical intervention targeting time perception.

Apart from these findings on individual differences in the date/delay effect, its mechanisms and its association with psychopathological symptom patterns, only very few correlates of the effect have been reported and, if so, their robustness is yet to be confirmed. A significant inverse correlation with the *past negative subscale* of the Zimbardo Time Perspective Inventory (Zimbardo and Boyd 1999), for instance, warrants replication in view of limited statistical power regarding between-subject correlations (Keidel et al. 2025a). Similarly, covariations with trait impulsivity constructs according to the UPPS-P Behavior Scale (Cyders et al. 2014; Lynam et al. 2006; Whiteside and Lynam 2001) appear not to be entirely consistent (Keidel et al. 2024a, 2025a) and should be investigated further.

Overall, there are still too few studies to draw definite conclusions regarding the cognitive and neural mechanisms of the date/delay effect. Based on the studies so far, it is very likely that different mechanisms play a role. Differential time estimation and reward-weighting in particular appear to be crucial factors, and individual differences in attentional strategies may explain individual differences in the magnitude of the date/delay effect. The importance of examining individual difference approaches to the date/delay effect is further emphasised by studies revealing correlations with transdiagnostic symptom dimensions, which suggest that the perception of future time may be a highly relevant target in clinical contexts.

11.4 Conclusions and Outlook

In this chapter, we have discussed intertemporal decisions in which individuals choose between a smaller, sooner and a larger, later reward. The phenomenon of temporal discounting that can be observed in this task is the consistent finding that, on average, people devalue future rewards, indicated by a tendency to choose the reward received sooner over a reward of the same size but received at a delay. The phenomenon is ubiquitous and observed across species (Vanderveldt et al. 2016). However, it also displays considerable malleability. One such form of malleability is the date/delay effect, viz., the observation that temporal discounting is less pronounced when the time until receipt of rewards is presented as dates rather than delay units. The date/delay effect is highly replicable, and whilst its cognitive and neural mechanisms are far from clear, testable hypotheses have been developed in the psychological literature that may inspire future experimental research. Individual differences studies have provided evidence of associations between the magnitude of the date/delay effect and measures of psychopathology, particularly in relation to symptom domains from externalizing disorders such as drug abuse, inattention, hyperactivity and impulsivity as well as positive schizotypy. However, the literature on the association between the date/delay effect and psychopathology is not very large yet, calling for further work.

In the following, we will briefly draw attention to some areas of research where further work may be particularly needed and fruitful.

First, despite promising experimental results and the availability of specific hypotheses, the cognitive mechanisms that underlie the date/delay effect are still unresolved. As argued above, it is unlikely that a single explanation will fully account for the data, and sophisticated experimental designs will be needed to further uncover the underlying processes. For example, combinations of date/delay framing with other experimental manipulations of temporal discounting may be instructive in this regard, particularly if combined with process tracing methods such as mouse-tracking or eye-tracking. Additionally, more studies of intra- and inter-individual reaction time distributions and computational modelling may be informative in this context (cf. Scherbaum et al. 2012).

A second major area of investigation concerns the neurophysiological processes that give rise to the date/delay effect. Only one functional magnetic resonance imaging (fMRI) study on the neural mechanisms of the effect and its individual differences is available to date (Keidel et al. 2024b). Apart from replicating the basic findings of that study, more data are sorely needed to provide a more complete picture of the neural mechanisms of the date/delay effect. Evidence from other cognitive neuroscience methods would be instructive as well, from neuroimaging approaches with higher temporal resolution, such as magnetoencephalography (MEG), to brain stimulation methods such as transcranial magnetic stimulation (TMS) or transcranial direct current stimulation (tDCS). Beyond providing more detailed knowledge on the macroscopic brain functional and structural correlates of the date/delay effect, its molecular basis remains completely uncharted. An important methodological approach in this context concerns pharmacological challenge studies, which allow drawing inferences regarding the neurotransmitter systems that are involved in specific cognitive functions based on observed behavioural effects (Friedman and Robbins 2022; Müller et al. 2006). Additionally, molecular genetic methods may be drawn upon to understand the molecular basis of inter-individual differences in the magnitude of the date/delay effect (see e.g., the chapter by Hogeterp et al. in this volume).

Third, there is exciting initial evidence of differential sensitivity to date/delay framing in individuals with substance use disorders and with high expressions of psychopathology-related traits. However, more empirical data are needed to replicate and extend these initial reports. Whilst future studies using both classic case–control and population-based dimensional designs are required, a focus on transdiagnostic approaches appears to be particularly promising (Keidel et al. 2024a). It will also be important to go beyond the level of describing and quantifying alterations in the date/delay effect in clinical populations; instead, the cognitive and affective as well as neural processes that underlie such alterations in individuals with high levels of psychopathology will need to be explicated. To do so, experimental psychological, neuroimaging and pharmacological approaches will be required, as outlined above. A particular promising area of work may be in the integration of date/delay framing with psychotherapeutic and psychoeducational approaches in clinical practice, particularly in disorders characterized by high levels of impulsivity, such as attention deficit and disruptive behaviour disorders.

References

Amasino DR, Sullivan NJ, Kranton RE, Huettel SA (2019) Amount and time exert independent influences on intertemporal choice. Nat Hum Behav 3(4):383–392. https://doi.org/10.1038/s41562-019-0537-2

Amlung M, Marsden E, Holshausen K, Morris V, Patel H, Vedelago L, Naish KR, Reed DD, McCabe RE (2019) Delay discounting as a transdiagnostic process in psychiatric disorders: a meta-analysis. JAMA Psychiat 76(11):1176–1186. https://doi.org/10.1001/jamapsychiatry.2019.2102

Audrain-McGovern J, Rodriguez D, Epstein LH, Cuevas J, Rodgers K, Wileyto EP (2009) Does delay discounting play an etiological role in smoking or is it a consequence of smoking? Drug Alcohol Depend 103(3):99–106. https://doi.org/10.1016/j.drugalcdep.2008.12.019

Axon RN, Bradford WD, Egan BM (2009) The role of individual time preferences in health behaviors among hypertensive adults: a pilot study. J Am Soc Hypertens 3(1):35–41. https://doi.org/10.1016/j.jash.2008.08.005

Bailey AJ, Romeu RJ, Finn PR (2021) The problems with delay discounting: a critical review of current practices and clinical applications. Psychol Med 51(11):1799–1806. https://doi.org/10.1017/S0033291721002282

Bickel WK, Athamneh LN, Basso JC, Mellis AM, DeHart WB, Craft WH, Pope D (2019) Excessive discounting of delayed reinforcers as a trans-disease process: Update on the state of the science. Curr Opin Psychol 30:59–64. https://doi.org/10.1016/j.copsyc.2019.01.005

Bradford WD (2010) The association between individual time preferences and health maintenance habits. Med Decis Making 30(1):99–112. https://doi.org/10.1177/0272989X09342276

Cyders MA, Littlefield AK, Coffey S, Karyadi KA (2014) Examination of a short English version of the UPPS-P impulsive behavior scale. Addict Behav 39(9):1372–1376. https://doi.org/10.1016/j.addbeh.2014.02.013

Dai J, Busemeyer JR (2014) A probabilistic, dynamic, and attribute-wise model of intertemporal choice. J Exp Psychol Gen 143(4):1489–1514. https://doi.org/10.1037/a0035976

de Ruijter MJT, Dahlén AD, Rukh G, Schiöth HB (2023) Job satisfaction has differential associations with delay discounting and risk-taking. Sci Rep 13(1), Article 754. https://doi.org/10.1038/s41598-023-27601-8

DeHart WB, Odum AL (2015) The effects of the framing of time on delay discounting. J Exp Anal Behav 103(1):10–21. https://doi.org/10.1002/jeab.125

Doyle JR (2013) Survey of time preference, delay discounting models. Judgm Decis Mak 8(2):116–135. https://doi.org/10.1017/s1930297500005052

Dshemuchadse M, Scherbaum S, Goschke T (2013) How decisions emerge: action dynamics in intertemporal decision making. J Exp Psychol Gen 142(1):93–100. https://doi.org/10.1037/a0028499

Edwards A, Elwyn G, Covey J, Matthews E, Pill R (2001) Presenting risk information—a review of the effects of "framing" and other manipulations on patient outcomes. J Health Commun 6(1):61–82. https://doi.org/10.1080/10810730150501413

Estle SJ, Green L, Myerson J, Holt DD (2006) Differential effects of amount on temporal and probability discounting of gains and losses. Mem Cognit 34(4):914–928. https://doi.org/10.3758/BF03193437

Ettinger U, Meyhöfer I, Steffens M, Wagner M, Koutsouleris N (2014) Genetics, cognition, and neurobiology of schizotypal personality: a review of the overlap with schizophrenia. Front Psychiatry 5, Article 18. https://doi.org/10.3389/fpsyt.2014.00018

Frederick S, Loewenstein G, O'Donoghue T (2002) Time discounting and time preference: a critical review. Journal of Economic Literature 40(2):351–401. https://doi.org/10.1257/002205102320161311

Freeney Y, O'Connell M (2010) Wait for it: delay-discounting and academic performance among an Irish adolescent sample. Learn Individ Differ 20(3):231–236. https://doi.org/10.1016/j.lindif.2009.12.009

Friedman NP, Robbins TW (2022) The role of prefrontal cortex in cognitive control and executive function. Neuropsychopharmacology 47(1):72–89. https://doi.org/10.1038/s41386-021-01132-0

Fröhner JH, Ripke S, Jurk S, Li SC, Banaschewski T, Bokde ALW, Quinlan EB, Desrivières S, Flor H, Grigis A, Garavan H, Heinz A, Brühl R, Martinot JL, Paillère Martinot ML, Artiges E, Nees F, Papadopoulos Orfanos D, Poustka L (2022) Associations of delay discounting and drinking

trajectories from ages 14 to 22. Alcohol: Clin Exp Res 46(4):667–681. https://doi.org/10.1111/acer.14799
Gallagher KM, Updegraff JA (2012) Health message framing effects on attitudes, intentions, and behavior: a meta-analytic review. Ann Behav Med 43(1):101–116. https://doi.org/10.1007/s12160-011-9308-7
Green L, Fristoe N, Myerson J (1994) Temporal discounting and preference reversals in choice between delayed outcomes. Psychon Bull Rev 1(3):383–389. https://doi.org/10.3758/BF03213979
Green L, Myerson J, McFadden E (1997) Rate of temporal discounting decreases with amount of reward. Mem Cognit 25(5):715–723. https://doi.org/10.3758/BF03211314
Hamilton KR, Mitchell MR, Wing VC, Balodis IM, Bickel WK, Fillmore M, Lane SD, Lejuez CW, Littlefield AK, Luijten M, Mathias CW, Mitchell SH, Napier TC, Reynolds B, Schütz CG, Setlow B, Sher KJ, Swann AC, Tedford SE, White MJ, Winstanley Catharine AY, Richard P, Marc N, Moeller, F. G. (2015) Choice impulsivity: Definitions, measurement issues, and clinical implications. Pers Disord: Theory Res Treat 6(2), 182–198.https://doi.org/10.1037/per0000099
Jackson JNS, MacKillop J (2016) Attention-deficit/hyperactivity disorder and monetary delay discounting: a meta-analysis of case-control studies. Biol Psychiatry: Cogn Neurosci Neuroimaging 1(4):316–325. https://doi.org/10.1016/j.bpsc.2016.01.007
Jiang J, Dai J (2021) Time and risk perceptions mediate the causal impact of objective delay on delay discounting: an experimental examination of the implicit-risk hypothesis. Psychon Bull Rev 28(4):1399–1412. https://doi.org/10.3758/s13423-021-01890-4
Kahneman D, Knetsch JL, Thaler RH (1991) Anomalies: the endowment effect, loss aversion, and status quo bias. J Econ Perspect 5(1):193–206. https://doi.org/10.1257/jep.5.1.193
Kahneman D, Tversky A (1979) Prospect theory: an analysis of decision under risk. Econometrica 47(2):263–292. https://doi.org/10.2307/1914185
Keidel K, Rramani Q, Weber B, Murawski C, Ettinger U (2021) Individual differences in intertemporal choice. Front Psychol 12, Article 643670. https://doi.org/10.3389/fpsyg.2021.643670
Keidel K, Lu X, Suzuki S, Murawski C, Ettinger U (2024a) Association of temporal discounting with transdiagnostic symptom dimensions. Npj Mental Health Res 3, Article 13. https://doi.org/10.1038/s44184-024-00060-3
Keidel K, Schröder R, Trautner P, Radbruch A, Murawski C, Ettinger U (2024b) The date/delay effect in intertemporal choice: a combined fMRI and eye-tracking study. Hum Brain Mapp 12(3), Article e26585. https://doi.org/10.1002/hbm.26585
Keidel K, Murawski C, Ettinger U (2025a) Attentional mechanisms of the date/delay effect in intertemporal choice: an eye-tracking study. J Exp Psychol: Learn Mem Cogn. 51(5):737–754. https://doi.org/10.1037/xlm0001363
Keidel K, Murawski C, Pantelis C, Ettinger U (2025b) The relationship between schizotypal personality traits and temporal discounting: the role of the date/delay effect. Schizophr Bull. 51(Supplement_2):S64–S73. https://doi.org/10.1093/schbul/sbad110
Kekic M, McClelland J, Bartholdy S, Chamali R, Campbell IC, Schmidt U (2020) Bad things come to those who do not wait: temporal discounting is associated with compulsive overeating, eating disorder psychopathology and food addiction. Front Psychiatry 10, Article 978. https://doi.org/10.3389/fpsyt.2019.00978
Kirby KN (2009) One-year temporal stability of delay-discount rates. Psychon Bull Rev 16(3):457–462. https://doi.org/10.3758/PBR.16.3.457
Kirby KN, Herrnstein RJ (1995) Preference reversals due to myopic discounting of delayed reward. Psychol Sci 6(2):83–89. https://doi.org/10.1111/j.1467-9280.1995.tb00311.x
Kirby KN, Petry NM, Bickel WK (1999) Heroin addicts have higher discount rates for delayed rewards than non-drug-using controls. J Exp Psychol Gen 128(1):78–87. https://doi.org/10.1037/0096-3445.128.1.78
Klapproth F (2012) The date-delay framing effect in temporal discounting depends on substance abuse. Behav Proc 90(3):420–423. https://doi.org/10.1016/j.beproc.2012.03.013

Koffarnus MN, Bickel WK (2014) A 5-trial adjusting delay discounting task: accurate discount rates in less than one minute. Exp Clin Psychopharmacol 22(3):222–228. https://doi.org/10.1037/a0035973

LeBoeuf RA (2006) Discount rates for time versus dates: the sensitivity of discounting to time-interval description. J Mark Res 43(1):59–72. https://doi.org/10.1509/jmkr.43.1.59

Lempert KM, Johnson E, Phelps EA (2016) Emotional arousal predicts intertemporal choice. Emotion 16(5):647–656. https://doi.org/10.1037/emo0000168

Lempert KM, Phelps EA (2016) The malleability of intertemporal choice. Trends Cogn Sci 20(1):64–74. https://doi.org/10.1016/j.tics.2015.09.005

Lempert KM, Steinglass JE, Pinto A, Kable JW, Simpson HB (2019) Can delay discounting deliver on the promise of RDoC? Psychol Med 49(2):190–199. https://doi.org/10.1017/S0033291718001770

Levitt EE, Oshri A, Amlung M, Ray LA, Sanchez-Roige S, Palmer AA, MacKillop J (2023) Evaluation of delay discounting as a transdiagnostic research domain criteria indicator in 1388 general community adults. Psychol Med 53(4):1649–1657. https://doi.org/10.1017/S0033291721005110

Loewenstein GF (1988) Frames of mind in intertemporal choice. Manage Sci 34(2):200–214. https://doi.org/10.1287/mnsc.34.2.200

Lynam DR, Smith GT, Whiteside SP, Cyders MA (2006) The UPPS-P: assessing five personality pathways to impulsive behavior (technical report). Purdue University

Magen E, Dweck CS, Gross JJ (2008) The hidden-zero effect: Representing a single choice as an extended sequence reduces impulsive choice. Psychol Sci 19(7):648–649. https://doi.org/10.1111/j.1467-9280.2008.02137.x

Magen E, Kim B, Dweck CS, Gross JJ, McClure SM (2014) Behavioral and neural correlates of increased self-control in the absence of increased willpower. Proc Natl Acad Sci 111(27):9786–9791. https://doi.org/10.1073/pnas.1408991111

Marzilli Ericson KM, White JM, Laibson D, Cohen JD (2015) Money earlier or later? Simple heuristics explain intertemporal choices better than delay discounting does. Psychol Sci 26(6):826–833. https://doi.org/10.1177/0956797615572232

Mazur JE (1987) An adjusting procedure for studying delayed reinforcement. In: Commons ML, Mazur JE, Nevin JA, Rachlin H (eds) Quantitative analyses of behavior: the effect of delay and of intervening events on reinforcement value, vol 5, pp 55–73. Lawrence Erlbaum Associates, Inc.

McCaffery EJ, Kahneman DJ, Spitzer ML (1995) Framing the jury: cognitive perspectives on pain and suffering awards. Virginia Law Rev 81(5):1341–1420. https://doi.org/10.2307/1073506

Müller U, Fletcher PC, Steinberg H (2006) The origin of pharmacopsychology: emil Kraepelin's experiments in Leipzig, Dorpat and Heidelberg (1882–1892). Psychopharmacology 184(2):131–138. https://doi.org/10.1007/s00213-005-0239-5

Munasinghe L, Sicherman N (2005) Why do dancers smoke? Smoking, time preference, and wage dynamics. SSRN Electron J 32(4):595–616. https://doi.org/10.2139/ssrn.203162

Myerson J, Baumann AA, Green L (2017) Individual differences in delay discounting: Differences are quantitative with gains, but qualitative with losses. J Behav Decis Mak 30(2):359–372. https://doi.org/10.1002/bdm.1947

Naudé GP, Kaplan BA, Reed DD, Henley AJ, DiGennaro Reed FD (2018) Temporal framing and the hidden-zero effect: Rate-dependent outcomes on delay discounting. J Exp Anal Behav 109(3):506–519. https://doi.org/10.1002/jeab.328

Nelson MT, Seal ML, Pantelis C, Phillips LJ (2013) Evidence of a dimensional relationship between schizotypy and schizophrenia: a systematic review. Neurosci Biobehav Rev 37(3):317–327. https://doi.org/10.1016/j.neubiorev.2013.01.004

Peters J, Büchel C (2010) Episodic future thinking reduces reward delay discounting through an enhancement of prefrontal-mediotemporal interactions. Neuron 66(1):138–148. https://doi.org/10.1016/j.neuron.2010.03.026

Peters J, Büchel C (2011) The neural mechanisms of inter-temporal decision-making: understanding variability. Trends Cogn Sci 15(5):227–239. https://doi.org/10.1016/j.tics.2011.03.002

Radu PT, Yi R, Bickel WK, Gross JJ, McClure SM (2011) A mechanism for reducing delay discounting by altering temporal attention. J Exp Anal Behav 96(3):363–385. https://doi.org/10.1901/jeab.2011.96-363

Read D, Frederick S, Orsel B, Rahman J (2005) Four score and seven years from now: the date/delay effect in temporal discounting. Manage Sci 51(9):1326–1335. https://doi.org/10.1287/mnsc.1050.0412

Read D, Frederick S, Scholten M (2013) DRIFT: an analysis of outcome framing in intertemporal choice. J Exp Psychol Learn Mem Cogn 39(2):573–588. https://doi.org/10.1037/a0029177

Reeck C, Wall D, Johnson EJ (2017) Search predicts and changes patience in intertemporal choice. Proc Natl Acad Sci USA 114(45):11890–11895. https://doi.org/10.1073/pnas.1707040114

Reyes-Huerta HE, Robles E, dos Santos CV (2023) Valuing the future at different temporal points: the role of time framing on discounting. J Exp Anal Behav 120(2):214–227. https://doi.org/10.1002/jeab.871

Roszkowski MJ, Snelbecker GE (1990) Effects of "framing" on measures of risk tolerance: financial planners are not immune. J Behav Econ 19(3):237–246. https://doi.org/10.1016/0090-5720(90)90029-7

Rung JM, Madden GJ (2018) Experimental reductions of delay discounting and impulsive choice: a systematic review and meta-analysis. J Exp Psychol Gen 147(9):1349–1381. https://doi.org/10.1037/xge0000462

Samuelson PA (1937) A note on measurement of utility. Rev Econ Stud 4(2):155–161. https://doi.org/10.2307/2967612

Scherbaum S, Dshemuchadse M, Goschke T (2012) Building a bridge into the future: dynamic connectionist modeling as an integrative tool for research on intertemporal choice. Front Psychol 3, Article 514. https://doi.org/10.3389/fpsyg.2012.00514

Schoemann M, Lüken M, Grage T, Kieslich PJ, Scherbaum S (2019) Validating mouse-tracking: How design factors influence action dynamics in intertemporal decision making. Behav Res Methods 51(5):2356–2377. https://doi.org/10.3758/s13428-018-1179-4

Scholten H, Scheres A, de Water E, Graf U, Granic I, Luijten M (2019) Behavioral trainings and manipulations to reduce delay discounting: a systematic review. Psychon Bull Rev 26(6):1803–1849. https://doi.org/10.3758/s13423-019-01629-2

Scholten M, Read D (2006) Discounting by intervals: a generalized model of intertemporal choice. Manage Sci 52(9):1424–1436. https://doi.org/10.1287/mnsc.1060.0534

Scholten M, Read D (2010) The psychology of intertemporal tradeoffs. Psychol Rev 117(3):925–944. https://doi.org/10.1037/a0019619

Sharma SN, Khan A (2022) Self-other differences in intertemporal decision making: An eye-tracking investigation. Conscious Cogn 102, Article 103356. https://doi.org/10.1016/j.concog.2022.103356

Strickland JC, Gelino BW, Rabinowitz JA, Ford MR, Dayton L, Latkin C, Reed DD (2023) Temporal reliability and stability of delay discounting: A 2-year repeated assessments study of the Monetary Choice Questionnaire. Exp Clin Psychopharmacol 31(5):902–907. https://doi.org/10.1037/pha0000651

Trope Y, Liberman N (2003) Temporal construal. Psychol Rev 110(3):403–421. https://doi.org/10.1037/0033-295X.110.3.403

Tversky A, Kahneman D (1981) The framing of decisions and the psychology of choice. Science 211(4481):453–458. https://doi.org/10.1126/science.7455683

Vanderveldt A, Oliveira L, Green L (2016) Delay discounting: Pigeon, rat, human—does it matter? J Exp Psychol: Anim Learn Cogn 42(2):141–162. https://doi.org/10.1037/xan0000097

Wang Y, Wang L, Keller LR (2015) Discounting over subjective time: subjective time perception helps explain multiple discounted utility anomalies. Int J Res Mark 32(4):445–448. https://doi.org/10.1016/j.ijresmar.2015.08.006

Weinsztok S, Brassard S, Balodis I, Martin LE, Amlung M (2021) Delay discounting in established and proposed behavioral addictions: a systematic review and meta-analysis. Front Behav Neurosci 15, Article 786358. https://doi.org/10.3389/fnbeh.2021.786358

Whiteside SP, Lynam DR (2001) The Five Factor Model and impulsivity: using a structural model of personality to understand impulsivity. Pers Individ Differ 30(4):669–689. https://doi.org/10.1016/S0191-8869(00)00064-7

Wu C-Y, He G-B (2012) The effects of time perspective and salience of possible monetary losses on intertemporal choice. Soc Behav Pers 40(10):1645–1654. https://doi.org/10.2224/sbp.2012.40.10.1645

Zauberman G, Kim BK, Malkoc SA, Bettman JR (2009) Discounting time and time discounting: subjective time perception and intertemporal preferences. J Mark Res 46(4):543–556. https://doi.org/10.1509/jmkr.46.4.543

Zimbardo PG, Boyd JN (1999) Putting time in perspective: a valid, reliable individual-differences metric. J Pers Soc Psychol 77(6):1271–1288. https://doi.org/10.1037/0022-3514.77.6.1271

Chapter 12
Neurogenetics of Human Decision-Making

Svea A. Hogeterp, Annabel Vetterlein, and Martin Reuter

Abstract The objective of this chapter is to review the existing literature on the genetic basis of human decision-making behavior. Identifying genetic influences has the potential to shed light on the underlying neurobiological mechanisms of decision-making, in that genes can convey information about which neurotransmitter and brain metabolism processes are relevant while making decisions. Most studies published to date have applied candidate gene approaches. In this type of study, researchers investigate the associations between distinct gene polymorphisms and certain aspects of decision-making behavior. However, since personality and behavior are the results of a complex interplay of different biological components, more recent research often considers multiple genes simultaneously, via haplotype analyses, gene–gene interaction studies, polygenetic risk scores or genome-wide analyses. A vast body of research has implicated genes related to dopaminergic, serotonergic, oxytocinergic and vasopressinergic neurotransmission in relation to decision-making, although the empirical evidence is often ambiguous. Therefore, studying interactions between biological and environmental factors, e.g., by means of epigenetics, might bridge the gap between 'nature and nurture' and might represent a starting point for future research approaches in the field.

Keywords Genetics · Polymorphism · Neurotransmitter · Neuroeconomics · Risk-taking · Social behavior

S. A. Hogeterp (✉) · A. Vetterlein · M. Reuter
Department of Psychology, University of Bonn, Bonn, Germany
e-mail: svea.hogeterp@uni-bonn-diff.de

A. Vetterlein
e-mail: annabel.vetterlein@uni-bonn-diff.de

M. Reuter
e-mail: martin.reuter@uni-bonn-diff.de

U. Ettinger et al. (eds.), *Decision Making*, Studies in Neuroscience, Psychology and Behavioral Economics, https://doi.org/10.1007/978-3-032-00880-0_12

12.1 General Remarks on the Genetics of Decision-Making

People are confronted with a multitude of decisions every day. These can be simple, such as what clothes to wear for the day, but can also be complex and far-reaching, as when deciding on a particular university program, a new job, or a change of residence. Decisions can depend on a variety of state factors, such as mood (e.g., Forgas 1989; Vinckier et al. 2018). Nevertheless, we can also observe stability in behavioral patterns in decision-relevant situations (e.g., Levin et al. 2007), which may be attributed to relatively stable personality traits like risk propensity (Josef et al. 2016). In this sense, both environmental and biological factors underlying such personality traits likely contribute to human decision-making. As genetic contributions to decision-making have been widely researched, the present chapter aims to provide a comprehensive overview of the field.

But what is the advantage of understanding the genetics of human decision-making behavior? If it is shown that certain genes related to, e.g., neurotransmitter functioning, play a role in decision-making, first conclusions can be drawn about the neurobiological underpinnings of this behavior. Next to basic scientific research this can also contribute to a better understanding of impaired decision-making or even risk-taking behavior in certain medical conditions (see below) and might thus provide novel starting points for treatment options.

The search for the underlying neuronal mechanisms in decision-making processes has given rise to the field of neuroeconomics, combining neuroscientific, psychological, and behavioral economic research (for an introduction see Reuter and Montag 2016). Hence, much of the below-cited research refers to decision-making in an economic context. To be able to study human decision-making under experimentally controlled conditions, behavioral economists have developed different paradigms, also known as economic games. As the following presentation of empirical evidence repeatedly refers to such paradigms, an overview of the games discussed here can be found in Table 12.1.

The heritability of various decision-relevant factors has been studied in twin studies. Research by Zyphur et al. (2009) indicated a heritability of 63% for economic risk preference with the remainder of variance explained by non-shared environment. In line with this, results by Zhong et al. (2009a, b, c) showed a heritability of economic risk attitudes of 57% and no significant influence of shared environments. In another twin study, the genetic components of investment portfolio decisions were examined, suggesting a heritability of around 25% (Cesarini et al. 2010). Apart from this, there are also studies directly investigating heritability estimates of behavior in specific economic games, e.g. a study by Wallace et al. (2007), which found that more than 40% of the variation in the participants' rejection behavior (indicating fairness preference) is explained by additive genetic effects in the so-called *Ultimatum Game* (UG, see Table 12.1).

Table 12.1 Overview of prominent economic paradigms

Economic game	Abbr	Original game structure	Outcome measures	Citations
Dictator game	DG	Player one (dictator) receives a certain amount of money and is asked to decide how much to share with player two (recipient). The recipient must accept their share.	Inter alia: Altruism/fairness/generosity versus competitiveness/greed in player one	Civai and Hawes (2016), Thielmann et al. (2021)
Ultimatum game	UG	Player one (proposer) receives a certain amount of money and is asked to decide how much to share with player two (recipient). The recipient can accept as well as reject the proposer's offer, affecting both of their payoffs (e.g., if the recipient rejects, both players receive nothing).	Inter alia: Altruism/fairness/generosity versus competitiveness/greed in the proposer; greed/social welfare in the recipient	Civai and Hawes (2016), Thielmann et al. (2021)
Trust game	TG	Player one (investor) receives a certain amount of money and is asked to decide how much to share with player two (trustee). The amount transferred is then multiplied by a constant and is added to the trustee's endowment. The trustee can then return any amount of money to the investor.	Inter alia: Altruism/social welfare/trust versus competitiveness/greed/spite in the investor and trustee	Civai and Hawes (2016), Thielmann et al. (2021)
Prisoner's dilemma game	PDG	Two players' payoffs depend on their combination of strategies (either cooperate or defect). The maximum will be yielded by unilateral defection, while the minimum payoff results from unilateral cooperation.	Inter alia: Altruism/social welfare/cooperative behavior versus competitiveness/greed/spite in both players	Civai and Hawes (2016), Thielmann et al. (2021)

(continued)

Table 12.1 (continued)

Economic game	Abbr	Original game structure	Outcome measures	Citations
Iowa gambling task	IGT	Participants receive a seed capital of play money, and are asked to take cards from four virtual decks of cards, with the goal of maximizing their capital. Turning some cards will have the participants gain money, turning others will have them lose money. The participants do not know that the decks differ in the distribution of reward and penalty cards. To win as much money as possible, two decks with smaller gains and losses are most advantageous.	Inter alia: sensitivity to future consequences, risk behavior	Bechara et al. (1994)
Balloon analogue risk task	BART	A virtual balloon is presented together with a pump. Each click on the pump yields a fixed amount of money while the balloon inflates a little. At each point in time, participants can collect the accumulated amount of money. Pumping the balloon past a varying point, it bursts and the participants are left with no money. Unbeknownst to the participants, balloon colors are matched with a certain probability of bursting. After money collection or explosion, a new balloon is presented with a total of 90 balloons.	Risk behavior	Lejuez et al. (2002)

Most of these studies aim to determine the overall heritability but do not yet identify the genes involved. In contrast, the candidate gene approach is frequently used to identify relevant gene loci by establishing associations between certain polymorphisms and specific personality or behavioral traits. However, the complexity of human cognition is reflected in that decision-making most likely is not only determined by single genes alone but rather by a complex interplay between different genes and the corresponding expression of neurotransmitters, enzymes, receptors and transporters affecting several neurotransmitter systems. Thus, newer approaches often investigate more than one gene, through haplotype analyses, the investigation of gene–gene interactions, polygenic risk scores or genome-wide association studies (GWAS). Finally, recent research also addresses the extent to which the genetic make-up interacts with the environment. For this purpose, gene-environment interactions are reviewed at the end of the chapter, following a general discussion of genetic research related to neurotransmitters such as dopamine and serotonin, as well as to the neuropeptides oxytocin and vasopressin. To reduce the abundance of polymorphisms mentioned in the scope of the chapter, Table 12.2 aims to provide the reader with an overview of those most prominently investigated. While it is the aim of this chapter to give an extensive review of studies in the field, due to the vast body of literature, it cannot claim to be exhaustive. As it is not possible to do justice to all aspects of decision-making behavior in this chapter, it is primarily focused on the domains of risk-taking and aspects of social behavior (e.g., fairness, trust, and altruism).

12.2 The Dopaminergic System

Dopamine is the most prevalent catecholamine in the brain, constituting about 80% of the catecholamine content (Vallone et al. 2000). Four different axonal pathways are differentiated and associated with several dopaminergic functions: (1) the nigrostriatal pathway involved in the control of movement, (2) the mesocortical pathway involved in learning and memory, (3) the mesolimbic pathway innervating parts of the limbic system and involved in affect and motivated behavior, and, lastly, (4) the tuberoinfundibular pathway where dopamine is responsible for the inhibition of prolactin release (Vallone et al. 2000).

Dopamine performs its physiological actions through five different G-protein-coupled receptors, namely D1, D2, D3, D4 and D5 (Vallone et al. 2000). Two distinct subtypes can be differentiated: D1-like receptor subtypes refer to D1 and D5, which exhibit excitatory effects, whereas the D2-like subtypes, comprising D2, D3 and D4 receptors, exhibit inhibitory actions (Missale et al. 1998). The dopaminergic receptors as well as the dopamine transporter (responsible for the termination of synaptic transmission by rapid reuptake of dopamine into presynaptic nerve endings; Jaber et al. 1997), are encoded by genes that are frequent targets for studies in the domain of decision-making behavior (see below).

One indication of the involvement of dopamine in decision-making processes stems from the finding that dysfunctions in the dopaminergic system are associated with alterations in this regard. For example, in Parkinson's disease, a neurodegenerative disorder associated with the progressive degeneration of dopaminergic neurons in the substantia nigra pars compacta of the basal ganglia, impaired processing in the final stages of decision-making (i.e., the evaluation of the outcome) can be observed (for a review see Ryterska et al. 2013). Deficits in decision-making can also be observed in individuals with attention deficit hyperactivity disorder (ADHD), another condition in which dopamine is involved in the pathogenesis (for a review see Mowinckel et al. 2015). Another indication is derived from normal dopaminergic functioning: Dopamine is involved in reward-seeking behavior (Arias-Carrión et al. 2010) as well as various other cognitive functions (e.g., cognitive control

Table 12.2 Prominently investigated polymorphisms and their associations with decision-making behavior

Neurotransmitter system	Gene	Polymorphism	Type	Alleles	Putative functional consequence	Decision-making related findings
Dopamine	DRD2	*C957T (rs6277)*	SNP	C/T	Increased DRD2 binding potential (C/C > C/T > T/T) (Hirvonen et al. 2009)	Higher impulsivity in intertemporal choice for delayed gain in T carriers (Kawamura et al. 2013).
		DRD2/ANKK1 Taq1A (rs1800497)	SNP	A1 (T)/A2 (C)	Association of A1 allele with reduced DRD2 receptor availability (Pohjalainen et al. 1998)	Association of A1 with substance disorders (Noble 2000); no association of A1 in experimental risk-behavior (Dreber et al. 2009). Higher generosity in A2 homozygotes (Reuter et al. 2013).
	DRD4	*DRD4 exon3 VNTR*	VNTR	Two to eleven 48-bp repeats (2R-11R)	7R may reduce DRD4 mRNA expression (Simpson et al. 2010)	Higher risk behavior in (particularly male) 7R carriers (Carpenter et al. 2011; Roussos et al. 2009; Kuhnen and Chiao 2009; Dreber et al. 2011) in addition to higher sensation seeking and lower performance in IGT, while problem solving skills are intact (Roussos et al. 2009). However, there is conflicting evidence (Anderson et al. 2015; Frydman et al. 2011; Muda et al. 2018; Dreber et al. 2012). Higher fairness preference in 4R homozygotes (Zhong et al. 2010; Reuter et al. 2013). Associations of selflessness with 4R allele (Bachner-Melman et al. 2005).
	DAT1	*DAT1 VNTR*	VNTR	Seven to eleven 40-bp repeats (7R-11R)	Striatal DAT expression might be higher in 9R carriers compared to 10R homozygotes (Faraone et al. 2014)	Evidence of higher risk behavior in 9R carriers (Zhong et al. 2009a, b, c) compared to 10R next to evidence of the reverse relationship (Mata et al. 2012; Guo et al. 2010).

(continued)

Table 12.2 (continued)

Neurotransmitter system	Gene	Polymorphism	Type	Alleles	Putative functional consequence	Decision-making related findings
Serotonin	5-HTT	*5-HTTPLR*	VNTR	Short (S) and long (L) allele	Higher 5-HTT expression in L than S allele (Lesch et al. 1996)	Sex-dependent effect of L/L on performance in the IGT under ambiguity (Stoltenberg and Vandever, 2010; He et al. 2010). Higher risk aversion in S homozygotes (Kuhnen and Chiao 2009) and S allele carriers (Crişan et al. 2009). However, there is conflicting evidence (Anderson et al. 2015; Frydman et al. 2011; Roiser et al. 2009). Higher fairness preference (Enge et al. 2017; Nishina et al. 2022) and prosocial behavior (Stoltenberg et al. 2013) in L/L carriers.
Oxytocin	OXTR	*rs53576*	SNP	A/G	Unclear	Higher trust behavior in G/G carriers (Krueger et al. 2012). No association with inequity aversion (Tanaka et al. 2023). Sex-dependent effect of G/G genotype on reciprocated cooperation (Feng et al. 2015).
		rs2268498	SNP	C/T	Higher OXTR expression in C allele (Reuter et al. 2017)	Higher trust is associated with T allele (Reuter et al. 2017). C allele carriers rate accidentally committed harm as more blameworthy than T allele homozygotes (Walter et al. 2012) and express more utilitarianism in response to moral dilemmas (Bernhard et al. 2016).
Vasopressin	AVPR1A	*RS3*	VNTR microsatellite	Short and long alleles	Higher AVPR1A expression in long repeat variant (Tansey et al. 2011)	Less generosity in long allele carriers (Avinun et al. 2011) and more prosocial preference in short allele homozygotes (Tanaka et al. 2023). In a male sample, the short allele was, however, associated with lower generosity (Knafo et al. 2008).

Note SNP Single nucleotide polymorphism. *VNTR* Variable number tandem repeats. *Bp* Base pair

and working memory, Cools and D'Esposito 2011), likely contributing to decision-making processes. Accordingly, a variety of genetic studies has already been carried out to identify the dopaminergic foundations with implications for different trait domains, which are briefly discussed below.

12.2.1 Risk-Taking

Due to its relevance in many decisions, risk-taking as personality trait, as well as state risk-taking (i.e., situational risk-taking) in economic games, are key targets for many studies in neuroeconomics.

Variation in the dopamine receptor D2 (DRD2) gene has been linked to differences in impulsive decision-making: the T allele in the SNP (single nucleotide polymorphism) *C957T* (see Table 12.2) is associated with higher impulsivity in choice of delayed gains (Kawamura et al. 2013). Interestingly, the T allele is associated with a lower DRD2 binding potential, an index for receptor availability and affinity (Hirvonen et al. 2009).

Regarding another polymorphism linked to the DRD2, the *rs1800497*, commonly referred to as *DRD2/ANKK1 Taq1A* polymorphism (see Table 12.2), similar evidence emerges. Drug dependencies can be interpreted as a proxy for risk-taking (Guo et al. 2010) and the A1 allele (putatively causing reduced DRD2 receptor availability; Pohjalainen et al. 1998) is commonly associated with different substance disorders, for example alcoholism (for a review see Noble 2000). In contrast, Dreber et al. (2009) did not find any association between the A1 allele and risk preferences in an experimental investment game with real monetary payoffs in a male sample.

Many studies have also investigated the *DRD4 Exon 3 VNTR* (variable number tandem repeats; see Table 12.2) in relation to risk-taking. Carpenter et al. (2011) found that carriers of the 7R (which may cause reduced DRD4 mRNA expression; Simpson et al. 2010) are more likely to exhibit risky behavior in a financial choice paradigm including losses, but not in a task that allows gains only. However, this finding was only marginally significant. Roussos et al. (2009) observed that individuals with at least one 7R allele had higher scores in novelty seeking, made more risky choices and won less money in the *Iowa Gambling Task* (IGT, see Table 12.1), which involves emotional processing of loss and reward, compared to persons without 7R allele. Notably, regarding the cognitive decision-making task *Stockings of Cambridge*, which assesses spatial planning, no differences were found, indicating intact planning for problem solving and decision-making (Roussos et al. 2009). Kuhnen and Chiao (2009) examined risk-taking behavior in investment decisions, showing that 7R carriers take greater financial risks. Similarly, Dreber et al. (2011) investigated a sample of serious tournament bridge players and found that male carriers of the 7R allele take more risks in a financial gamble than men without this allele. For women, no difference in risk-taking due to genetic influences were observed (Dreber et al. 2011). This confirms results of an earlier study, in which Dreber et al. (2009) found more risk proneness in male 7R allele carriers than in men without, although it should be noted that no women were included in the study. The authors further showed that the *DRD4 Exon 3 VNTR* seems to account for about 20% of the genetic variation in financial risk-taking in men. Hence, evidence from multiple empirical studies indicates an effect of genetic variation in the DRD4 gene on risk-taking. Yet, at the same time, the literature provides a multitude of conflicting results, showing no association between different measures of risk-taking behavior and the polymorphism (Anderson et al. 2015; Frydman et al. 2011; Muda et al. 2018). What is more, Dreber et al. (2012) even found a negative relationship between the 7R genotype and social risk-taking (i. e., trust behavior).

In addition to the dopaminergic receptors, the dopamine transporter (DAT) has been examined frequently in relation to its association with risk-taking. Individuals carrying the 9R allele in the 40-base-pair (bp) variable number of tandem repeats (VNTR) polymorphism in the DAT1 gene exhibit more risk tolerance over gains in a gain-based version of a lottery game compared to carriers of the 10R allele (Zhong et al. 2009a, b, c). In contrast, in a study by Mata et al. (2012), 9R carriers showed less risk-taking compared to 10R homozygotes or 11R carriers and thus earned less money in the *Balloon Analogue Risk Task* (BART, see Table 12.1). In line with this, Guo et al. (2010) identified being homozygous for the 9R allele as a protective factor against risk-taking behavior in the form of delinquency, variety of sexual partners, drug consumption and not wearing a seatbelt in young adults and adolescents. With regard to the functionality of the polymorphism, a meta-analysis of human molecular imaging studies suggest that striatal DAT expression might be higher in 9R carriers compared to 10R homozygotes (Faraone et al. 2014). This trend was also seen in an earlier meta-analysis, although the results did not reach statistical significance (Costa et al. 2011).

12.2.2 Social Behavior

As decision-making often occurs in interpersonal contexts, differences in social behavior (e.g., fairness, altruism) can cause interindividual variation.

Zhong et al. (2010) found that a *DRD4 Exon 3 VNTR* containing a 48 bp repeat is associated with a preference for fairness. Participants with the 4R/4R genotype reported a higher minimal acceptable offer than participants with 2R/4R or 2R/2R genotypes in the UG, indicating a higher fairness preference expressed through an insistence on a more equitable wealth distribution. In line with this, 4R-homozygotes also showed a higher minimal acceptable offer than participants with any other genotype in a study by Reuter et al. (2013), but no association with the first player (proposer) behavior (which in the UG can be interpreted as a combination of strategic consideration and prosocial perspective taking). Additionally, the authors identified a DRD2-haplotype-block spanning 15 kb of the DRD2/ANKK1 region and containing the SNPs *rs1800497* (*DRD2/ANKK1 Taq1A*, see above) and *rs2283265*, that was significantly associated with the offer of player one (proposer) but not with the behavior of player two (the recipient) in the UG. Thus, according to their results, a dissociation can be made between DRD2 influencing proposer and DRD4 affecting recipient behavior (Reuter et al. 2013).

Further examining the *DRD4 Exon 3 VNTR*, Enge et al. (2017) found larger feedback-related negativity amplitudes (FRN, an EEG measure) for unfair than for fair trials in 7R carriers but not in 7R non carriers in a *Dictator Game* (DG; see Table 12.1) from a first and a third party perspective, indicating DRD4 genotype-dependent processing of fairness.

With respect to altruistic tendencies, Bachner-Melman et al. (2005) observed significant multivariate associations between self-reported selflessness and the *DRD4 Exon 3 VNTR* (4R allele) and a *DRD5 polymorphism* allele, while no association was found for neither three DRD4 promoter region polymorphisms nor for any haplotype investigated. The authors conclude that the *DRD4 Exon 3 VNTR* 4R allele is associated with more altruistic and prosocial behavior than the 7R allele that represents a more aggressive and novelty seeking type of behavior (Bachner-Melman et al. 2005).

12.3 The Serotonergic System

Due to its distribution across many brain areas related to perception, cognition, and emotion, including the cerebral cortex and limbic structures (Waider et al. 2011), a role of the serotonergic system in decision-making appears plausible. The most prominently examined serotonergic polymorphism in the context of decision-making is the *5-HTTLPR (5-hydroxytryptamine transporter gene-linked polymorphic region)*, affecting the transcriptional activity of the serotonin transporter gene (Lesch et al. 1996; see Table 12.2). The short allele (also called S allele, which is less active according to Lesch et al. 1996) is associated with several markers of negative emotionality such as neuroticism and harm avoidance (Lesch et al. 1996), sensitivity to negative emotional cues (Papousek et al. 2013) and enhanced social learning of fear (Crişan et al. 2009).

12.3.1 Risk-Taking

The associations of the *5-HTTLPR* and of the tryptophan hydroxylase-2 (TPH-2, an enzyme critical in the synthesis of serotonin)-related SNP *rs1386438* with performance in the IGT was studied by Stoltenberg and Vandever (2010). While *rs1386438* had no effect, a significant sex by *5-HTTLPR* interaction effect was detected for decision-making under ambiguity (the first block of 20 choices in the IGT). Men with at least one S allele made more advantageous choices than men homozygous for the L allele, whereas women with L/L genotype made more beneficial choices than female S carriers. For decision-making under risk (IGT blocks 2–5), this effect was not observed. In line with this, He et al. (2010) also found an influence of the

5-HTTLPR on decision-making under ambiguity in the first trials of the IGT, with carriers of at least one L allele yielding higher scores than individuals homozygous for the S allele. In line with Stoltenberg and Vandever (2010), they also observed an interaction with sex under ambiguity, to the extent that male L allele carriers decided more advantageously than S allele homozygotes. Similarly, no *5-HTTLPR* effects for decisions under risk in the IGT were seen; but in a loss aversion experiment, carriers of the L allele showed less loss aversion than the S/S genotype. The modulation of loss aversion via the serotonergic system is also reflected in the study results of Zhong et al. (2009a, b, c), whose findings concerning dopaminergic associations have already been presented above. They also found associations to the *5-HTT Stin2 VNTR* (another polymorphism in the 5-HTT gene) in that individuals with the 10R allele were more risk-tolerant over losses in a loss-based version of a lottery game than 12R allele carriers. Although carriers of the L allele of the *5-HTTLPR* descriptively showed more risk-tolerance than carriers of the S allele, this difference did not reach significance. The authors interpret their results as evidence for the hypothesis that dopamine tone modulates sensitivity towards valuation of gains while serotonin tone modulates sensitivity towards valuation of losses (Zhong et al. 2009a, b, c). Fitting to the aforementioned results, Kuhnen and Chiao (2009) demonstrated that, besides their findings of dopaminergic associations mentioned earlier, S allele homozygotes are significantly more risk averse than persons with at least one L allele in investment decisions. For both self-report risk-taking as well as performance in the BART, carriers of at least one S allele also showed less financial risk-taking than homozygous L allele carriers (Crişan et al. 2009). However, contradicting results can also be found, reporting no association between the *5-HTTLPR* and risk-taking (Anderson et al. 2015; Frydman et al. 2011; Roiser et al. 2009).

By and large, the serotonergic system seems to be of importance in studying loss aversion and risk tolerance as well as decision-making under ambiguous circumstances.

12.3.2 Social Behavior

With regard to fairness, Enge et al. (2017) found involvement of the *5-HTTLPR* by observing that homozygous L allele carriers showed larger FRN amplitudes to unfair relative to fair monetary assignments in a DG, implicating a stronger responsiveness to unfair behavior of others, which could be not seen for S allele carriers. Fittingly, in another study, a combination of alleles (the S allele of the *5-HTTLPR* and the G allele of the *rs4570625*, another TPH2 gene polymorphism) was associated with less altruistic punishment in response to an unfair offer in the UG (Gärtner et al. 2018). Nishina et al. (2022) found no relationship between the *5-HTTLPR* and punishment for unfair proposals in a third party punishment paradigm, but demonstrated that individuals with at least one L allele were more likely to reject unfair offers in the UG compared to persons homozygous for the S allele. Inspired by the role of serotonin in emotional aversion to harmful actions, Yang et al. (2019) investigated the relationship of the *5-HTTLPR* with moral permissibility of impersonal and personal harmful behaviors with a moral dilemma task. They found that S/S carriers rate impersonal harms as more permissible, with stability over two waves of data collection (with a six months interval in between), while this was not the case for personal action (Yang et al. 2019). These results highlight the role of serotonin in considerations of fairness, although the overall evidence remains rather inconclusive.

Lastly, Stoltenberg et al. (2013) showed that the *5-HTTLPR* triallelic genotype (a combination of the alleles in the *5-HTTLPR* and in a SNP in the same gene, *rs25531*) was significantly associated with prosocial behavior (helping others) and that this effect was partially mediated by social anxiety. Participants homozygous for the L'allele (Lallele of the *5-HTTLPR* combined with the A allele of the *rs25531*) had higher scores on measures of prosocial behavior and less anxiety in terms of agoraphobic and sociophobic tendencies than individuals with at least one S' allele (comprising persons with the S allele of the *5-HTTLPR* as well as persons with the L allele of the *5-HTTLPR* in combination with the G allele of the *rs25531*). Self-reported avoidance of places and social situations showed strong negative correlations of self-reports of engagements in prosocial behavior (Stoltenberg et al. 2013), suggesting the mediation effect of anxiety between the serotonergic marker and prosociality.

12.4 Neuropeptides

Two neurochemicals crucial for modulating social interactions are the neuropeptides oxytocin and vasopressin (Caldwell 2017). Accordingly, their role in decision-making processes based on social interactions are frequently studied, including associations to trust, altruism and moral decision-making.

12.4.1 Oxytocin

Oxytocin is a neuropeptide synthesized in the hypothalamus, primarily in neurons of the supraoptic nucleus and the paraventricular nucleus. High levels of mRNA expression of oxytocinergic genes have been found in subcortical and temporal structures as well as in the olfactory region of the human brain (Quintana et al. 2019). Oxytocin is of particular relevance for decision-making behavior in the context of social interactions as it is related to a range of cognitive and behavioral functions associated with pair bonding, affiliation, and prosocial behavior in general (de Dreu 2012). Accordingly, it is involved in promoting trust in the ingroup and modulating cooperation, but also in defending the ingroup against an outgroup. Consequently, depending on the context, oxytocin can also stimulate aggression (de Dreu 2012). However, oxytocin does not appear to be exclusively linked to social decisions, as findings show that non-social decision-making behavior can also be influenced by the neuropeptide, e.g., by observing an increase in delay of gratification (i. e., higher patience) under oxytocin (Kapetaniou et al. 2021).

Plenty of research has investigated the effect of oxytocin on decision-making behavior by direct elevation of oxytocin through intranasal application (e.g., modulation of decision-making in the IGT by Zebhauser et al. 2022; modulation of decision-making in the *Trust Game* (TG, see Table 12.1) by Baumgartner et al. 2008). Yet, there is also evidence that variation in genetic markers of the oxytocinergic system contributes to differential decision-making behavior. These studies mainly focus on polymorphisms in the oxytocin receptor (OXTR) gene.

12.4.1.1 Social Behavior

As research most prominently indicates that oxytocin is associated with trust (e.g., Kosfeld et al. 2005; Mikolajczak et al. 2010), oxytocinergic genes are often studied in relation to trust-related economic games. Apicella et al. (2010) studied nine different SNPs in the OXTR gene, and found no significant associations between any of the polymorphisms, neither to behavior in the TG nor in the DG. In contrast to this, Krueger et al. (2012) showed that genetic variation in the OXTR indeed appears to promote trust. They found that the *rs53576* (see Table 12.2) was related to trust behavior, with G/G carriers demonstrating higher trust behavior than individuals with at least one A allele. However, the polymorphism was neither associated with trustworthiness nor with risk-taking behavior in general (Krueger et al. 2012). In line with this, findings by Reuter et al. (2009) found two haplotypes significantly related to trust, with the SNP *rs2268498* (see Table 12.2), located in the promoter region of the OXTR, being the main driver of the relationship in both haplotype blocks. Functionality analyses revealed that the genetic variant associated with high trust (T allele) showed a 50% reduction in OXTR mRNA expression compared to the low trust variant (C, Reuter et al. 2017). Also in line with the aforementioned associations, Tabak et al. (2014) examined ten SNPs in the OXTR gene in association to behavior and emotional reactions after a betrayal of trust by using an iterated *Prisoner's Dilemma Game* (PDG; see Table 12.1). Different polymorphisms were associated with a range of emotional reactions: the C/T haplotype (*rs9840864*, *rs2268494*) was related to higher levels of anger and quicker retaliation after a betrayal, the A/C haplotype (*rs237887*, *rs2268490*) was linked to higher satisfaction after a betrayal, and the G/C haplotype (*rs237887*, *rs2268490*) exhibited lower satisfaction post-betrayal. Thus, it can be deduced that oxytocin is not only associated with the giving of trust, but also with emotional behaviors in response to breaches of trust. By contrast, Tanaka et al. (2023) observed no association between the OXTR *rs53576* and inequity aversion.

Regarding altruism, Israel et al. (2009) studied multiple SNPs as well as haplotypes, finding associations of variation in the OXTR gene with performance in the DG as well as in a social value orientation paradigm (Israel et al. 2009).

Moreover, studies have been conducted with respect to the OXTR gene and moral judgments, showing a genetic contribution to complex moral decisions: Walter et al.

(2012) studied the functional polymorphism *rs2268498* in the promotor region of the OXTR gene in the evaluation of moral dilemmas, in which a protagonist intentionally or accidentally committed harm or attempted but failed to commit harm. Carriers of at least one C allele rated accidentally committed harms as more blameworthy than persons homozygous for the T allele (Walter et al. 2012). A different polymorphism, *rs237889*, was also linked to moral decision-making as a linear trend of C/C > C/T > T/T with respect to utilitarian responses to moral dilemmas was observed (Bernhard et al. 2016).

12.4.1.2 Interaction Between Intranasal Oxytocin Application and Genetics

Given that the application of oxytocin could exert different effects depending on the genetic make-up, the interaction between these parameters is also an interesting subject of investigation. For instance, if a genetic variant leads to the oxytocin receptor being expressed more frequently, this could also modulate the effects of intranasal oxytocin on behavior, as oxytocin can bind to more receptors and could thereby perform stronger neurobiological actions. In a PDG, the *rs53576* OXTR polymorphism modulated intranasal oxytocin effects on the neuronal response to social cooperation. In carriers of the G/G genotype, the application of oxytocin increased activation in the left ventral caudate nucleus in the basal ganglia to reciprocated cooperation in men but decreased this activation in women (Feng et al. 2015). According to the authors, the regulation of the response of the left ventral caudate nucleus by oxytocin may represent alterations in the reward or salience of reciprocated cooperation with an increase for G allele homozygous men but a decrease for homozygous women (Feng et al. 2015). Similarly, Bozorgmehr et al. (2019) showed that application of intranasal oxytocin in carriers of the T/T genotype of the *rs2254295* leads to improved performance in the IGT compared with oxytocin application in C/T and C/C carriers. The authors assume that the C allele of this polymorphism could constitute a risk allele for uncertain decision-making behavior (Bozorgmehr et al. 2019). In a more recent study, Bozorgmehr et al. (2020) investigated the effects of another SNP, *rs2254298*, in combination with oxytocin application on impulsivity in a go/no-go task. Their results showed that oxytocin reduces impulsivity, and participants with the G/G genotype demonstrated a lower degree of impulsivity compared to G/A or A/A genotypes, suggesting that the G allele may increase the expression of the OXTR gene. However, it should be pointed out that the only SNP for which functionality has been demonstrated so far is the *rs2268498* (Reuter et al. 2017).

Based on the pharmacological studies cited above, it is likely that the application of oxytocin results in gene-dependent actions, so that genotyping could be a useful addition in studying the effects of intranasal oxytocin, as it may provide additional explanations.

In conclusion, it can be convincingly derived from the studies presented above that genetic markers of the oxytocin system are associated with a wide range of prosocial behaviors and can thus exert a crucial influence on decision-making processes. Importantly, the functionality of polymorphisms in the OXTR gene should be more thoroughly investigated in future studies, in order to further promote the neurobiological understanding.

12.4.2 Vasopressin

Arginine vasopressin, an antidiuretic hormone secreted from the posterior pituitary, regulates diverse physiological functions in the human body: Fluid balance, blood osmolarity, reproduction, as well as social behavior, memory, and learning (Hammock et al. 2005; Sparapani et al. 2021). Although the neuropeptide is also currently under investigation in terms of its role in decision-making by means of nasal application (e.g., Brunnlieb et al. 2016; Patel et al. 2015; Rilling et al. 2012) and in animal studies (e.g., Smith et al. 2017), the study of the underlying genetic mechanisms in humans is less elaborated in contrast to oxytocin.

Vasopressin exerts its actions inter alia at the arginine vasopressin 1a receptor (AVPR1A; Hammock et al. 2005), a target for studying genetic effects. The gene contains promotor region VNTR microsatellites that are simple sequence repeats in a non-coding region (Avinun et al. 2011; Tansey et al. 2011). Of these, the *AVPR1a RS3* (see Table 12.2) was examined by Avinun et al. (2011) in association with altruism in preschoolers, showing that carriers of at least one copy of the 327 bp allele (i.e., long allele) were less generous in the DG. This fits to the results of Tanaka et al. (2023), who found that participants with two short alleles (i.e., less than 330 bp) had higher advantageous inequity aversion (where the target player receives more than others, interpreted as prosocial preference) in five economic game experiments compared to heterozygous or homozygous long allele (greater or equal to 330 bp) genotypes. No association to disadvantageous inequity aversion (where others receive more than the target player) was found. In contrast, a reversed association with altruism was seen in a study by Knafo et al. (2008), who investigated a male sample. Participants with short alleles (308–325 bp repeats) of the *AVPR1a RS3* were less generous to the other player in a DG than participants with long alleles (327–343 bp). This relationship between the polymorphism and altruism was further confirmed by self-report scales (Knafo et al. 2008). Of interest, the long repeat variant has been related to higher AVPR1a expression (Tansey et al. 2011).

Finally, no association between measures of either social or economic risk-taking and the *RS3* was found by Dreber et al. (2012).

12.5 Other Genetic Associations

Next to investigations of genes related to dopamine, serotonin, oxytocin and vasopressin, other neurotransmitter systems, neurotrophins, and enzymes have been discussed in the context of decision-making, although to a lesser extent, which is why they are omitted from the present discussion. These comprise the cholinergic system (e.g., Roe et al. 2009), brain-derived neurotrophic factor (BDNF; e.g., Kang et al. 2010), the catecholamine-degrading enzyme catechol-O-methyltransferase (COMT; e.g., Amstadter et al. 2012; Kang et al. 2010; Reuter et al. 2011; Roussos et al. 2008) and the monoamine-degrading enzyme monoaminoxidase A (MAO-A; e.g., Frydman et al. 2011; Zhong et al. 2009a, b, c).

12.6 Interactions Between Different Neurotransmitter Systems

In the previous sections, we focused primarily on single neurotransmitter systems and their corresponding candidate genes, sometimes complemented by haplotype analyses. However, the wealth of research on various neurotransmitters clearly indicates that decision-making processes are based on a complex interplay of several transmitter systems. This issue has been taken into account in previous research by studies on gene–gene-interactions, polygenic risk scores or GWAS, some of which will be briefly discuss below.

For example, Ha et al. (2009) examined the effects of *5-HTTLPR* (triallelic) and DRD4 gene polymorphisms on performance in the IGT. While they observed no main effects of these polymorphisms, an interaction between them became evident. Accordingly, the performance of persons with the S'/S' genotype was better for the group who simultaneously carried a *DRD4 Exon 3 VNTR* 2R allele than for the group without the 2R allele. In contrast, not carrying the 2R allele was beneficial for the group without the S'/S' genotype. This effect was stronger for decisions under risk (second half of the IGT) than for decisions under ambiguity (first half). Their results suggest that DRD4 genotypes affect IGT performance differently depending on the background genotypes of *5-HTTLPR*, implicating an interaction of the dopaminergic and serotonergic systems in emotion-based decision-making.

In a study by Voigt et al. (2015), an interaction between the *DRD2/ANKK1 Taq1A* and the *BDNF Val66Met* polymorphism on loss aversion was seen. Lowest scores in loss aversion were seen for carriers of the Met+/A1+ combination in contrast to all other allelic groups.

Using a polygenic approach, Hogeterp et al. (2023) showed the dopaminergic modulation of decision-making by an interplay of different polymorphisms in male participants of a stock market simulation. The configuration of carrying no 9R allele of the *DAT1 VNTR*, carrying the Met/Met genotype of the *COMT Val158Met* polymorphism (*rs4680*), no A1 allele of the *DRD2/ANKK1 Taq1A* polymorphism as well as carrying at least one 7R allele of the *DRD4 Exon 3 VNTR* was beneficial for achieving a high financial success throughout the simulation.

By assessing GWAS data and using gene-based principal components regression method, Gao et al. (2017) found that genetic variations in the COMT and 5-HTT genes was associated with a framing effect in decision-making. This effect is evident when an individual exhibits a tendency to be risk-averse when options are presented positively but a tendency to be risk-seeking when the same options are presented negatively. Moreover, individual differences in susceptibility to framing were also based on genetic variations of the aromatic-L-amino-acid decarboxylase gene, involved in the synthesis of serotonin and dopamine. Karlsson Linnér et al. (2019) also performed multiple GWAS of general risk tolerance, adventurousness and risky behaviors in driving, drinking, smoking, and sexual domains. Although their analyses found no evidence for the main biological pathways previously hypothesized to influence risk tolerance (inter alia, dopaminergic and serotonergic genes), their results implicate the involvement of glutamatergic and GABAergic neurotransmission. Furthermore, the authors found substantial genetic overlap between different measures of risk tolerance and behavior (Karlsson Linnér et al. 2019). Ye et al. (2020) also investigated different risky behaviors (automobile speeding, alcoholic drinks per week, ever-smoker, number of sexual partners) in a GWAS. They identified several common genes associated with these risky behaviors which have previously been related to cognitive impairment as well as Alzheimer's disease. Based on a single question measuring risk-taking behavior ("Would you describe yourself as someone who takes risks?"—yes/no), Strawbridge et al. (2018) also identified two gene loci in a GWAS. These are located on chromosome 3 (*rs13084531*) and on chromosome 6 (*rs9379971*), although merely the first association is confirmed by their replication and meta-analytic endeavors. This SNP is associated with expression levels of cell adhesion molecule 2 (CADM2), likely playing a role in brain plasticity (Strawbridge et al. 2018).

12.7 Gene-Environment Associations

Quantitative genetic studies have demonstrated that genetic as well as environmental factors influence decision-making processes and it is conceivable that both factors-genes and environment, interact with each other (Cesarini et al. 2008). A few of these studies are presented in the following to demonstrate the relevance of such interactions.

He et al. (2012) found that IGT performance was influenced by the *COMT Val158Met* polymorphism in interaction with environmental factors. Among Met allele carriers, the experience of more stressful life events was linked to higher reward sensitivity. In Val/Val carriers, parental warmth acted as a protective factor for IGT performance. Interestingly, no main effect of any of the environmental or genetic factors on IGT performance was observed (He et al. 2012).

Even more fundamental environmental influences, such as the season in which one was born, have been linked to differential genetic influences on decision-making behavior. Roussos et al. (2010) observed a different performance in the IGT dependent on the *DRD4 Exon 3 VNTR* and season of birth: the performance of winter-born 7R carriers was significantly worse compared to 4R carriers. An interaction between season and birth and this polymorphism was also observed by Zhong et al. (2010) with regard to fairness preference in the UG.

While Poulin et al. (2012) observed no direct association between OXTR SNP *rs53576* and prosocial behavior in form of charitable activities (e.g., blood donation, working for a charity) and civic duty, an interaction with perceived threat (seeing the world as malevolent and unsafe) was observed, in that greater perceived threat predicted fewer charitable activities for individuals with at least one A allele, but not for individuals with a G/G genotype. Similarly, in the same study, no main effect of the *AVPR1a RS1* polymorphism on prosocial behavior was found, however, an interaction of perceived threat with a heterozygous genotype: Greater perceived threat predicted lower commitment to civic duty in the S/L genotype. No interaction or main effects were found for *AVPR1a RS3.*

Finally, it should be mentioned that while some epigenetic studies have been conducted with respect to decision-making behavior and the neurotransmitters of interest to the present chapter (e.g., DRD4 methylation and physical aggression; Cecil et al. 2018; OXTR methylation and discounting behavior; Anzani et al. 2022), epigenetic approaches in the field are generally in their infancy. However, due to the numerous gene-environment interactions demonstrated in previous studies, epigenetic analyses certainly offer great potential for further research.

12.8 Conclusion

The studies presented in this chapter show that there are differences in decision-making behavior that are to a substantial extent attributable to genetic variation. Most of the research has focused on the dopaminergic and the serotonergic system as well as interactions between these, likely indicating an involvement of the two in the neurobiological foundations of human decision-making. Although quantitative genetic studies unequivocally demonstrate the impact of genetic factors on decision-making, the empirical evidence regarding the specific associations often remains inconclusive, which may be partly due to inconsistencies between investigations (e.g., inconsistent definitions of alleles and/or inconsistent implementations of economic games). Many studies also point to sex differences in the genetic modulation of decision-making that require further clarification. As decision-making likely is the result of polygenic interactions, future research is also needed to further promote the neurobiological understanding of the underlying processes. Finally, epigenetic studies should be conducted in order to resolve nurture vs. nature controversies.

Despite the ascribed heterogeneity in the field and the associated challenges, ongoing research into the genetic basis of decision-making behavior offers great potential to one day obtain an overarching picture of the biological basis of how we all form decisions every day.

References

Anderson A, Dreber A, Vestman R (2015) Risk taking, behavioral biases, and genes: results from 149 active investors. SSRN Electron J. Advance online publication. https://doi.org/10.2139/ssrn.2598483

Amstadter AB, MacPherson L, Wang F, Banducci AN, Reynolds EK, Potenza MN, Gelernter J, Lejuez CW (2012) The relationship between risk-taking propensity and the COMT Val158Met polymorphism among early adolescents as a function of sex. J Psychiatr Res 46(7):940–945

Apicella CL, Cesarini D, Johannesson M, Dawes CT, Lichtenstein P, Wallace B, Beauchamp J, Westberg L (2010) No association between oxytocin receptor (OXTR) gene polymorphisms and experimentally elicited social preferences. PLoS ONE 5(6):e11153. https://doi.org/10.1371/journal.pone.0011153

Arias-Carrión O, Stamelou M, Murillo-Rodríguez E, Menéndez-González M, Pöppel E (2010) Dopaminergic reward system: a short integrative review. Int Arch Med 3(1):24. https://doi.org/10.1186/1755-7682-3-24

Avinun R, Israel S, Shalev I, Gritsenko I, Bornstein G, Ebstein RP, Knafo A (2011) Avpr1a variant associated with preschoolers' lower altruistic behavior. PLoS ONE 6(9):e25274. https://doi.org/10.1371/journal.pone.0025274

Bachner-Melman R, Gritsenko I, Nemanov L, Zohar AH, Dina C, Ebstein RP (2005) Dopaminergic polymorphisms associated with self-report measures of human altruism: a fresh phenotype for the dopamine D4 receptor. Mol Psychiatry 10(4):333–335. https://doi.org/10.1038/sj.mp.4001635

Baumgartner T, Heinrichs M, Vonlanthen A, Fischbacher U, Fehr E (2008) Oxytocin shapes the neural circuitry of trust and trust adaptation in humans. Neuron 58(4):639–650. https://doi.org/10.1016/j.neuron.2008.04.009

Bechara A, Damasio AR, Damasio H, Anderson SW (1994) Insensitivity to future consequences following damage to human prefrontal cortex. Cognition 50(1–3):7–15. https://doi.org/10.1016/0010-0277(94)90018-3

Bernhard RM, Chaponis J, Siburian R, Gallagher P, Ransohoff K, Wikler D, Perlis RH, Greene JD (2016) Variation in the oxytocin receptor gene (OXTR) is associated with differences in moral judgment. Soc Cogn Affect Neurosci 11(12):1872–1881. https://doi.org/10.1093/scan/nsw103

Bozorgmehr A, Alizadeh F, Sadeghi B, Shahbazi A, Norouzi Ofogh S, Joghataei MT, Razian S, Heydari F, Ghadirivasfi M (2019) Oxytocin moderates risky decision-making during the Iowa gambling task: a new insight based on the role of oxytocin receptor gene polymorphisms and interventional cognitive study. Neurosci Lett 708:134328. https://doi.org/10.1016/j.neulet.2019.134328

Bozorgmehr A, Moayedi R, Sadeghi B, Ghadirivasfi M, Joghataei MT, Shahbazi A (2020) A novel link between the oxytocin receptor gene and impulsivity. Neuroscience 444:196–208. https://doi.org/10.1016/j.neuroscience.2020.07.033

Brunnlieb C, Nave G, Camerer CF, Schosser S, Vogt B, Münte TF, Heldmann M (2016) Vasopressin increases human risky cooperative behavior. Proc Natl Acad Sci 113(8):2051–2056. https://doi.org/10.1073/pnas.1518825113

Caldwell HK (2017) Oxytocin and vasopressin: powerful regulators of social behavior. Neurosci: Rev J Bringing Neurobiol, Neurol Psychiatry 23(5):517–528. https://doi.org/10.1177/1073858417708284

Carpenter JP, Garcia JR, Lum JK (2011) Dopamine receptor genes predict risk preferences, time preferences, and related economic choices. J Risk Uncertain 42(3):233–261. https://doi.org/10.1007/s11166-011-9115-3

Cesarini D, Dawes CT, Fowler JH, Johannesson M, Lichtenstein P, Wallace B (2008) Heritability of cooperative behavior in the trust game. Proc Natl Acad Sci 105(10):3721–3726. https://doi.org/10.1073/pnas.0710069105

Cesarini D, Johanesson M, Lichtenstein P, Sandewall Ö, Wallace B (2010). Genetic variation in financial decision-making. J Financ 65(5):1725–1754.https://doi.org/10.1111/j.1540-6261.2010.01592.x

Civai C, Hawes DR (2016) Game theory in neuroeconomics. Neuroeconomics 13–37.https://doi.org/10.1007/978-3-642-35923-1_2

Cools R, D'Esposito M (2011) Inverted-U-shaped dopamine actions on human working memory and cognitive control. Biol Psychiat 69(12):e113–e125. https://doi.org/10.1016/j.biopsych.2011.03.028

Costa A, Riedel M, Müller U, Möller HJ, Ettinger U (2011) Relationship between SLC6A3 genotype and striatal dopamine transporter availability: a meta-analysis of human single photon emission computed tomography studies. Synapse 65(10):998–1005

Crişan LG, Pana S, Vulturar R, Heilman RM, Szekely R, Druğa B, Dragoş N, Miu AC (2009) Genetic contributions of the serotonin transporter to social learning of fear and economic decision-making. Soc Cogn Affect Neurosci 4(4):399–408. https://doi.org/10.1093/scan/nsp019

de Dreu CKW (2012) Oxytocin modulates cooperation within and competition between groups: an integrative review and research agenda. Horm Behav 61(3):419–428. https://doi.org/10.1016/j.yhbeh.2011.12.009

Dreber A, Apicella CL, Eisenberg DT, Garcia JR, Zamore RS, Lum JK, Campbell B (2009) The 7R polymorphism in the dopamine receptor D4 gene (DRD4) is associated with financial risk taking in men. Evol Hum Behav 30(2):85–92. https://doi.org/10.1016/j.evolhumbehav.2008.11.001

Dreber A, Rand DG, Wernerfelt NC, Garcia J, Lum JK, Zeckhauser RJ (2011) The dopamine receptor D4 Gene (DRD4) and self-reported risk taking in the economic domain. HKS Faculty Research Working Paper Series. https://dash.harvard.edu/handle/1/5347066

Dreber A, Rand DG, Wernerfelt N, Montgomery C, Malhotra DK (2012) Genetic correlates of economic and social risk taking.https://doi.org/10.2139/ssrn.2141601

Enge S, Mothes H, Fleischhauer M, Reif A, Strobel A (2017) Genetic variation of dopamine and serotonin function modulates the feedback-related negativity during altruistic punishment. Sci Rep 7(1):2996. https://doi.org/10.1038/s41598-017-02594-3

Faraone SV, Spencer TJ, Madras BK, Zhang-James Y, Biederman J (2014) Functional effects of dopamine transporter gene genotypes on in vivo dopamine transporter functioning: a meta-analysis. Mol Psychiatry 19(8):880–889

Feng C, Lori A, Waldman ID, Binder EB, Haroon E, Rilling JK (2015) A common oxytocin receptor gene (OXTR) polymorphism modulates intranasal oxytocin effects on the neural response to social cooperation in humans. Genes Brain Behav 14(7):516–525. https://doi.org/10.1111/gbb.12234

Forgas JP (1989) Mood effects on decision-making strategies. Aust J Psychol 41(2):197–214. https://doi.org/10.1080/00049538908260083

Frydman C, Camerer C, Bossaerts P, Rangel A (2011) Maoa-L carriers are better at making optimal financial decisions under risk. Proc Biol Sci 278(1714):2053–2059. https://doi.org/10.1098/rspb.2010.2304

Gao X, Liu J, Gong P, Wang J, Fang W, Yan H, Zhu L, Zhou X (2017) Identifying new susceptibility genes on dopaminergic and serotonergic pathways for the framing effect in decision-making. Soc Cogn Affect Neurosci 12(9):1534–1544. https://doi.org/10.1093/scan/nsx062

Gärtner A, Strobel A, Reif A, Lesch K-P, Enge S (2018) Genetic variation in serotonin function impacts on altruistic punishment in the ultimatum game: a longitudinal approach. Brain Cogn 125:37–44. https://doi.org/10.1016/j.bandc.2018.05.007

Guo G, Cai T, Guo R, Wang H, Harris KM (2010) The dopamine transporter gene, a spectrum of most common risky behaviors, and the legal status of the behaviors. PLoS ONE 5(2):e9352. https://doi.org/10.1371/journal.pone.0009352

Ha RY, Namkoong K, Kang J, Kim YT, Kim SJ (2009) Interaction between serotonin transporter promoter and dopamine receptor D4 polymorphisms on decision-making. Prog Neuro-Psychopharmacol Biol Psychiatry 33(7):1217–1222.https://doi.org/10.1016/j.pnpbp.2009.07.009

Hammock EAD, Lim MM, Nair HP, Young LJ (2005) Association of vasopressin 1a receptor levels with a regulatory microsatellite and behavior. Genes Brain Behav 4(5):289–301. https://doi.org/10.1111/j.1601-183X.2005.00119.x

He Q, Xue G, Chen C, Lu Z-L, Chen C, Lei X, Liu Y, Li J, Zhu B, Moyzis RK, Dong Q, Bechara A (2012) Comt Val158Met polymorphism interacts with stressful life events and parental warmth to influence decision-making. Sci Rep 2(1):677. https://doi.org/10.1038/srep00677

He Q, Xue G, Chen C, Lu Z, Dong Q, Lei X, Ding N, Li J, Li H, Chen C, Li J, Moyzis RK, Bechara A (2010) Serotonin transporter gene-linked polymorphic region (5-HTTLPR) influences decision-making under ambiguity and risk in a large Chinese sample. Neuropharmacology 59(6):518–526. https://doi.org/10.1016/j.neuropharm.2010.07.008

Hirvonen MM, Laakso A, Någren K, Rinne JO, Pohjalainen T, Hietala J (2009) C957t polymorphism of dopamine D2 receptor gene affects striatal DRD2 in vivo availability by changing the receptor affinity. Synapse 63(10):907–912. https://doi.org/10.1002/syn.20672

Hogeterp SA, Plieger T, Monzel M, Vetterlein A, Reuter M (2023) From genes to performance: dopaminergic modulation of decision-making in a stock market simulation. J Neurosci Psychol Econ 16(4):194–215. https://doi.org/10.1037/npe0000181

Israel S, Lerer E, Shalev I, Uzefovsky F, Riebold M, Laiba E, Bachner-Melman R, Maril A, Bornstein G, Knafo A, Ebstein RP (2009) The oxytocin receptor (OXTR) contributes to prosocial fund allocations in the dictator game and the social value orientations task. PLoS ONE 4(5):e5535. https://doi.org/10.1371/journal.pone.0005535

Jaber M, Jones S, Giros B, Caron MG (1997) The dopamine transporter: a crucial component regulating dopamine transmission. Mov Disord: Off J Mov Disord Soc 12(5):629–633. https://doi.org/10.1002/mds.870120502

Josef AK, Richter D, Samanez-Larkin GR, Wagner GG, Hertwig R, Mata R (2016) Stability and change in risk-taking propensity across the adult life span. J Pers Soc Psychol 111(3):430–450. https://doi.org/10.1037/pspp0000090

Kang J, Namkoong K, Ha RY, Jhung K, Kim YT, Kim SJ (2010) Influence of BDNF and COMT polymorphisms on emotional decision-making. Neuropharmacology 58(7):1109–1113.https://doi.org/10.1016/j.neuropharm.2010.02.001

Kapetaniou GE, Reinhard MA, Christian P, Jobst A, Tobler PN, Padberg F, Soutschek A (2021) The role of oxytocin in delay of gratification and flexibility in non-social decision-making. ELife 10.https://doi.org/10.7554/eLife.61844

Karlsson Linnér R, Biroli P, Kong E, Meddens SFW, Wedow R, Fontana MA, Lebreton M, Tino SP, Abdellaoui A, Hammerschlag AR, Nivard MG, Okbay A, Rietveld CA, Timshel PN, Trzaskowski M, de Vlaming R, Zünd CL, Bao Y, Buzdugan L, Beauchamp JP (2019) Genome-wide association analyses of risk tolerance and risky behaviors in over 1 million individuals identify hundreds of loci and shared genetic influences. Nat Genet 51(2):245–257. https://doi.org/10.1038/s41588-018-0309-3

Kawamura Y, Takahashi T, Liu X, Nishida N, Noda Y, Yoshikawa A, Umekage T, Sasaki T (2013) Variation in the DRD2 gene affects impulsivity in intertemporal choice. Open J Psychiatry 03(01):26–31. https://doi.org/10.4236/ojpsych.2013.31005

Knafo A, Israel S, Darvasi A, Bachner-Melman R, Uzefovsky F, Cohen L, Feldman E, Lerer E, Laiba E, Raz Y, Nemanov L, Gritsenko I, Dina C, Agam G, Dean B, Bornstein G, Ebstein RP (2008) Individual differences in allocation of funds in the dictator game associated with length of the arginine vasopressin 1a receptor RS3 promoter region and correlation between RS3 length and hippocampal mRNA. Genes Brain Behav 7(3):266–275. https://doi.org/10.1111/j.1601-183X.2007.00341.x

Kosfeld M, Heinrichs M, Zak PJ, Fischbacher U, Fehr E (2005) Oxytocin increases trust in humans. Nature 435(7042):673–676. https://doi.org/10.1038/nature03701

Krueger F, Parasuraman R, Iyengar V, Thornburg M, Weel J, Lin M, Clarke E, McCabe K, Lipsky RH (2012) Oxytocin receptor genetic variation promotes human trust behavior. Front Hum Neurosci 6:4. https://doi.org/10.3389/fnhum.2012.00004

Kuhnen CM, Chiao JY (2009) Genetic determinants of financial risk taking. PLoS ONE 4(2):e4362. https://doi.org/10.1371/journal.pone.0004362

Lejuez CW, Read JP, Kahler CW, Richards JB, Ramsey SE, Stuart GL, Strong DR, Brown RA (2002). Evaluation of a behavioral measure of risk taking: the Balloon Analogue Risk Task (BART). J Experiml Psychol Appl 8(2):75–84. https://doi.org/10.1037//1076-898x.8.2.75

Lesch K-P, Bengel D, Heils A, Sabol SZ, Greenberg BD, Petri S, Benjamin J, Müller CR, Hamer DH, Murphy DL (1996) Association of anxiety-related traits with a polymorphism in the serotonin transporter gene regulatory region. Science 274(5292):1527–1531

Levin IP, Hart SS, Weller JA, Harshman LA (2007) Stability of choices in a risky decision-making task: a 3-year longitudinal study with children and adults. J Behav Decis-Mak 20(3):241–252. https://doi.org/10.1002/bdm.552

Mata R, Hau R, Papassotiropoulos A, Hertwig R (2012) Dat1 polymorphism is associated with risk taking in the Balloon Analogue Risk Task (BART). PLoS ONE 7(6):e39135. https://doi.org/10.1371/journal.pone.0039135

Mikolajczak M, Gross JJ, Lane A, Corneille O, de Timary P, Luminet O (2010) Oxytocin makes people trusting, not gullible. Psychol Sci 21(8):1072–1074. https://doi.org/10.1177/0956797610377343

Missale C, Nash SR, Robinson SW, Jaber M, Caron MG (1998) Dopamine receptors: from structure to function. Physiol Rev 78(1):189–225. https://doi.org/10.1152/physrev.1998.78.1.189

Mowinckel AM, Pedersen ML, Eilertsen E, Biele G (2015) A meta-analysis of decision-making and attention in adults with ADHD. J Atten Disord 19(5):355–367. https://doi.org/10.1177/1087054714558872

Muda R, Kicia M, Michalak-Wojnowska M, Ginszt M, Filip A, Gawda P, Majcher P (2018) The dopamine receptor D4 Gene (DRD4) and financial risk-taking: stimulating and instrumental risk-taking propensity and motivation to engage in investment activity. Front Behav Neurosci 12:320020. https://doi.org/10.3389/fnbeh.2018.00034

Nishina K, Shou Q, Takahashi H, Sakagami M, Inoue-Murayama M, Takagishi H (2022) Association between polymorphism (5-HTTLPR) of the serotonin transporter gene and behavioral response to unfair distribution. Front Behav Neurosci 16:762092. https://doi.org/10.3389/fnbeh.2022.762092

Noble EP (2000) Addiction and its reward process through polymorphisms of the D2 dopamine receptor gene: a review. Eur Psychiatry: J Assoc Eur Psychiatr 15(2):79–89. https://doi.org/10.1016/S0924-9338(00)00208-X

Papousek I, Reiser EM, Schulter G, Fink A, Holmes EA, Niederstätter H, Nagl S, Parson W, Weiss EM (2013) Serotonin transporter genotype (5-HTTLPR) and electrocortical responses indicating the sensitivity to negative emotional cues. Emotion 13(6):1173–1181. https://doi.org/10.1037/a0033997

Patel N, Grillon C, Pavletic N, Rosen D, Pine DS, Ernst M (2015) Oxytocin and vasopressin modulate risk-taking. Physiol Behav 139:254–260. https://doi.org/10.1016/j.physbeh.2014.11.018

Pohjalainen T, Rinne JO, Någren K, Lehikoinen P, Anttila K, Syvälahti EK, Hietala J (1998) The A1 allele of the human D2 dopamine receptor gene predicts low D2 receptor availability in healthy volunteers. Mol Psychiatry 3(3):256–260. https://doi.org/10.1038/sj.mp.4000350

Poulin MJ, Holman EA, Buffone A (2012) The neurogenetics of nice: receptor genes for oxytocin and vasopressin interact with threat to predict prosocial behavior. Psychol Sci 23(5):446–452. https://doi.org/10.1177/0956797611428471

Quintana DS, Rokicki J, van der Meer D, Alnæs D, Kaufmann T, Córdova-Palomera A, Dieset I, Andreassen OA, Westlye LT (2019) Oxytocin pathway gene networks in the human brain. Nat Commun 10(1):668. https://doi.org/10.1038/s41467-019-08503-8

Reuter M, Frenzel C, Walter NT, Markett S, Montag C (2011) Investigating the genetic basis of altruism: the role of the COMT Val158Met polymorphism. Soc Cogn Affect Neurosci 6(5):662–668. https://doi.org/10.1093/scan/nsq083

Reuter M, Felten A, Penz S, Mainzer A, Markett S, Montag C (2013) The influence of dopaminergic gene variants on decision-making in the ultimatum game. Front Hum Neurosci 7:242. https://doi.org/10.3389/fnhum.2013.00242

Reuter M, Montag C (eds) (2016) Neuroeconomics. Springer, Berlin Heidelberg

Reuter M, Montag C, Altmann S, Bendlow F, Elger C, Kirsch P, Becker A, Schoch-McGovern S, Simon M, Weber B, Felten A (2017) Functional characterization of an oxytocin receptor gene variant (rs2268498) previously associated with social cognition by expression analysis in vitro and in human brain biopsy. Soc Neurosci 12(5):604–611. https://doi.org/10.1080/17470919.2016.1214174

Rilling JK, DeMarco AC, Hackett PD, Thompson R, Ditzen B, Patel R, Pagnoni G (2012) Effects of intranasal oxytocin and vasopressin on cooperative behavior and associated brain activity in men. Psychoneuroendocrinology 37(4):447–461. https://doi.org/10.1016/j.psyneuen.2011.07.013

Roe BE, Tilley MR, Gu HH, Beversdorf DQ, Sadee W, Haab TC, Papp AC (2009) Financial and psychological risk attitudes associated with two single nucleotide polymorphisms in the nicotine receptor (CHRNA4) gene. PLoS ONE 4(8):e6704. https://doi.org/10.1371/journal.pone.0006704

Roiser JP, de Martino B, Tan GCY, Kumaran D, Seymour B, Wood NW, Dolan RJ (2009) A genetically mediated bias in decision-making driven by failure of amygdala control. J Neurosci 29(18):5985–5991. https://doi.org/10.1523/JNEUROSCI.0407-09.2009

Roussos P, Giakoumaki SG, Pavlakis S, Bitsios P (2008) Planning, decision-making and the COMT rs4818 polymorphism in healthy males. Neuropsychologia 46(2):757–763

Roussos P, Giakoumaki SG, Bitsios P (2009) Cognitive and emotional processing in high novelty seeking associated with the L-DRD4 genotype. Neuropsychologia 47(7):1654–1659. https://doi.org/10.1016/j.neuropsychologia.2009.02.005

Roussos P, Giakoumaki SG, Bitsios P (2010) Cognitive and emotional processing associated with the Season of Birth and dopamine D4 receptor gene. Neuropsychologia 48(13):3926–3933. https://doi.org/10.1016/j.neuropsychologia.2010.09.021

Ryterska A, Jahanshahi M, Osman M (2013) What are people with Parkinson's disease really impaired on when it comes to making decisions? A meta-analysis of the evidence. Neurosci Biobehav Rev 37(10 Pt 2):2836–2846. https://doi.org/10.1016/j.neubiorev.2013.10.005

Simpson J, Vetuz G, Wilson M, Brookes KJ, Kent L (2010) The DRD4 receptor Exon 3 VNTR and 5' SNP variants and mRNA expression in human post-mortem brain tissue. Am J Med Genet B Neuropsychiatr Genet 153B(6):1228–1233. https://doi.org/10.1002/ajmg.b.31084

Smith CJW, Poehlmann ML, Li S, Ratnaseelan AM, Bredewold R, Veenema AH (2017) Age and sex differences in oxytocin and vasopressin V1a receptor binding densities in the rat brain: focus on the social decision-making network. Brain Struct Funct 222(2):981–1006. https://doi.org/10.1007/s00429-016-1260-7

Sparapani S, Millet-Boureima C, Oliver J, Mu K, Hadavi P, Kalostian T, Ali N, Avelar CM, Bardies M, Barrow B, Benedikt M, Biancardi G, Bindra R, Bui L, Chihab Z, Cossitt A, Costa J, Daigneault T, Dault J, Gamberi C (2021) The biology of vasopressin. Biomedicines 9(1):89. https://doi.org/10.3390/biomedicines9010089

Stoltenberg SF, Christ CC, Carlo G (2013) Afraid to help: social anxiety partially mediates the association between 5-HTTLPR triallelic genotype and prosocial behavior. Soc Neurosci 8(5):400–406. https://doi.org/10.1080/17470919.2013.807874

Stoltenberg SF, Vandever JM (2010) Gender moderates the association between 5-HTTLPR and decision-making under ambiguity but not under risk. Neuropharmacology 58(2):423–428. https://doi.org/10.1016/j.neuropharm.2009.09.010

Strawbridge RJ, Ward J, Cullen B, Tunbridge EM, Hartz S, Bierut L, Horton A, Bailey MES, Graham N, Ferguson A, Lyall DM, Mackay D, Pidgeon LM, Cavanagh J, Pell JP, O'Donovan M, Escott-Price V, Harrison PJ, Smith DJ (2018) Genome-wide analysis of self-reported risk-taking behaviour and cross-disorder genetic correlations in the UK Biobank cohort. Transl Psychiatry 8(1):1–11. https://doi.org/10.1038/s41398-017-0079-1

Tabak BA, McCullough ME, Carver CS, Pedersen EJ, Cuccaro ML (2014) Variation in oxytocin receptor gene (OXTR) polymorphisms is associated with emotional and behavioral reactions to betrayal. Soc Cogn Affect Neurosci 9(6):810–816. https://doi.org/10.1093/scan/nst042

Tanaka H, Nishina K, Shou Q, Takahashi H, Sakagami M, Matsuda T, Inoue-Murayama M, Takagishi H (2023) Association between arginine vasopressin receptor 1A (AVPR1A) polymorphism and inequity aversion. Proc R Soc B(290):20230378

Tansey KE, Hill MJ, Cochrane LE, Gill M, Anney RJ, Gallagher L (2011) Functionality of promoter microsatellites of arginine vasopressin receptor 1A (AVPR1A): implications for autism. Molecular Autism 2(1):3. https://doi.org/10.1186/2040-2392-2-3

Thielmann I, Böhm R, Ott M, Hilbig BE (2021) Economic games: an Introduction and guide for research. Collabra: Psychol 7(1), Article 19004. https://doi.org/10.1525/collabra.19004

Vallone D, Picetti R, Borrelli E (2000) Structure and function of dopamine receptors. Neurosci Biobehav Rev 24(1):125–132. https://doi.org/10.1016/S0149-7634(99)00063-9

Vinckier F, Rigoux L, Oudiette D, Pessiglione M (2018) Neuro-computational account of how mood fluctuations arise and affect decision-making. Nat Commun 9(1):1708. https://doi.org/10.1038/s41467-018-03774-z

Voigt G, Montag C, Markett S, Reuter M (2015) On the genetics of loss aversion: an interaction effect of BDNF Val66Met and DRD2/ANKK1 Taq1a. Behav Neurosci 129(6):801–811. https://doi.org/10.1037/bne0000102

Waider J, Araragi N, Gutknecht L, Lesch K-P (2011) Tryptophan hydroxylase-2 (TPH2) in disorders of cognitive control and emotion regulation: a perspective. Psychoneuroendocrinology 36(3):393–405. https://doi.org/10.1016/j.psyneuen.2010.12.012

Wallace B, Cesarini D, Lichtenstein P, Johannesson M (2007) Heritability of ultimatum game responder behavior. Proc Natl Acad Sci USA 104(40):15631–15634. https://doi.org/10.1073/pnas.0706642104

Walter NT, Montag C, Markett S, Felten A, Voigt G, Reuter M (2012) Ignorance is no excuse: moral judgments are influenced by a genetic variation on the oxytocin receptor gene. Brain Cogn 78(3):268–273. https://doi.org/10.1016/j.bandc.2012.01.003

Yang Y, Wang C, Li X, Yu R, Zhang M, Xue M, Guo W, He L, Gao X, Gong P (2019) The 5-HTTLPR polymorphism impacts moral permissibility of impersonal harmful behaviors. Soc Cogn Affect Neurosci 14(8):911–918. https://doi.org/10.1093/scan/nsz060

Ye J, Liu L, Xu X, Wen Y, Li P, Cheng B, Cheng S, Zhang L, Ma M, Qi X, Liang C, Kafle OP, Wu C, Wang S, Wang X, Ning Y, Chu X, Niu L, Zhang F (2020) A genome-wide multiphenotypic association analysis identified candidate genes and gene ontology shared by four common risky behaviors. Aging (Albany NY) 12(4):3287–3297. https://doi.org/10.18632/aging.102812

Zebhauser PT, Macchia A, Gold E, Salcedo S, Burum B, Alonso-Alonso M, Gilbert DT, Pascual-Leone A, Brem A-K (2022) Intranasal oxytocin modulates decision-making depending on outcome predictability-a randomized within-subject controlled trial in healthy males. Biomedicines 10(12):3230. https://doi.org/10.3390/biomedicines10123230

Zhong S, Chew SH, Set E, Zhang J, Xue H, Sham PC, Ebstein RP, Israel S (2009a) The heritability of attitude toward economic risk. Twin Res Hum Genet 12(1):103–107. https://doi.org/10.1375/twin.12.1.103

Zhong S, Israel S, Xue H, Ebstein RP, Chew SH (2009b) Monoamine oxidase A gene (MAOA) associated with attitude towards longshot risks. PLoS ONE 4(12):e8516. https://doi.org/10.1371/journal.pone.0008516

Zhong S, Israel S, Xue H, Sham PC, Ebstein RP, Chew SH (2009c) A neurochemical approach to valuation sensitivity over gains and losses. Proceedings Biol Sci 276(1676):4181–4188. https://doi.org/10.1098/rspb.2009.1312

Zhong S, Israel S, Shalev I, Xue H, Ebstein RP, Chew SH (2010) Dopamine D4 receptor gene associated with fairness preference in ultimatum game. PLoS ONE 5(11):e13765. https://doi.org/10.1371/journal.pone.0013765

Zyphur MJ, Narayanan J, Arvey RD, Alexander GJ (2009) The genetics of economic risk preferences. J Behav Decis-Mak 22(4):367–377. https://doi.org/10.1002/bdm.643

Implications and Applications

Chapter 13
Decision Making in Major Depressive Disorder

Laura Asché, Julian Pakos, Annika Wiebe, Benjamin Selaskowski, Christine Leonards, Ben Harrison, Alexandra Philipsen, Chris Davey, and Niclas Braun

Abstract Impaired cognitive processing such as adverse decision-making is characteristic of major depressive disorder and can have considerable negative psychosocial consequences. This chapter presents and discusses existing theories and therapeutic approaches relevant to decision-making in depression. We show that, despite significant progress in fundamental research, there are still major gaps in our current understanding of adverse decision-making in major depressive disorder. Moreover, existing psychotherapeutic treatment approaches do not directly target adverse decision-making behaviors but do so inadvertently. Currently, only few treatment approaches (e.g., problem-solving and self-management skills) have been developed that primarily focus on treating adverse decision-making behavior. Therefore, a better understanding of altered decision-making processes in this clinical population is needed in order to better target adverse decision-making and associated problems, and to develop better therapeutic interventions.

Keywords Affective disorders · Depression · Major depressive disorder · Decision-making · Cognitive neuroscience

13.1 Introduction

Adverse decision-making (DM) is a distinctive feature of major depressive disorder (MDD) that can lead to significant negative psychosocial outcomes. People with depression tend to be more prone to engage in risky behavior (Lu et al. 2024), experience decisional conflicts (van Randenborgh et al. 2010), and tend to blame themselves more for their poor DM (Kraines et al. 2017). Despite broad research interest into the nature of cognitive impairment in depression (Gotlib and Joormann 2010),

L. Asché · J. Pakos · A. Wiebe · B. Selaskowski · A. Philipsen · N. Braun (✉)
Department of Psychiatry and Psychotherapy, University Hospital Bonn, Bonn, Germany
e-mail: niclas.braun@ukbonn.de

C. Leonards · B. Harrison · C. Davey
Department of Psychiatry, University of Melbourne, Melbourne, Australia

U. Ettinger et al. (eds.), *Decision Making*, Studies in Neuroscience, Psychology and Behavioral Economics, https://doi.org/10.1007/978-3-032-00880-0_13

there is less work that focused on adverse DM and unfavorable behavioral outcomes. Gaining a deeper understanding of the neurobiological underlying of adverse DM in MDD is crucial if we are to address DM-related problems and develop more effective therapeutic interventions. In this chapter, we will first provide a brief overview of MDD before delving into the existing theories and therapeutic strategies concerning adverse DM in the context of MDD. Finally, we will conclude the chapter by offering a critical analysis, and recommendations for future directions.

13.2 Clinical Presentation

13.2.1 Symptoms and Clinical Presentation

According to the Diagnostic and Statistical Manual of Mental Disorders Fifth Edition (DSM-5), MDD is one of the most common affective disorders. The hallmark characteristics of MDD are depressed mood and/or reduced interest or pleasure in all or almost all activities over a period of at least two weeks (American Psychiatric Association 2013). In addition to having at least one of these two symptoms, a diagnosis of MDD requires that five out of the nine following symptoms are present over the same period: appetite and sleep disturbance; psychomotoric agitation or slowness; loss of energy; feelings of worthlessness or guilt; reduced ability to concentrate or make decisions; and recurrent thoughts of death or suicidal ideation, planning, or even attempts. Further, for a positive clinical diagnosis to be reached, symptoms need to cause clinically significant distress or impairment in social, occupational, or other important areas of functioning.

13.2.2 Epidemiology

MDD typically emerges in early adulthood (Solmi et al. 2022) and its prevalence remains relatively stable across adult age groups. The estimated occurrence of MDD varies significantly, depending on the employed metrics and study designs (Marx et al. 2023). For instance, the World Mental Health Survey (World Health Organization 2022), published by the World Health Organization, reported a ~ 10% lifetime prevalence of MDD. However, a prospective study suggested a > 30% lifetime prevalence (Moffitt et al. 2010). The 12-month prevalence is estimated to be around 4–5% (Ferrari et al. 2013; World Health Organization 2017). The cumulative recurrence of MDD after remission is estimated to be 13% after 5 years and 42% after 20 years (Hardeveld et al. 2013). The genetic contribution to MDD is estimated to be between 30 and 40% (Sullivan et al. 2000). The prevalence of MDD is almost twice as high in women compared to men, suggesting gender differences may be a contributing factor (Kuehner 2017).

Cognitive difficulties are frequently observed in MDD, with 20–30% of people with depression showing objective evidence of cognitive impairment (McIntyre et al. 2013). However, we are not aware of any demographic studies that have specifically investigated the prevalence rates of adverse DM in MDD. And yet, research suggests that depression is associated with a higher probability for risky behavior, such as risky sexual behavior and drug use, which may in turn be driven by adverse DM (cf. next sections).

13.2.3 Adverse DM in MDD and Its Personal and Social Implications

Depression, depressive symptomatology, and subsequent alterations in DM are associated with numerous negative implications for the affected person and their social environment. For example, alterations in DM were apparent in psychiatric patients that attempted suicide, including those with depression (Richard-Devantoy et al. 2016; Sastre-Buades et al. 2021). Additionally, a meta-analysis reported a significant difference in risky DM, as measured by the Iowa Gambling Task, in persons with MDD with suicidal behavior compared to persons with MDD without suicidal behavior and healthy controls (Richard-Devantoy et al. 2016).

Alterations in risky DM are not restricted to people with MDD who exhibit suicidal behavior but are also generally present in the disorder. In fact, a recent meta-analysis reported that people with MDD showed a significant increase in risky DM measured by a variety of neuropsychological DM tasks compared to healthy controls (Lu et al. 2024). Interestingly, while overall risk-taking behavior was increased, the authors also reported significant risk aversion in one of the tasks compared to healthy controls. According to the authors, this might be caused by the different operationalization or underlying mechanisms of different DM tasks.

A similar picture of risk preferences and risk-taking behavior has also been found in a study by Cobb-Clark et al. (2022). In this study, a large, representative household panel survey data set from Germany (SOEP) was used to compare individuals vulnerable to depression with those who are not. While the vulnerable group reported a lower preference to take risks in general, they also reported a greater willingness to take health risks. In addition, the authors reported a small increase in risky financial and health behaviors and a decreased likelihood to lend their belongings in their investigated population. Further studies exploring risky behavior in adolescents report associations between depressive symptoms and risky behaviors, such as risky sexual behavior, substance use, delinquency, self-harm, and suicidal behavior in adolescents in low- and middle-income countries (for a meta-analysis, see Pozuelo et al. 2022) as well as high-income countries (Katon et al. 2010; Lehrer et al. 2006).

Moreover, it has been argued that negative beliefs about oneself and the surrounding world that are frequent in depression may have a negative causal effect on economic DM (Ridley et al. 2020). Choung et al. (2022) using a large longitudinal cohort data set, supported this idea. They reported a shorter financial planning horizon of patients with MDD compared to controls, which appears to be partially driven by the individual's behavioral traits such as optimism/pessimism, and perceived control over the outcomes in their lives. The authors concluded that the differences in financial behavior might be explained by a diminished locus of control, a pessimistic outlook and the underestimation of life years left. Similarly, Pulcu et al. (2015) found that a higher discount rate of delayed rewards in a monetary delay discounting task in patients with MDD compared to HC correlated with MDD symptoms such as hopelessness.

Studies have also reported deficits in social DM in MDD which are associated with the impairment in social functioning characteristic of the disorder (Kupferberg et al. 2016). For example, Destoop et al. (2012) reported that patients accepted fair and unfair offers in the Ultimatum Game similar to healthy controls, while offering more money as proposers than healthy controls. As reported by Pulcu et al. (2015), this alteration might be dependent on the specific social interaction paradigm and on whether or not the patient is currently depressed or remitted. They reported less altruistic decisions in currently depressed patients compared to controls and non-symptomatic MDD patients in the Prisoner's Dilemma Game.

It has also been discussed whether MDD can impair the capacity to make treatment or research participation decisions. In their review, Hindmarch et al. (2013) tentatively concluded that the literature primarily suggests an impairment of appreciation, i.e., "the ability to appreciate the significance of [...] information for one's own situation", primarily regarding one's illness and potential consequences of treatment options.

In conclusion, DM in patients with MDD is altered in multiple contexts such as risk-taking behavior, social interactions, and financial planning.

13.3 Explanatory Approaches

There are two distinct approaches that propose possible mechanisms and/or explanations for adverse DM in MDD: computational theories and clinical theories. Whereas computational theories attempt to apply computational models of DM in the context of MDD, clinical models of depression implicitly consider adverse DM as an important component within maladaptive cognitive processes of MDD more broadly. Moreover, computational theories restrict themselves to DM in the narrower sense and have developed concrete experimental paradigms, by means of which general theories of DM are developed and then applied to mechanisms of depression. Clinical theories, on the other hand, are based on the examination of the disorder's pathomechanisms, taking phenomenological and behavioral observations as their starting point for theory development. The following section will present how general DM models can be applied to MDD, and the DM-relevant clinical theories of MDD.

13.3.1 Computational Theories and Models

Computational theories of DM are frameworks that describe how individuals make decisions in different situations. These theories often include mathematical or computational formulations to detail the underlying processes of DM.

13.3.1.1 Prospect Theory

Prospect theory describes how individuals evaluate potential outcomes in relation to a subjective reference point (Kahneman and Tversky 1979). Specifically, it states that people tend to make decisions based on perceived gains and losses, thereby being more risk-averse towards gains and more risk-taking towards losses. While prospect theory is one of the most influential behavioral models, particularly for risk-taking behavior, contextual elements alone do not appear to adequately describe the variability of individual DM (Kühberger 1998). More recently, the influence of affective states has been increasingly considered (Hodgkinson and Healey 2011). With respect to prospect theory, emotions can alter the value of prospects, the probability weighting, and the reference point that determines whether a particular prospect is considered a potential gain or loss (Prietzel 2020). Therefore, not only variations in the accuracy of parameter estimation, but also differences in parameter weighting could characterize the DM of individuals with affective disorders such as depression (Bishop and Gagne 2018).

Findings based on data from healthy individuals have linked positively valanced emotions to increased loss aversion, while negatively valanced emotions appear to be more diverse in their outcome (Prietzel 2020). The effects of sadness indicate a tendency to increase the subjective value of immediate and high positive outcomes in high-risk decisions (Lerner et al. 2013), as well as a preference for high-risk, high-reward options, while anxiety is more strongly associated with low-risk, low-reward decisions (Raghunathan and Pham 1999). At a clinical level, depression has been linked with lower expectation of reward value and therefore a positive prediction error in reward receipt (for details, see 3.1.3), as well as with difficulties updating expected reward value over time (Bishop and Gagne 2018). In line with these findings, preliminary exploratory neuroimaging studies have demonstrated impaired reward learning in patients with depression (Kumar et al. 2018).

In conclusion, prospect theory may offer insights into adverse DM in depression, including reduced expectation of reward value and impaired reward learning, illustrating the complex interplay between emotions and DM processes.

13.3.1.2 Dual-Process Theory

Dual-process theory assumes that there are two separate cognitive systems involved in DM: An intuitive, associative system and an analytical, reflective system (Evans and Stanovich 2013). The interaction between these systems influences our DM, with the former primarily using fast and automatic processes, while the latter uses slower and more deliberate thinking (Wason and Evans 1974). A further development of the model distinguishes between heuristic mechanisms, which identify relevant task information, and analytical mechanisms, which process the identified information to formulate conclusions or judgments (Evans 1984). While factors such as perceptions and semantic connections influence the heuristic selection, analytical processes are primarily guided by context information whereby individuals reason on the basis of previous experience rather than predefined inference rules.

In mood regulation, dual-process theory has been applied in the sense that individuals use both associative and reflective processing mechanisms. Associative processing serves to maintain the current mood state by retrieving mood-congruent information, while reflective processing is activated at certain intensities of affect and consequently attempts to alter the imbalanced state by accessing mood-incongruent information (Forgas 2000).

A dual-process perspective can be used to explain vulnerability to depression. Specifically, dual-process theory proposes that the interaction between associative and reflective processing may be rooted in a negative association bias in the processing of self-related information as suggested by Beevers (2005). While reflective processing can potentially attenuate this bias, there are some cases in which associative processing remains uncorrected. This includes cases where biased associative processing contradicts expectations but the person lacks the cognitive resources for reflective processing; cases where biased associative processing is consistent with expectations and therefore does not initiate reflective processing; and cases where biased associative processing contradicts expectations, but reflective processing does not adequately compensate for the negativity bias. Consequently, vulnerability to depression may persist due to the inadequate correction of biased associative processing.

13.3.1.3 Model-Based and Model-Free Learning

Depression-associated characteristics are also found in the context of reinforcement learning. Similar to the dual-process theory, in the computational approach to reinforcement learning a general differentiation is made between two distinct, but interacting, learning strategies: Model-free and model-based learning. In model-free (MF) learning, decisions are made solely based on past trial-and-error feedback. Specifically, for each action, the difference between expected and actual outcomes (reward prediction error) is calculated—a positive prediction error increases the expected value of an action, and a negative prediction error decreases it (Daw et al. 2005). The rate at which action values are adapted thereby depends on the individual learning rate (Bishop and Gagne 2018). In contrast, in model-based (MB) DM, the individual engages in mental simulation and recall of the consequences of potential decision alternatives, creating a model of the world, and choosing the alternative with the best expected long-term outcome (Daw et al. 2005).

Interestingly, evidence has been presented for depression-associated deficits in both MF learning and MB learning. Regarding the former, a meta-analysis suggests that patients with MDD demonstrate impairments in reward processing tasks (Halahakoon et al. 2020). At the neural level, altered activations in reward-associated brain areas during reward processing have been found, such as in the ventral striatum and orbitofrontal cortex, including blunted reward prediction error signalling (see reviews by Borsini et al. 2020; Chen et al. 2015; and meta-analysis by Ng et al. 2019). Further, a recent meta-analysis (Pike and Robinson 2022) found increased punishment learning rates and slightly reduced reward learning rates in patients with anxiety and mood disorders. Taken together, these results point towards a depression-associated deficit in MF learning.

In contrast, Huys et al. (2015) argue that depression is characterized by a deficit in MB rather than in MF learning. As illustrated in Sect. 3.1.2, depression has been linked to inadequate correction of mood-congruent negative association bias that may affect attention, memory, and interpretation of situations. This, in turn, influences the internal model of the world due to a reduced ability to simulate future positive events and increased recall of negative events. In addition to these negative affective biases, depression also appears to be associated with an increased valuation of effort (Bishop and Gagne 2018; Horne et al. 2021).

Attempting to explain how model-based and model-free processes interact in maintaining the depressed state, Bishop and Gagne (2018) developed the altered computations in DM (ACDM) framework. According to this framework, the combination of reduced simulation of positive events coupled with increased recall of negative events and altered effort-reward tradeoff leads to a decrease in decisions to engage in potentially rewarding activities. This combination also reduces the possibilities for model-free trial-and-error learning. As a result, action values are no longer updated and become increasingly dependent on model-based simulation processes, which are, however, influenced by the negative affective biases. Over time, this vicious cycle perpetuates or even exacerbates the symptoms of depression.

13.3.2 Clinical Theories

The following clinical theories attempt to understand the key factors underlying the development and maintenance of depression. However, conclusions can also be drawn from these theories regarding adverse DM associated with the disorder.

13.3.2.1 Cognitive Theory

Cognitive theory was formulated by Beck in the 1960s and 1970s and has since been supported by extensive empirical research (e.g., in a review by Haaga et al. 1991). This theory is widely accepted as a way of explaining depression and altered cognitive processes, including impaired DM behavior, that are characteristic of the disorder. Similar to the model-building process that is described in computational theories (cf. 3.1.3), the basic premise of the cognitive approach is that individuals with depression have an entrenched negative cognitive bias that influences their information processing. This cognitive bias amplifies negative experiences and interpretations, while positive experiences and memories are unnoticed or suppressed (Beck 2002). According to Beck, the experience of early adverse life events creates a cognitive vulnerability to depression, as a negative attitude arises, which is activated with every further negative experience and solidifies into a dysfunctional attitude (schema) (Beck 2008) – an information process comparable to the experience-based development of a model in model-based DM. This negative processing system thereby encompasses three core negative beliefs: about the self, the world and the future (known as the "cognitive triad"). These negative biases create a cognitive vulnerability to depression.

Furthermore, schemas produce cognitive distortions such as arbitrary conclusion, selective abstraction, overgeneralization, magnification and minimization, and absolutist (all-or-nothing) thinking that maintain depressive symptoms (Beck 1963). As a consequence of cognitive schemas, behavioral and motivational schemas are triggered, which promote the symptoms of depression such as sleep, appetite, or concentration disturbances (Beck 2008). Information processing is also controlled by the network of schemas and resources for adaptive processes such as coping and problem solving are diminished. The internalization of a schema can also have far-reaching behavioral and decisional consequences by creating negative automatic thoughts. For example, individuals with a core belief that they are unlovable, may assume that nobody will talk to them at a party. As a result of this assumption, triggered by automatic negative thoughts, they may choose to not attend the party (Hyland and Boduszek 2012). In the context of prospect theory (cf. 3.1.1), these automated negative thoughts might also be regarded as the reason why possible outcomes are evaluated differently than in a stable emotional situation, which in turn influences DM. Considering dual-process theory, the automated negative thoughts could be, in turn, interpreted as the result of uncorrected biased associative processing (cf. 3.1.2).

13.3.2.2 Rational-Emotive Behavior Theory

Another cognitive model of psychopathology is Ellis' rational-emotive behavior theory (REBT). According to Ellis, rational beliefs can be turned into rigid and extreme irrational beliefs that significantly influence emotional experience (Ellis 1989). Consequently, Ellis originally referred to his theory as "rational-emotive theory". In his later work, however, Ellis increasingly acknowledged the significant influence of these beliefs on a person's DM and behavior and renamed his theory "rational-emotive behavior theory" (Ellis 1999).

Ellis's theory is based on the ABC model (Ellis 1991), which states that an activating event (A) triggers a reaction with a person's beliefs (B). Depending on the beliefs being either rational or irrational, the cognitive, emotional, and behavioral consequences (C) arise in a functional and healthy or a dysfunctional and maladaptive way. According to Ellis, the notions "rational" and "irrational" describe whether a belief is based on empirical evidence as well as on logically sound and pragmatic reasoning, or not. Rational beliefs are flexible and non-extreme, while irrational beliefs are rigid and extreme. Hence, the basic thesis is that it is not the event itself but the beliefs about it that influence a person's emotional reactions and resulting behavior. This might be linked to dual-process theory in that irrational beliefs might be similar to uncorrected biased associative processing (cf. 3.1.2).

Furthermore, Ellis stated that people have the disposition to want to achieve their goals and ambitions, and that there is a tendency for these initially flexible desires to develop into rigid, absolute, and dogmatic "demands". According to Ellis, this is the primary irrational mechanism that forms the core of psychological disturbances, called demandingness (David et al. 2009). This underscores secondary irrational appraisal beliefs such as catastrophizing, low frustration tolerance, and depreciation. In response to an activating event, the primary and secondary beliefs create cognitive distortions, negative emotions, and maladaptive behaviors, which are considered symptoms of various psychological disorders. Hence, according to REBT, adverse DM is primarily caused by the existence of unreflected irrational beliefs.

While cognitive theory and REBT share many similarities, as noted by Ellis (2003), e.g., they are both framed around the ABC model (Hyland and Boduszek 2012), there are a few key differences between the theories. First, it is argued that Beck's cognitive theory pursues an empirical approach and is based on a more trial-guided foundation while REBT is rooted in philosophical reflections (Padesky and Beck 2003). Moreover, while REBT is transdiagnostic and avoids detailed explanations of disorder-specific psychopathologies, Beck has applied his general theory to individual disorders and developed disorder-specific adaptations of his general theory on the basis of empirical findings (Matweychuk et al. 2019).

Another key difference between the two theories can be seen by using the approach of cold and hot cognition first postulated by Abelson and Rosenberg (1958). Cold cognition describes the subjective knowledge of an event, i.e., the description, interferences and attributes, while hot cognition describes the evaluation of this mental representation, i.e., the way in which a person further processes the cold cognition (David and Szentagotai 2006). According to Beck's theory, cold cognition is what influences the dysfunctionality of cognitive, emotional and behavioral responses, while according to Ellis, these responses only arise through the persons appraisal of the event, the hot cognition (Hyland and Boduszek 2012).

Another theoretical distinction pointed out by Hyland and Boduszek (2012) lies in the way irrational beliefs seem to be structured in the two models. In cognitive theory, the core of dysfunctional responses lies in depreciation. In other words, negative schemas lead to distorted information processing that reinforce the negative beliefs and can thus cause psychopathology. According to REBT, psychopathology originally arises from a person's entrenched ambitions, demandingness, which are mediated by secondary factors such as depreciation and can cause dysfunctional responses in reaction to an event. Both theories form the basis of the most commonly used method for the treatment of mental disorders, cognitive behavioral therapy (Hyland and Boduszek 2012).

13.3.2.3 Learned Helplessness

Learned helplessness is a psychological concept that refers to the conviction that, as a result of negative experiences, one has lost the ability to change one's own situation (Maier and Seligman 2016). The term was first coined after animal experiments by Martin Seligman and Steven Maier who showed that prior conditioning, in which dogs had no influence on a repeated aversive outcome (e.g., an electroshock), resulted in lethargy and passive behavior in a new experimental setup (Seligman 1975). Previous learning appeared to prevent the search for a solution in the new situation, in which an escape of the aversive environment would have been possible. This finding was replicated in analogous experimental setups with humans that involved, e.g., loud noises or unsolvable puzzles (Hiroto and Seligman 1975). Seligman used this as a laboratory model of clinical depression and found evidence for most of the nine DSM-5 symptoms in the combination of animal and human experiments (Maier and Seligman 2016). It was assumed that a lack of contingency between action and outcome can lead to a loss of control and the conviction that nothing can be changed in a situation anyway, resulting in passive behavior. While the idea that adverse life events favor the development of depression is similar between Seligman's and Beck's cognitive theories, this is the central difference between the theories, as Beck would argue that the self-attribution of a negative event leads to depression and not the lack of control over it (Blaney 1977). The hypothesis was later altered, since the passive behavior displayed in the experiments was only momentary (Abramson et al. 1978). The extent of helplessness seemed to be determined by the explanations that an individual finds for the cause of an aversive event. That is, only when the inability to find a solution is globally, internally, and permanently attributed, an individual will display helplessness in the long-term and give up trying across different situations. This view more closely aligns with the assumptions of cognitive theory and the development of schemas. Countering learned helplessness is the approach of depressive realism which states that most people are prone to optimistic bias and that depressed individuals have a more accurate perception of their personal control (Alloy and Abramson 1979). However, this offers mixed empirical support (Dev et al. 2022). Nowadays, after researching the neural circuits behind learned helplessness, Seligman concluded that shock and passivity are not learned but rather the default reaction to ongoing adverse events (Maier and Seligman 2016).

13.4 Possible Interventions and Treatment Strategies

While adverse DM is clearly a concomitant problem or in some cases even a maintaining factor in depression, only a few treatment approaches have been developed that specifically address the treatment of adverse DM in depression. Instead, adverse DM, in the context of depression, has been implicitly addressed as part of general therapeutic interventions and broader cognitive impairments. Due to their stronger orientation towards MDD pathomechanisms and their greater empirical basis in this area in contrast to computational theories, most therapeutic approaches are based on clinical theories. In the following section, some of these therapeutic interventions are presented, whereby most of them can be assigned to cognitive behavioral therapy.

13.4.1 General Recommendations for Decision-Making Behavior During a Depressive Episode

As already described, in the context of MDD, adverse decisions (e.g., premature termination of a relationship) are often made during an acute MDD phase which are then regretted after the acute episode has subsided. Unfortunately, these adverse decisions are often accompanied by considerable personal, social or economic damage (e.g., loss of a key stabilizing factor). Accordingly, clinicians recommend that, if possible, no critical or life-altering decisions should be made during an acute depressive episode (Leykin et al. 2011) and, in particular, impulsive, emotion-driven actions are discouraged. Instead, if decisions are required, they should not be made on one's own, but be discussed with a trusted person or therapist. Additionally, while emotions are, of course, an integral part of DM, the rational basis of the decision should be reviewed and ensured. For this purpose, it may also be advisable to use appropriate DM aids to put the DM process on a rational foundation (Leykin et al. 2011).

13.4.2 Practice of Problem-Solving and Self-management Skills

As with other mental illnesses, the problem-solving and self-management skills of people with MDD are often severely impaired, which in many cases exacerbates or maintains the depressive symptoms. However, various interventions have been developed to address these skill deficits, with which the patient can learn to improve their problem-solving and self-control skills.

With regards to problem solving, individual concepts from solution-oriented therapy (Gingerich and Peterson 2013), for example, have proven their merit. This form of therapy places particular emphasis on recognizing and using a person's resources and strengths. The focus is less on analyzing the causes of problems and more on identifying and implementing solutions to existing problems. Furthermore, problem-solving therapy (PST) has been shown to be an effective treatment for MDD (Cuijpers et al. 2018). This intervention involves training on specific problem-solving behavior and developing an adaptive worldview towards problems encountered in life (e.g., optimism, positive self-efficacy) (D'Zurilla and Nezu 2010). Importantly, the problem-solving skills are thought to be general and applicable to all problems. PST consists of 14 modules with various topics, including "Decision Making", concerned with identifying consequences of decisions and conducting cost–benefit analyses of actions/decisions (D'Zurilla and Nezu 2010).

In order to improve a patient's self-control skills, self-management therapy (Rehm and Adams 2004) is a cognitive-behavioral treatment program for MDD primarily delivered in a group format. It focuses on poor self-control skills, such as heightened attention to negative stimuli, maladaptive attribution styles, and the preference for working towards immediate instead of long-term consequences of behavior (Rehm 1984). It includes weekly exercises and homework, such as self-monitoring of positive activities and self-statements, investigating negative short- and positive long-term consequences of behavior, illuminating depressogenic attribution styles, as well as setting goals and how to use rewards to influence one's behavior (Rehm 1984).

In addition, there are various non-clinical coaching tools for both competence domains, which can also be used to improve patients' self-management and problem-solving skills. One example would be the S.M.A.R.T. rule introduced by George T. Doran in project management (Doran 1981), according to which objectives should be formulated in such a way that they are specific, measurable, achievable, reasonable, and time-bound.

13.4.3 Behavioral Plans

Behavioral plans or contracts are written agreements between a patient and therapist, in which the patient makes a commitment to themselves and/or the therapist to behave in a clearly defined way in a defined life situation (Hautzinger 2015). A typical example would be a behavioral activation intervention in which the patient sets specific goals to increase their activity (Hopko et al. 2003), for instance, to go out for at least 1 h every day regardless of their mental state. Another example would be a crisis-response-plan in suicide prevention that, in a step-by-step format, specifies precisely what the patient will do in a crisis, for instance, how to identify warning-signs, which pre-defined self-coping mechanisms to use, and under which conditions to seek professional help (Bryan et al. 2017; Stanley et al. 2018). The implicit background assumption underlying behavioral contracts is that patients in mental crisis, such as when experiencing an acute depressive episode, are not able to make decisions appropriate to the situation and that it is therefore helpful if the patient and therapist already work out and define possible behaviors in a healthy state, which can then be used in the crisis situation.

13.4.4 Cognitive Restructuring

Cognitive restructuring was developed on the basis of the aforementioned cognitive theory according to Beck, the rational-emotive theory according to Ellis as well as other theoretical approaches (Beck 1979; Ellis 2003). Depending on the underlying theoretical approach, cognitive restructuring is carried out by slightly different methods. Most cognitive restructuring methods, however, share the goal of identifying irrational, dysfunctional thought patterns and replacing them by more functional ones, in order to promote positive changes in emotional experiences and behavior (Clark 2013). This implies an improvement of the patient's DM behavior by dissolving cognitive distortions and irrational beliefs and replacing them with more situation-appropriate, functional cognition that drive behavior (Ellis 2003). An example based on clinical practice would be a person with MDD that is convinced that other people reject him or her, although in reality this is not the case. Through cognitive restructuring (e.g., by means of a socratic dialogue, for a review see Clark and Egan 2015), the person could now learn to recognize the irrational nature of this belief and to replace the belief with a more realistic view and to acknowledge positive social interactions and supportive relationships. As a consequence, this could lead to the person becoming more open to social interactions and making more positive choices regarding interpersonal relationships. Other typical cognitive distortions that might be worked on are, for example, negative self-referent schemas that can lead to excessive self-criticism, pessimistic expectations of the future, the feeling of hopelessness, and the tendency to overgeneralize negative and systematically ignore positive events (Clark 2013).

13.4.5 Improvement of Self-efficacy Expectations

An important therapeutic intervention derived from Seligman's above-described learned helplessness model is the promotion of the MDD patient's own self-efficacy expectations. This involves regaining or strengthening a person's conviction that they are able to influence their environment and bring about positive changes to their life (Bandura 1977). Self-efficacy expectations can thereby be improved through various techniques. Typical examples based on clinical practice include setting realistic and clearly verifiable goals, practicing positive self-instructions or consciously recognizing small successes. Behavioral activation, calling in social support, and exposing oneself to increasingly challenging tasks also play a decisive role.

13.5 Conclusion

The comparison of the different models and theories discussed in this chapter illustrates that there are several pathways through which deficits in DM can affect the lives of people with MDD. As a result, symptoms can be exacerbated, which may have a significant impact on the personal and social aspects of those affected.

The explanatory approaches to understanding these deficits are broadly divided into computational theories and clinical theories. Computational theories offer insights into the cognitive processes underlying DM in the context of MDD. Clinical theories, on the other hand, focus on psychological and emotional aspects that may contribute to the negative effects of DM.

Therapeutic treatments for depression may not yet adequately address strategies to improve DM capacities. Such interventions and treatment strategies may include general recommendations on DM behavior during depressive episodes, practicing self-management and problem-solving skills, behavioral contracts, cognitive restructuring, and improving self-efficacy expectations. A stronger therapeutic focus on these aspects could ultimately contribute towards better management of the symptoms of depression in the future. Furthermore, findings from computational theories have not yet been taken into account in the development of therapeutic interventions for MDD, which could lead to innovative approaches on a deeper cognitive level.

References

Abelson RP, Rosenberg MJ (1958) Symbolic psychologic: a model of attitudinal cognition. Behav Sci 3:1–13

Abramson LY, Seligman ME, Teasdale JD (1978) Learned helplessness in humans: critique and reformulation. J Abnorm Psychol 87(1):49–74. https://doi.org/10.1037/0021-843X.87.1.49

Alloy LB, Abramson LY (1979) Judgment of contingency in depressed and nondepressed students: sadder but wiser? J Exp Psychol Gen 108(4):441–485. https://doi.org/10.1037/0096-3445.108.4.441

American psychiatric association (2013) Diagnostic and statistical manual of mental disorders (DSM-5®). American Psychiatric Publishing

Bandura A (1977) Self-efficacy: toward a unifying theory of behavioral change. Psychol Rev 84(2):191–215. https://doi.org/10.1037/0033-295X.84.2.191

Beck AT (2002) Cognitive models of depression. Clin Adv Cogn Psychother: Theory Appl 14(1):29–61

Beck AT (2008) The evolution of the cognitive model of depression and its neurobiological correlates. Am J Psychiatry 165(8):969–977. https://doi.org/10.1176/appi.ajp.2008.08050721

Beck AT (1963) Thinking and depression: I. idiosyncratic content and cognitive distortions. Arch Gen Psychiatry 9(4):324–333. https://doi.org/10.1001/archpsyc.1963.01720160014002

Beck AT (1979) Cognitive therapy of depression. Guilford press

Beevers CG (2005) Cognitive vulnerability to depression: a dual process model. Clin Psychol Rev 25(7):975–1002. https://doi.org/10.1016/j.cpr.2005.03.003

Bishop SJ, Gagne C (2018) Anxiety, depression, and decision making: a computational perspective. Annu Rev Neurosci 41(1):371–388. https://doi.org/10.1146/annurev-neuro-080317-062007

Blaney PH (1977) Contemporary theories of depression: critique and comparison. J Abnorm Psychol 86(3):203–203

Borsini A, Wallis ASJ, Zunszain P, Pariante CM, Kempton MJ (2020) Characterizing anhedonia: a systematic review of neuroimaging across the subtypes of reward processing deficits in depression. Cogn Affect Behav Neurosci 20(4):816–841. https://doi.org/10.3758/s13415-020-00804-6

Bryan CJ, Mintz J, Clemans TA, Leeson B, Burch TS, Williams SR, Maney E, Rudd MD (2017) Effect of crisis response planning vs. contracts for safety on suicide risk in U.S. army soldiers: a randomized clinical trial. J Affect Disord 212:64–72. https://doi.org/10.1016/j.jad.2017.01.028

Chen C, Takahashi T, Nakagawa S, Inoue T, Kusumi I (2015) Reinforcement learning in depression: a review of computational research. Neurosci Biobehav Rev 55:247–267. https://doi.org/10.1016/j.neubiorev.2015.05.005

Choung Y, Chatterjee S, Pak T-Y (2022) Depression and financial planning horizon. J Behav Exp Econ 98:101877. https://doi.org/10.1016/j.socec.2022.101877

Clark GI, Egan SJ (2015) The Socratic method in cognitive behavioural therapy: a narrative review. Cogn Ther Res 39(6):863–879. https://doi.org/10.1007/s10608-015-9707-3

Clark DA (2013) Cognitive restructuring. In: The Wiley handbook of cognitive behavioral therapy (S. 1–22). John Wiley & Sons, Ltd. https://doi.org/10.1002/9781118528563.wbcbt02

Cobb-Clark DA, Dahmann SC, Kettlewell N (2022) Depression, risk preferences, and risk-taking behavior. J Hum Resour 57(5):1566–1604. https://doi.org/10.3368/jhr.58.1.0419-10183R1

Cuijpers P, de Wit L, Kleiboer A, Karyotaki E, Ebert DD (2018) Problem-solving therapy for adult depression: an updated meta-analysis. Eur Psychiatry: J Assoc Eur Psychiatr 48:27–37. https://doi.org/10.1016/j.eurpsy.2017.11.006

D'Zurilla TJ, Nezu AM (2010) Problem-solving therapy. Handbook of cognitive-behavioral therapies 3:197–225

David D, Szentagotai A (2006) Cognitions in cognitive-behavioral psychotherapies; toward an integrative model. Clin Psychol Rev 26(3):284–298. https://doi.org/10.1016/j.cpr.2005.09.003

David D, Lynn SJ, Ellis A (2009) Rational and irrational beliefs: research, theory, and clinical practice. Oxford University Press

Daw ND, Niv Y, Dayan P (2005) Uncertainty-based competition between prefrontal and dorsolateral striatal systems for behavioral control. Nat Neurosci 8(12). https://doi.org/10.1038/nn1560

Destoop M, Schrijvers D, De Grave C, Sabbe B, De Bruijn ERA (2012) Better to give than to take? Interactive social decision-making in severe major depressive disorder. J Affect Disord 137(1–3):98–105. https://doi.org/10.1016/j.jad.2011.12.010

Dev AS, Moore DA, Johnson SL, Garrett KT (2022) Sadder ≠ wiser: depressive realism is not robust to replication. Collabra: Psychol 8(1):38529. https://doi.org/10.1525/collabra.38529

Doran GT (1981) There's a SMART way to write management's goals and objectives. Manage Rev 70(11):35–36

Ellis A (1989) Rational psychotherapy. TACD J 17(1):67–80. https://doi.org/10.1080/1046171X.1989.12034348

Ellis A (1991) The revised ABC's of rational-emotive therapy (RET). J Rational-Emot Cognitive Behav Ther 9(3):139–172. https://doi.org/10.1007/BF01061227

Ellis A (1999) Why rational-emotive therapy to rational emotive behavior therapy? Psychotherapy: Theory. Res, Pract, Train 36(2):154–154

Ellis A (2003) Cognitive restructuring of the disputing of irrational beliefs. Cogn Behav Ther: Appl EmpirLy Support Tech Your Pract, pp 79–83

Evans JSBT (1984) Heuristic and analytic processes in reasoning. Br J Psychol 75(4):451–468. https://doi.org/10.1111/j.2044-8295.1984.tb01915.x

Evans JSt. BT, Stanovich KE (2013) Dual-process theories of higher cognition: advancing the debate. Perspect Psychol Sci 8(3):223–241. https://doi.org/10.1177/1745691612460685

Ferrari AJ, Somerville AJ, Baxter AJ, Norman R, Patten SB, Vos T, Whiteford HA (2013) Global variation in the prevalence and incidence of major depressive disorder: a systematic review of

the epidemiological literature. Psychol Med 43(3):471–481. https://doi.org/10.1017/S0033291712001511
Forgas JP (2000) Managing moods: toward a dual-process theory of spontaneous mood regulation. Psychol Inq 11(3):172–177
Gingerich WJ, Peterson LT (2013) Effectiveness of solution-focused brief therapy: a systematic qualitative review of controlled outcome studies. Res Soc Work Pract 23(3):266–283. https://doi.org/10.1177/1049731512470859
Gotlib IH, Joormann J (2010) Cognition and depression: current status and future directions. Annu Rev Clin Psychol 6:285–312
Haaga DA, Dyck MJ, Ernst D (1991) Empirical status of cognitive theory of depression. Psychol Bull 110(2):215–215
Halahakoon DC, Kieslich K, O'Driscoll C, Nair A, Lewis G, Roiser JP (2020) Reward-processing behavior in depressed participants relative to healthy volunteers: a systematic review and meta-analysis. JAMA Psychiat 77(12):1286–1295. https://doi.org/10.1001/jamapsychiatry.2020.2139
Hardeveld F, Spijker J, De Graaf R, Nolen WA, Beekman ATF (2013) Recurrence of major depressive disorder and its predictors in the general population: results from the Netherlands mental health survey and incidence study (NEMESIS). Psychol Med 43(1):39–48. https://doi.org/10.1017/S0033291712002395
Hautzinger M (2015) Verhaltensverträge. In: Linden M, Hautzinger M (Hrsg.), Verhaltenstherapiemanual (S. 299–301), Springer. https://doi.org/10.1007/978-3-642-55210-6_63
Hindmarch T, Hotopf M, Owen GS (2013) Depression and decision-making capacity for treatment or research: a systematic review. BMC Med Ethics 14(1):54. https://doi.org/10.1186/1472-6939-14-54
Hiroto DS, Seligman ME (1975) Generality of learned helplessness in man. J Pers Soc Psychol 31(2):311–327. https://doi.org/10.1037/h0076270
Hodgkinson GP, Healey MP (2011) Psychological foundations of dynamic capabilities: reflexion and reflection in strategic management. Strateg Manag J 32(13):1500–1516. https://doi.org/10.1002/smj.964
Hopko DR, Lejuez CW, Ruggiero KJ, Eifert GH (2003) Contemporary behavioral activation treatments for depression: procedures, principles, and progress. Clin Psychol Rev 23(5):699–717. https://doi.org/10.1016/S0272-7358(03)00070-9
Horne SJ, Topp TE, Quigley L (2021) Depression and the willingness to expend cognitive and physical effort for rewards: a systematic review. Clin Psychol Rev 88:102065. https://doi.org/10.1016/j.cpr.2021.102065
Huys QJM, Daw ND, Dayan P (2015) Depression: a decision-theoretic analysis. Annu Rev Neurosci 38:1–23. https://doi.org/10.1146/annurev-neuro-071714-033928
Hyland P, Boduszek D (2012) Resolving a difference between cognitive therapy and rational emotive behaviour therapy: towards the development of an integrated CBT model of psychopathology. Ment Health Rev J 17(2):104–116. https://doi.org/10.1108/13619321211270425
Kahneman D, Tversky A (1979) Prospect theory: an analysis of decisions under risk. Econometrica 47:278
Katon W, Richardson L, Russo J, McCarty CA, Rockhill C, McCauley E, Richards J, Grossman DC (2010) Depressive symptoms in adolescence: the association with multiple health risk behaviors. Gen Hosp Psychiatry 32(3):233–239. https://doi.org/10.1016/j.genhosppsych.2010.01.008
Kraines MA, Krug CP, Wells TT (2017) Decision justification theory in depression: regret and self-blame. Cogn Ther Res 41(4):556–561. https://doi.org/10.1007/s10608-017-9836-y
Kuehner C (2017) Why is depression more common among women than among men? Lancet Psychiatry 4(2):146–158. https://doi.org/10.1016/S2215-0366(16)30263-2
Kühberger A (1998) The influence of framing on risky decisions: a meta-analysis. Organ Behav Hum Decis Process 75(1):23–55. https://doi.org/10.1006/obhd.1998.2781
Kumar P, Goer F, Murray L, Dillon DG, Beltzer ML, Cohen AL, Brooks NH, Pizzagalli DA (2018) Impaired reward prediction error encoding and striatal-midbrain connectivity in depression.

Neuropsychopharmacol: Off Publ Am Coll Neuropsychopharmacol 43(7):1581–1588. https://doi.org/10.1038/s41386-018-0032-x

Kupferberg A, Bicks L, Hasler G (2016) Social functioning in major depressive disorder. Neurosci Biobehav Rev 69:313–332. https://doi.org/10.1016/j.neubiorev.2016.07.002

Lehrer JA, Shrier LA, Gortmaker S, Buka S (2006) Depressive symptoms as a longitudinal predictor of sexual risk behaviors among US middle and high school students. Pediatrics 118(1):189–200. https://doi.org/10.1542/peds.2005-1320

Lerner JS, Li Y, Weber EU (2013) The financial costs of sadness. Psychol Sci 24(1):72–79. https://doi.org/10.1177/0956797612450302

Leykin Y, Roberts CS, DeRubeis RJ (2011) Decision-making and depressive symptomatology. Cogn Ther Res 35(4):333–341. https://doi.org/10.1007/s10608-010-9308-0

Lu J, Zhao X, Wei X, He G (2024) Risky decision-making in major depressive disorder: a three-level meta-analysis. Int J Clin Health Psychol 24(1):100417. https://doi.org/10.1016/j.ijchp.2023.100417

Maier SF, Seligman MEP (2016) Learned helplessness at fifty: insights from neuroscience. Psychol Rev 123(4):349–367. https://doi.org/10.1037/rev0000033

Marx W, Penninx BWJH, Solmi M, Furukawa TA, Firth J, Carvalho AF, Berk M (2023) Major depressive disorder. Nat Rev Dis Prim 9(1). https://doi.org/10.1038/s41572-023-00454-1

Matweychuk W, DiGiuseppe R, Gulyayeva O (2019) A comparison of REBT with other cognitive behavior therapies. Adv REBT: Theory, Pract, Res, Meas, Prev Promot, pp 47–77

McIntyre RS, Cha DS, Soczynska JK, Woldeyohannes HO, Gallaugher LA, Kudlow P, Alsuwaidan M, Baskaran A (2013) Cognitive deficits and functional outcomes in major depressive disorder: determinants, substrates, and treatment interventions. Depress Anxiety 30(6):515–527

Moffitt TE, Caspi A, Taylor A, Kokaua J, Milne BJ, Polanczyk G, Poulton R (2010) How common are common mental disorders? Evidence that lifetime prevalence rates are doubled by prospective versus retrospective ascertainment. Psychol Med 40(6):899–909. https://doi.org/10.1017/S0033291709991036

Ng TH, Alloy LB, Smith DV (2019) Meta-analysis of reward processing in major depressive disorder reveals distinct abnormalities within the reward circuit. Transl Psychiatry 9(1). https://doi.org/10.1038/s41398-019-0644-x

Padesky CA, Beck AT (2003) Science and philosophy: comparison of cognitive therapy and rational emotive behavior therapy. J Cogn Psychother 17(3):211–224. https://doi.org/10.1891/jcop.17.3.211.52536

Pike AC, Robinson OJ (2022) Reinforcement learning in patients with mood and anxiety disorders vs control individuals: a systematic review and meta-analysis. JAMA Psychiat 79(4):313–322. https://doi.org/10.1001/jamapsychiatry.2022.0051

Pozuelo JR, Desborough L, Stein A, Cipriani A (2022) Systematic review and meta-analysis: depressive symptoms and risky behaviors among adolescents in low- and middle-income countries. J Am Acad Child Adolesc Psychiatry 61(2):255–276. https://doi.org/10.1016/j.jaac.2021.05.005

Prietzel TT (2020) The effect of emotion on risky decision making in the context of prospect theory: a comprehensive literature review. Manag Rev Q 70(3):313–353. https://doi.org/10.1007/s11301-019-00169-2

Pulcu E, Thomas EJ, Trotter PD, McFarquhar M, Juhasz G, Sahakian BJ, Deakin JFW, Anderson IM, Zahn R, Elliott R (2015) Social-economical decision making in current and remitted major depression. Psychol Med 45(6):1301–1313. https://doi.org/10.1017/S0033291714002414

Raghunathan R, Pham MT (1999) All negative moods are not equal: motivational influences of anxiety and sadness on decision making. Organ Behav Hum Decis Process 79(1):56–77. https://doi.org/10.1006/obhd.1999.2838

Rehm LP (1984) A self-management therapy program for depression. Int J Ment Health 13(3–4):34–53. https://doi.org/10.1080/00207411.1984.11448975

Rehm LP, Adams JH (2004) 58 Self-management. Cogn Behav Ther 466

Richard-Devantoy S, Olié E, Guillaume S, Courtet P (2016) Decision-making in unipolar or bipolar suicide attempters. J Affect Disord 190:128–136. https://doi.org/10.1016/j.jad.2015.10.001

Ridley M, Rao G, Schilbach F, Patel V (2020) Poverty, depression, and anxiety: Causal evidence and mechanisms. Science 370(6522):eaay0214. https://doi.org/10.1126/science.aay0214

Sastre-Buades A, Alacreu-Crespo A, Courtet P, Baca-Garcia E, Barrigon ML (2021) Decision-making in suicidal behavior: a systematic review and meta-analysis. Neurosci Biobehav Rev 131:642–662. https://doi.org/10.1016/j.neubiorev.2021.10.005

Seligman ME (1975) Helplessness: on depression, development, and health. WH Freeman

Solmi M, Radua J, Olivola M, Croce E, Soardo L, Salazar de Pablo G, Il Shin J, Kirkbride JB, Jones P, Kim JH, Kim JY, Carvalho AF, Seeman MV, Correll CU, Fusar-Poli P (2022) Age at onset of mental disorders worldwide: Large-scale meta-analysis of 192 epidemiological studies. Mol Psychiatry 27(1). https://doi.org/10.1038/s41380-021-01161-7

Stanley B, Brown GK, Brenner LA, Galfalvy HC, Currier GW, Knox KL, Chaudhury SR, Bush AL, Green KL (2018) Comparison of the safety planning intervention with follow-up vs usual care of suicidal patients treated in the emergency department. JAMA Psychiat 75(9):894–900. https://doi.org/10.1001/jamapsychiatry.2018.1776

Sullivan PF, Neale MC, Kendler KS (2000) Genetic epidemiology of major depression: review and meta-analysis. Am J Psychiatry 157(10):1552–1562. https://doi.org/10.1176/appi.ajp.157.10.1552

van Randenborgh A, de Jong-Meyer R, Hüffmeier J (2010) Decision making in depression: differences in decisional conflict between healthy and depressed individuals. Clin Psychol Psychother 17(4):285–298. https://doi.org/10.1002/cpp.651

Wason PC, Evans JSTBT (1974) Dual processes in reasoning? Cognition 3(2):141–154. https://doi.org/10.1016/0010-0277(74)90017-1

World health organization (2017) Depression and other common mental disorders: Global health estimates: Bd. No. WHO/MSD/MER/2017.2. World Health Organization

World health organization (2022) World mental health report: transforming mental health for all: executive summary

Chapter 14
Applying Reinforcement Learning to the Psychopathology of Obsessive–Compulsive and Gambling Disorders: Practices and Pitfalls in Computational Model Fitting

Shinsuke Suzuki and Kentaro Katahira

Abstract Computational modeling is gaining increasing popularity in the field of psychiatry. By applying a machine learning framework known as reinforcement learning (RL), researchers in the domain of computational psychiatry have begun to shed light on the latent decision-making processes that are altered in mental disorders. In this chapter, we first explain how RL can be used to characterize the neurocomputational processes of decision-making and learning. To illustrate this approach, we focus on a set of studies that employ a basic class of RL, known as model-free RL, to gain insights into the psychopathology of obsessive–compulsive and gambling disorders. Lastly, we review a common model-fitting procedure used in computational psychiatry research and offer a practical guide, also considering potential pitfalls. This encompasses discussions on estimating model parameters (parameter estimation), comparing multiple models based on their goodness of fits (model comparison), and testing the robustness of model-fitting procedures using synthesized data (model and parameter recovery analyses). Given the rapid growth of this area of study, we believe that these discussions will be of great value to authors, reviewers, and readers of computational psychiatry literature.

Keywords Computational psychiatry · Reinforcement learning · Obsessive-compulsive disorder · Gambling disorder · Model fitting · Functional MRI

S. Suzuki (✉)
Faculty of Social Data Science, Hitotsubashi University, Tokyo, Japan
e-mail: shinsuke.szk@gmail.com

HIAS Brain Research Center, Hitotsubashi University, Tokyo, Japan

Centre for Brain, Mind and Markets, The University of Melbourne, Melbourne, Australia

K. Katahira
Human Informatics and Interaction Research Institute, National Institute of Advanced Industrial Science and Technology (AIST), Tokyo, Japan

U. Ettinger et al. (eds.), *Decision Making*, Studies in Neuroscience, Psychology and Behavioral Economics, https://doi.org/10.1007/978-3-032-00880-0_14

14.1 Introduction

Learning to make good decisions is a process vital for our quality of life and survival. Let us consider a simple scenario. Imagine that you have relocated to Australia from another country. Close to your new residence, you find two major supermarket chains: Coles and Woolworths. Which would you select for your regular grocery shopping? Initially, you might not know which supermarket to visit, but with experience, you will learn which one caters to your shopping needs best. This kind of learning is prevalent in diverse scenarios, from animal foraging to human financial decisions, highlighting its crucial role in our life.

A machine learning framework, known as reinforcement learning (RL), offers an explanation for the process of learning in sequential decision-making (Dayan and Daw 2008; Rangel et al. 2008; Sutton and Barto 1998). The RL framework posits that an individual assigns a "value" to each of the available options and selects one based on that value. Revisiting the example presented above through the lens of this framework, you would compute the values of Coles and Woolworths and select the supermarket with the higher value. After making a choice, you would update the value of the chosen option based on your experience. For instance, if you choose Coles and have a positive experience, you would increase the value assigned to Coles. Specifically, the learning process is driven by a reward prediction error: the discrepancy between the amount of reward obtained and the current value of the chosen option. The prediction error is positive and negative for better- and worse-than-expected outcomes, respectively, and the error is zero for outcomes that are exactly equal to the expected outcome (see Sect. 14.2 for the mathematical description). Crucially, such a learning process enables individuals to estimate the true values of the available options, thus leading to optimal decision-making.

Accumulating evidence in neuroscience suggests that reward prediction errors are encoded in dopaminergic neural activities in the basal ganglia and other brain regions (Glimcher 2011; Schultz et al. 1997). In humans, functional neuroimaging studies have consistently shown covariation between reward prediction errors and blood-oxygenation-level-dependent (BOLD) signals in the ventral striatum, a major projection site of dopaminergic neurons in the midbrain, as well as other brain loci including the dorsal striatum, medial prefrontal cortex, and anterior insula in diverse contexts (Corlett et al. 2022; O'Doherty et al. 2004; Suzuki et al. 2012).

By relating the formal RL framework to behavior, researchers can infer the latent characteristics of decision-makers that are not directly observable (Daw 2011; Wilson and Collins 2019). Examples of these characteristics include how new information influences value updating and future behavior, the level of noise in decision-making, and the extent to which past choices (regardless of their outcomes) affect behavior. Specifically, fitting RL models to behavioral data allows for the estimation of model parameters that capture those latent characteristics of decision-makers: for example, the parameters *learning rate*, *inverse temperature*, and *perseverance* reflect the extent of value updating in response to new information, the degree of noise, and the stickiness of choices, respectively (see Sect. 14.2 for details). Recent studies in the field of computational psychiatry have shed light on the latent processes of decision-making that are affected by mental disorders, utilizing RL models (Huys et al. 2016; Montague et al. 2012). For instance, patients with mood and anxiety disorders have been shown to display a greater degree of noise in decision-making, indicated by a lower inverse temperature (equivalent to reduced reward sensitivity) (Huys et al. 2013; Kunisato et al. 2012), as well as higher and lower learning rates for reward and punishment outcomes, respectively (Aylward et al. 2019; Pike and Robinson 2022). These findings offer computational explanations for certain symptoms of mood and anxiety disorders, such as anhedonia.

In this chapter, we review the procedure of model fitting in computational psychiatry research that utilizes the RL framework to gain a deeper understanding of the psychopathology of obsessive–compulsive disorder (OCD) and gambling disorder (GD). We provide a practical guide to conducting model fitting in computational psychiatry research, including discussions on estimating parameters of the models (parameter estimation), comparing each model's goodness of fit (model comparison), and verifying the robustness of fitting procedures. Given the rapid expansion of this research domain, we believe that these discussions will benefit authors, reviewers, and readers of computational psychiatry literature.

OCD is characterized by intrusive and involuntary thoughts (i.e., obsessions) that lead to repetitive behavioral rituals (i.e., compulsions). GD, previously known as pathological gambling, is characterized by persistent and recurrent problematic gambling behavior that leads to clinically significant impairment or distress. Remarkably, both OCD and GD exhibit behavioral inflexibility, that is, impairments in the ability to alter behaviors (Figee et al. 2016; Fontenelle et al. 2011; Robbins et al. 2019; Timmeren et al. 2018). Here, we focus on computational psychiatry research on OCD and GD using simple model-free RL models, while acknowledging prior research using model-based RL models (Gillan and Robbins 2014; Gillan et al. 2011; Wyckmans et al. 2019), RL models incorporating internal states (Sakai et al. 2022; Sato et al. 2023), and Bayesian RL (Kalman filter) models (Wiehler et al. 2021). These prior studies have offered considerable insights into the psychopathology of OCD and GD.

The remainder of this chapter is structured as follows. Initially, we present formal descriptions of the standard model-free RL model and its variants in Sect. 14.2. Subsequently, in Sect. 14.3, we review the computational psychiatry literature on studies that have applied model-free RL models to examine the psychopathology of OCD and GD. We then discuss common practices and pitfalls associated with RL model fitting in Sect. 14.4. The chapter concludes with a summary of the key points.

14.2 Reinforcement Learning as a Model of Decision-Making

Consider a simple experimental task known as the two-armed bandit task. In this task, participants repeatedly choose between two options, receiving a reward for each choice based on a pre-assigned hidden probability corresponding to the selected option. To maximize their rewards, participants need to learn the reward probability associated with each option based on their experiences. The RL framework is believed to offer a reasonable explanation for the latent computational processes underpinning learning and decision-making in such sequential decision-making tasks.

Conventional RL. The conventional model-free RL model supposes that an individual represents the values of the two options, $Q(A)$ and $Q(B)$, and selects the option with the higher value using a probabilistic rule. The probability of one option, for instance A, being chosen is often given by the *softmax* function:

$$P(A) = \frac{\exp(\beta Q(A))}{\exp(\beta Q(A)) + \exp(\beta Q(B))} \tag{14.1}$$

where the parameter $\beta \geq 0$ governs the degree of noise or random exploration, called *inverse temperature*. The individual then updates the value of the chosen option based on the information about reward delivery. Suppose an option A is chosen and the reward R is delivered, then the value of A is updated by the reward prediction error, δ, as follows:

$$Q(A) \leftarrow Q(A) + \alpha\delta \tag{14.2}$$

where $\delta = R - Q(A)$. Here, $0 \leq \alpha \leq 1$ denotes the *learning rate* that controls for how much the value is updated in proportion to the reward prediction error. By fitting this model to the behavioral data, we can estimate the latent characteristics of the individual: that is, learning rate (α) reflecting the extent of value update in response to new information and inverse temperature (β) representing the degree of noise in decision-making.

RL with choice perseverance. A straightforward extension of the conventional RL model that accounts for behavioral inflexibility is a model that includes "choice perseverance", which is also called "choice stickiness" or "decision inertia." RL models with "long-term perseverance" assume that an individual, when making a decision, considers not only the value of each option but also the history of choices, that is, how often s/he has chosen the corresponding option in the past. That is, the probability of A being chosen is given by

$$P(A) = \frac{\exp(\beta Q(A) + \gamma C(A))}{\exp(\beta Q(A) + \gamma C(A)) + \exp(\beta Q(B) + \gamma C(B))} \tag{14.3}$$

where $C(\cdot)$ denotes the choice trace, and $\gamma \in (-\infty, \infty)$ represents the *perseverance* controlling the degree to repeat the same choice. The choice trace for one option, for instance A, is initially set at 0 and updated as follows (same for the other option B):

$$C(A) \leftarrow C(A) + \eta(I(A) - C(A)), \tag{14.4}$$

where $I(A)$ is 1 if the option A is chosen and 0 otherwise, and $0 \leq \eta \leq 1$ is the *choice-trace decay rate*. Individuals with positive perseverance, γ, are likely to repeat a recently selected choice, leading to behavioral inflexibility.

Some studies have employed a simpler version, termed as "short-term perseverance." RL models with short-term perseverance assume that the choice trace is derived based solely on a choice in the previous trial (not over the past multiple trials: i.e., $\eta = 1$). That is, the choice trace of one option is 1 if that option was chosen in the previous trial and 0 otherwise.

RL with asymmetric learning rates. Besides choice perseverance, behavioral inflexibility can also be captured by RL models with asymmetric learning from outcomes above and below expectations. That is, differential learning rates control for the speed of learning from positive and negative reward prediction errors:

$$\alpha = \begin{cases} \alpha_{(+)} \text{if } \delta \geq 0 \\ \alpha_{(-)} \text{if } \delta < 0 \end{cases}. \tag{14.5}$$

Notably, a higher learning rate for positive reward prediction errors and/or a lower learning rate for negative reward prediction errors can lead to repetitive choice behaviors, indicative of behavioral inflexibility. A higher learning rate for positive prediction errors causes an overestimation of value from positive outcomes, resulting in the chosen option being overvalued after positive experiences. A lower learning rate for negative prediction errors leads to an under-devaluation from negative outcomes, meaning that the chosen option's value is not sufficiently decreased despite negative experiences. These dynamics promote the repetition of current choice behaviors.

RL with choice perseverance and asymmetric learning rates. We can consider RL models that incorporate both choice perseverance and asymmetric learning rates. According to a theoretical study (Katahira 2018), to disentangle the computational processes underlying behavioral inflexibility, it is necessary to compare the RL model with perseverance and that with asymmetric learning rates as well as the full model with both perseverance and asymmetric learning rates.

By fitting these RL models to behavioral data from experiments such as a two-armed bandit task, we can estimate parameters (e.g., α, $\alpha_{(+)}$, $\alpha_{(-)}$, β, γ, η) that reflect the latent characteristics of the decision-maker. In computational psychiatry research, these estimated characteristics are compared between patient groups and healthy controls, or correlated with individual differences in continuous measurements of psychiatric symptoms (e.g., psychiatric questionnaire scores).

14.3 Reinforcement Learning in OCD and GD

In a typical research program in computational psychiatry, researchers develop various RL models to analyze behavioral and neural data from patients with mental disorders, such as OCD and GD, alongside data from healthy controls. During data collection, samples of patients and healthy controls perform a sequential decision-making task (e.g., multi-armed bandit task), often while additionally undergoing functional magnetic resonance imaging (fMRI) scans. The researchers then fit the RL models to the behavioral data to compare the model fits and estimate the parameters (e.g., learning rates, perseverance). Subsequent comparisons of the parameter estimates between patients and controls provide insights into the computational mechanisms of mental disorders. Further, by associating fMRI data with latent variables (e.g., reward prediction error) derived from the best-fitting model and its parameters, the researchers can pinpoint brain regions that encode these variables differently in patients and healthy controls.

A pioneering study on RL in OCD patients (Hauser et al. 2017) evaluated four models in a two-armed bandit task: the conventional model-free RL, RL with short-term perseverance, RL with asymmetric learning rates, and RL with both perseverance and asymmetric learning rates. The findings from this study indicated that the RL model incorporating perseverance provided a better explanation of the behavioral data for both OCD patients and healthy controls compared to other models. Notably, the researchers identified a significant difference in the perseverance parameter (i.e., γ in Eq. (14.3)) between OCD patients and healthy controls, with the OCD patients exhibiting less perseverance. This finding appears paradoxical considering that behavioral inflexibility is a hallmark of OCD. Hauser and colleagues speculated that the reduced perseverance might be linked to the compulsive rechecking behaviors, a key characteristic of OCD, reflecting a pronounced urge to eliminate uncertainty by verifying outcomes of unselected alternatives. While this speculation warrants further investigation, the observed reduction of perseverance in OCD patients has been replicated in subsequent studies (Kanen et al. 2019; Marzuki et al. 2021). They also revealed that across the groups of OCD patients and healthy controls, fMRI signals in the dorsal anterior cingulate cortex (dACC) inversely correlated with reward prediction errors—higher prediction errors corresponded with diminished fMRI signals in this region. Furthermore, this inverse relationship between reward prediction errors and fMRI signals in the dACC was notably diminished in OCD patients.

Suzuki et al. (2023) explored neural computations in RL among OCD patients using a modified two-armed bandit task, where decisions to seek rewards and avoid losses were randomly interleaved across trials. Their findings demonstrated that the RL model with asymmetric learning rates more accurately predicted behavior in reward-seeking contexts for both OCD patients and healthy controls, compared to the conventional RL and RL with perseverance models. Conversely, in loss-avoidance scenarios, the RL model featuring perseverance was superior. Remarkably, the authors observed a reduced learning rate from negative reward prediction errors (i.e., $\alpha_{(-)}$ in Eq. (14.5)) in OCD patients during reward-seeking decisions, aligning partially with a previous study (Marzuki et al. 2021) but not with others (Hauser et al. 2017; Kanen et al. 2019). No significant differences in the parameters were identified in loss-avoidance decision-making between OCD patients and healthy controls. Furthermore, corresponding to the diminished learning rate from negative reward prediction errors, neural representations of the negative prediction errors in reward-seeking decision-making were found to be diminished in the dorsal striatum and dorsomedial prefrontal cortex (dmPFC)—adjacent to the dACC reported in the study by Hauser et al. (2017). Based on these results, Suzuki and colleagues speculated that OCD patients might exhibit a blunted response to negative reward prediction errors in the dorsal striatum and dmPFC, potentially contributing to the behavioral inflexibility characteristic of the disorder.

Suzuki et al. (2023) also examined GD patients using the same experimental task and analyzed the data with the RL models. They found that in reward-seeking scenarios, GD patients exhibited altered learning rates for positive and negative reward prediction errors ($\alpha_{(+)}$ and $\alpha_{(-)}$ respectively in Eq. (14.5)) compared to healthy controls. Specifically, GD patients showed increased learning rates for positive reward prediction errors and decreased rates for negative errors. The heightened learning rate for positive reward prediction errors corresponded with enhanced neural representations of positive prediction errors in the anterior insula (AI) in GD patients (compared to healthy controls). However, neural responses to negative reward prediction errors showed no significant difference between GD patients and healthy controls. The enhanced neural response to positive prediction errors in the AI broadly aligns with existing literature showing that cravings for gambling in pathological gamblers are associated with an increased AI response to gambling cues (Limbrick-Oldfield et al. 2017) and that individuals with insula damage show reduced gambling-related cognitive distortions (Clark et al. 2014). Collectively, these findings suggest that GD patients may be hypersensitive to positive reward prediction errors compared to healthy controls, which could contribute to the behavioral inflexibility observed in GD.

14.4 Practices and Pitfalls of Model Fitting

As described above, in the field of computational psychiatry, researchers fit RL models to behavioral data and specify the best-fit model and its parameter estimates. They then compare the parameter estimates between patients with mental disorders and healthy controls. In this section, we review the procedures of model fitting, also presenting some guidelines and highlighting some pitfalls.

14.4.1 Parameter Estimation

Researchers compare parameter values (e.g., perseverance and learning rates) in RL models between patients and healthy controls to better understand the latent computational mechanisms behind mental disorders. In this context, an important question emerges: How are parameter values estimated from behavioral data?

A basic concept in parameter estimation is to find values of the parameters that enable the model to accurately represent the observed data, a process known as *Maximum Likelihood Estimation* (MLE). MLE is achieved by maximizing the likelihood function, which quantifies how well the model, with a given set of parameters, explains the observed data: $P(\boldsymbol{D}|\boldsymbol{\theta})$ where $\boldsymbol{D}$ denotes the data and $\boldsymbol{\theta}$ represents the set of model parameters (see below for an example case of RL). The parameter estimates can be obtained by

$$\boldsymbol{\theta}^* = \arg\max_{\boldsymbol{\theta}} P(\boldsymbol{D}|\boldsymbol{\theta}). \tag{14.6}$$

In actual implementations, we often minimize the negative log-likelihood, which is effectively the same as maximizing the likelihood, by using functions that are built into many programming languages (e.g., *fminsearch* and *fmincon* in Matlab, *optim* in R, and *scipy.optimize.minimize* in Python).

In Bayesian estimation, the posterior probability distribution over the parameters is computed by combining the likelihood function and the prior belief about the parameters, $P(\boldsymbol{\theta})$, as follows:

$$P(\boldsymbol{\theta}|\boldsymbol{D}) = \frac{P(\boldsymbol{D}|\boldsymbol{\theta})\,P(\boldsymbol{\theta})}{\int P(\boldsymbol{D}|\boldsymbol{\theta})\,P(\boldsymbol{\theta})d\boldsymbol{\theta}}. \tag{14.7}$$

While the estimation of the posterior distribution is often computationally intensive, the software package *Stan* offers a practical solution to numerically approximate this distribution using Markov Chain Monte Carlo (MCMC) sampling methods (https://mc-stan.org/).

Individual modeling approach. Let us consider an example of the standard computational psychiatry experiments, where patients with mental disorders and healthy controls engage in a sequential decision-making task with two choice options (i.e., two-armed bandit task) over T trials. How do we estimate each participant's learning rate (α) and inverse temperature (β) in the conventional model-free RL model (see Sect. 14.2 for details of the model and its parameters)?

A straightforward method is to estimate parameters for each participant *individually*, known as the "individual modeling approach," using either MLE or Bayesian estimation. This approach posits that each participant possesses their own parameters, independent of others, which generate the choice data (see Fig. 14.1a). Suppose participant i chooses one option A (over the other option B) at trial t. Then the likelihood of the RL model on that choice $c_t^{(i)}$ is given by Eq. (14.1):

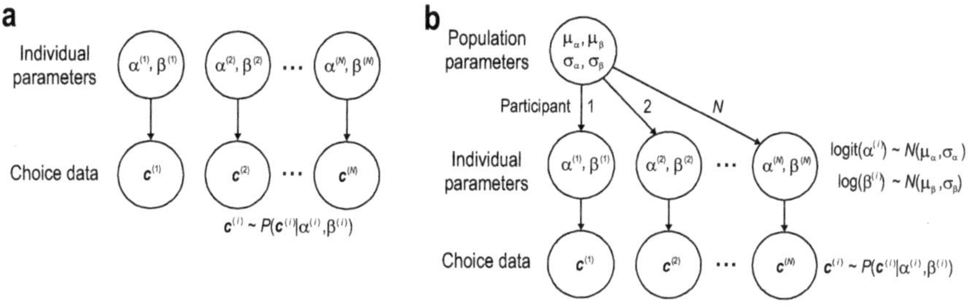

Fig. 14.1 Schematic illustration of model fitting. **a** Individual modeling approach. Each participant has an independent set of parameters (e.g., $\alpha^{(i)}$ and $\beta^{(i)}$) that generates their choice data, $\boldsymbol{c}^{(i)}$. N, the number of participants. **b** Hierarchical modeling approach. Participants' individual-level parameters ($\alpha^{(i)}$ and $\beta^{(i)}$) are drawn from common distributions with the population-level parameters (μ_α, σ_α, μ_β, and σ_β)

$$P\left(c_t^{(i)}\right) = \frac{\exp\left(\beta^{(i)} Q_t^{(i)}(A)\right)}{\exp\left(\beta^{(i)} Q_t^{(i)}(A)\right) + \exp\left(\beta^{(i)} Q_t^{(i)}(B)\right)}. \tag{14.8}$$

Note that the values of the two options $Q_t^{(i)}(A)$ and $Q_t^{(i)}(A)$ are derived from the sequence of previous choices and rewards (see Eq. (14.2)). The full likelihood over the trials $t = 1, \ldots, T$ for participant i is given by

$$P\left(\boldsymbol{c}^{(i)} | \alpha^{(i)}, \beta^{(i)}\right) = \prod_{t=1}^{T} P\left(c_t^{(i)}\right). \tag{14.9}$$

By maximizing the likelihood function, we derive the point estimation of the parameters:

$$\left(\hat{\alpha}^{(i)}, \hat{\beta}^{(i)}\right) = \underset{\alpha^{(i)}, \beta^{(i)}}{\operatorname{argmax}} P\left(\boldsymbol{c}^{(i)} | \alpha^{(i)}, \beta^{(i)}\right). \tag{14.10}$$

In the Bayesian estimation, the posterior probability distribution of the parameters is derived by the likelihood function and the prior belief, $P\left(\alpha^{(i)}, \beta^{(i)}\right)$, as follows:

$$P\left(\alpha^{(i)}, \beta^{(i)} | \boldsymbol{c}^{(i)}\right) = \frac{P\left(\boldsymbol{c}^{(i)} | \alpha^{(i)}, \beta^{(i)}\right) P\left(\alpha^{(i)}, \beta^{(i)}\right)}{\iint P\left(\boldsymbol{c}^{(i)} | \alpha^{(i)}, \beta^{(i)}\right) P\left(\alpha^{(i)}, \beta^{(i)}\right) d\alpha^{(i)} d\beta^{(i)}}. \tag{14.11}$$

By repeating this procedure for all the participants, it is possible to estimate the individual parameters for each one.

The individual modeling approach is widely used in computational psychiatry research. For instance, Hauser et al. (2017) fitted RL models to each participant separately and then compared the mean values of the individual estimated parameter of interest—such as learning rate and perseverance—between OCD patients and healthy controls using two-sample t-tests. However, the individual modeling approach often fails to reliably estimate parameter values, especially with a small number of trials (Katahira 2016). This poses a significant challenge in computational psychiatry studies, as patients with mental disorders often find it difficult to engage in experimental tasks for extended periods, resulting in a limited number of trials (i.e., data points).

Hierarchical modeling approach. To overcome this issue, researchers have begun to employ a new method, known as "hierarchical modeling approach" (Daw 2011; Katahira 2016). The hierarchical approach posits that each participant's parameters are drawn from common population-level distributions (Fig. 14.1b). For example, each participant's learning rate, $0 \leq \alpha \leq 1$, is often assumed to be sampled from the population-level normal distribution with the logistic (inverse logit) function transformation: $\alpha^{(i)} = 1/\left(1 + \exp\left(-\alpha\prime^{(i)}\right)\right)$ where $\alpha\prime^{(i)} \sim N(\mu_\alpha, \sigma_\alpha)$. Likewise, each participant's $\beta \geq 0$ is often assumed to be drawn from the population-level distribution as follows: $\beta^{(i)} = \exp\left(\beta\prime^{(i)}\right)$ where $\beta\prime^{(i)} \sim N\left(\mu_\beta, \sigma_\beta\right)$.

In this hierarchical model, the likelihood of the RL model for participant i given the population-level parameters, $\mu_\alpha, \sigma_\alpha, \mu_\beta, \sigma_\beta$, can be derived as follows:

$$\begin{aligned} &P\left(\boldsymbol{c}^{(i)} | \mu_\alpha, \sigma_\alpha, \mu_\beta, \sigma_\beta\right) \\ &= \iint P\left(\boldsymbol{c}^{(i)} | \alpha^{(i)}, \beta^{(i)}\right) P\left(\alpha^{(i)} | \mu_\alpha, \sigma_\alpha\right) P\left(\beta^{(i)} | \mu_\beta, \sigma_\beta\right) d\alpha^{(i)} d\beta^{(i)}. \end{aligned} \tag{14.12}$$

In other words, the likelihood can be obtained by marginalization of the likelihood given the individual-level parameters, $P\left(\boldsymbol{c}^{(i)} | \alpha^{(i)}, \beta^{(i)}\right)$ (see Eq. (14.9)), over the probabilities of the individual-level parameters given the population-level parameters. Then, the full likelihood is given by the product over the participants 1, …, N:

$$P\left(\boldsymbol{c}^{(1)}, \ldots, \boldsymbol{c}^{(N)} | \mu_\alpha, \sigma_\alpha, \mu_\beta, \sigma_\beta\right) = \prod_{i=1}^{N} P\left(\boldsymbol{c}^{(i)} | \mu_\alpha, \sigma_\alpha, \mu_\beta, \sigma_\beta\right). \tag{14.13}$$

Given the full likelihood function, the posterior probability distribution of the population-level parameters, $\mu_\alpha, \sigma_\alpha, \mu_\beta, \sigma_\beta$, can be derived by Bayes' rule:

$$\begin{aligned} &P\left(\mu_\alpha, \sigma_\alpha, \mu_\beta, \sigma_\beta | \boldsymbol{c}^{(1)}, \ldots, \boldsymbol{c}^{(N)}\right) \\ &\propto P\left(\boldsymbol{c}^{(1)}, \ldots, \boldsymbol{c}^{(N)} | \mu_\alpha, \sigma_\alpha, \mu_\beta, \sigma_\beta\right) P\left(\mu_\alpha, \sigma_\alpha, \mu_\beta, \sigma_\beta\right). \end{aligned} \tag{14.14}$$

Once the population-level parameters are estimated, the individual-level parameters can be restored as follows:

$$\begin{aligned} &P\left(\alpha^{(i)}, \beta^{(i)} | \boldsymbol{c}^{(i)}, \mu_\alpha, \sigma_\alpha, \mu_\beta, \sigma_\beta\right) \\ &\propto P\left(\boldsymbol{c}^{(i)} | \alpha^{(i)}, \beta^{(i)}\right) P\left(\alpha^{(i)}, \beta^{(i)} | \mu_\alpha, \sigma_\alpha, \mu_\beta, \sigma_\beta\right). \end{aligned} \tag{14.15}$$

The hierarchical modeling approach is becoming increasingly popular in computational psychiatry (Kanen et al. 2019; Marzuki et al. 2021; Suzuki et al. 2023; Wiehler et al. 2021). For instance, using the hierarchical modeling approach, Suzuki et al. (2023) demonstrated that patients with OCD exhibited a lower learning rate for negative reward prediction errors compared to healthy controls (see Sect. 14.3 for details). They assumed that the individual-level parameters for patients and healthy controls were drawn from separate population-level parameters (see Fig. 14.2a) and compared the population-level mean parameters of the learning rate between the two groups (e.g., $\mu_\alpha^{(G1)}$ versus $\mu_\alpha^{(G2)}$ in Fig. 14.2a). Specifically, they evaluated the Bayes Factor based on the posterior distribution of the difference between the population-level parameters in the two groups (e.g., $\mu_\alpha^{(G1)} - \mu_\alpha^{(G2)}$ in Fig. 14.2a). It is important to note that in such hierarchical models, comparing individual-level parameters between groups (e.g., $\alpha^{(i)}$ in Group 1 versus $\alpha^{(j)}$ in Group 2 in Fig. 14.2a) is not appropriate: the estimated individual-level parameters for each group are shrunk by their respective population-level distributions, which potentially increases the likelihood of detecting false positive differences.

Other studies estimated the individual-level parameters of two groups of participants, under the assumption that their individual-level parameters were drawn from common population-level distributions regardless of their group assignment (see Fig. 14.2b) (Mkrtchian et al. 2017; Oba et al. 2019). In this scenario, comparing individual-level parameters between the two groups (e.g., $\alpha^{(i)}$ in Group 1 versus $\alpha^{(j)}$ in Group 2 in Fig. 14.2b) is appropriate since the estimates for both groups are influenced by the same population-level parameters. Through such comparisons, Oba and colleagues found that participants with high-psychopathic traits exhibited a lower learning rate for positive reward prediction errors during loss-avoidance decision-making tasks, compared to those with low-psychopathic traits (Oba et al. 2019).

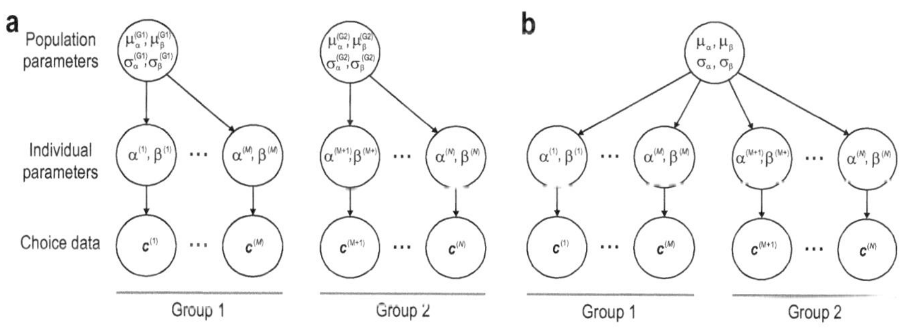

Fig. 14.2 Schematic illustration of the two ways of hierarchical model fitting. **a** For each of the groups, participants' individual-level parameters are drawn from the respective common distributions with population-level parameters. For instance, the individual-level parameters of participants who belong to Group 1 are sampled from the distributions with the population-level parameters, $\mu_\alpha^{(G1)}$, $\mu_\beta^{(G1)}$, $\sigma_\alpha^{(G1)}$, and $\sigma_\beta^{(G1)}$. Meanwhile, for those who belong to Group 2, individual-level parameters are drawn from the other distributions with the population-level parameters, $\mu_\alpha^{(G2)}$, $\mu_\beta^{(G2)}$, $\sigma_\alpha^{(G2)}$, and $\sigma_\beta^{(G2)}$. M, the number of participants in Group 1. N, the total number of participants. $\boldsymbol{c}^{(i)}$, participant i's choice data. **b** Participants' individual-level parameters are drawn from a common set of the population-level distributions regardless of their group assignment

Which approach should researchers use, individual or hierarchical modeling? The individual modeling approach is known to work fairly well when the number of trials is sufficiently large. In contrast, when the number of trials is not sufficiently large, which is often the case in computational psychiatry studies, it is recommended that researchers employ a hierarchical modeling approach. Importantly, the validity of parameter estimation procedures can be evaluated using simulated data, as discussed in the "parameter recovery analysis" subsection of Sect. 4.3. This evaluation hinges on the premise that well-functioning procedures yield parameter estimates that align with the true parameters used to generate the simulated data. For example, a parameter recovery analysis has shown that individual modeling is effective even for complicated model-free RL models with five parameters (β, $\alpha_{(+)}$, $\alpha_{(-)}$, γ, andη; see Sect. 14.2), with data from 500 trials in a three-armed bandit task (Suzuki et al. 2021). Adopting a model-fitting approach informed by parameter recovery analysis is currently considered the best practice, as it systematically assesses the performance of the parameter estimation procedures.

14.4.2 Model Comparison

Researchers often compare multiple RL models based on their goodness-of-fit metrics to identify the most suitable one. For instance, in computational psychiatry studies on OCD, Hauser et al. (2017) evaluated the conventional model-free RL, RL model with short-term perseverance, RL model with asymmetric learning rates, and the full RL model incorporating both perseverance and asymmetric learning rates. Suzuki et al. (2023) did not test the full model but instead assessed other advanced models, such as the volatility Kalman filter and Pearce-Hall models (Li et al. 2011; Piray and Daw 2020).

How to evaluate the goodness of fit of a model? In the framework of MLE, the model can be evaluated by the maximum likelihood (see Eq. (14.6)):

$$L^* = \max_{\boldsymbol{\theta}} P(\boldsymbol{D}|\boldsymbol{\theta}). \tag{14.16}$$

However, this is not appropriate as it leads to overfitting. That is, a model with a higher degree of complexity (i.e., larger number of free parameters) leads to a larger maximum likelihood due to overfitting (Bishop 2006). To evaluate the goodness of fit of a model based on maximum likelihood, we thus need to have a penalty for free parameters. Among the widely used metrics for the evaluation of each model's goodness of fit with penalizing additional free parameters are the Akaike information criterion (AIC) (Akaike 1992) and Bayesian information criterion (BIC) (Schwarz 1978), which can be easily computed from the maximum likelihood (see Eq. (14.16)):

$$AIC \equiv -2\ln(L^*) + 2K, \tag{14.17}$$

$$BIC \equiv -2\ln(L^{*}) + K\ln(n), \tag{14.18}$$

where K and n denotes the number of free parameters in the model and the number of data points, respectively. Note that while AIC and BIC have similar formulations, they are based on different principles: AIC evaluates the *generalization loss* (i.e., how well the model predicts new data), while BIC approximates the *log model evidence* (i.e., how likely the model is true one that generates the observed data). It is also important to note that AIC and BIC are not suitable for hierarchical modeling approaches (see Sect. 4.1), where the posterior distribution of parameters is not well approximated by a normal distribution. For singular models, including hierarchical ones, the use of the widely applicable AIC (WAIC) (Watanabe 2010) or widely applicable BIC (WBIC) (Watanabe 2012) is advised. These metrics can be approximated using the "*loo*" package in *R* (Vehtari et al. 2017) in conjunction with sampling methods implemented in software such as *Stan*.

In the Bayesian framework, the goodness of fit of a model is evaluated by *model evidence* (also called *marginal likelihood* or *free energy*):

$$P(\boldsymbol{D}) = \int P(\boldsymbol{D}|\boldsymbol{\theta})P(\boldsymbol{\theta})\mathrm{d}\boldsymbol{\theta}. \tag{14.19}$$

As it is generally difficult to obtain the exact value, model evidence can be approximated by using BIC, WBIC, and *Laplace approximation* as well as *Bridge Sampling* and *Variational Bayes* (Daw 2011). Comparing the metrics of competing models, researchers can specify the most probable one given the data.

Note that, when using metrics such as AIC, BIC, WAIC, WBIC, and model evidence for model comparison, it is typically assumed that all participants can be modeled with the same underlying process, meaning that the model is treated as a fixed variable. In contrast, another framework for model comparison, known as Bayesian model selection (BMS), treats the model as a random variable with the assumption that it is sampled from a higher-level Dirichlet distribution for each participant (Stephan et al. 2009). In BMS, the goodness of fit for a model, given a set of competing models, is evaluated by *model probability*—that is, the likelihood that a particular model generated the data of a randomly chosen participant—and *exceedance probability*, which reflects the likelihood that a particular model is more probable than any competing model.

In previous computational psychiatry research on OCD and GD, Kanen et al. (2019) and Marzuki et al. (2021) fitted RL models using a hierarchical modeling approach and compared their model evidence as estimated by Bridge Sampling. Similarly, Suzuki et al. (2023) and Wiehler et al. (2021) also employed hierarchical modeling for model fitting, but they used WAIC to compare candidate RL models. Conversely, Hauser et al. (2017) adopted an individual modeling approach, obtaining log model evidence for each participant and model through Variational Bayes approximation, and compared the goodness of fits of the competing models using BMS.

An important consideration in model comparison is determining which models should be included. Ideally, all models that could potentially explain the data and computational processes of interest should be considered. However, it is impractical to test every possible model. Indeed, computational psychiatry studies on OCD and GD (Hauser et al. 2017; Kanen et al. 2019; Marzuki et al. 2021; Suzuki et al. 2023) have not examined RL models that incorporate long-term perseverance (see Sect. 14.2 for an explanation of the distinction between short-term and long-term perseverance), despite the theoretical demonstration of its importance in unraveling the latent computational processes behind behavioral inflexibility (Katahira 2018). Notably, similar to parameter estimation, procedures of model comparison (including the metrics of goodness of fit, set of candidate models, and setting of the experimental task) can be evaluated through simulation. By simulating the actual experiment—where the true model that generates the data is known—it is possible to assess if a model comparison procedure is effective. If the procedure is successful, it should correctly identify the true model in the simulated data as the best fit (see the subsection "model recovery analysis" in Sect. 4.3). For example, one study (Suzuki et al. 2023) established a set of candidate models for comparison based on simulation data analysis. It is now widely regarded as best practice to base decisions about model comparison procedures on such simulation analyses.

14.4.3 Validation of the Model-Fitting Procedure

How do we validate our model-fitting procedure? Researchers may wish to verify to what extent model comparison procedures can identify the true model and how accurately parameter estimation captures the true parameter values of the model within a given experimental setup. However, this task is challenging because the true model and its parameters are unknown in real experimental data. Here, we review the way to assess and validate model-fitting procedures using simulated data (Wilson and Collins 2019). The basic idea involves applying the model-fitting procedure to simulated data for which the true generative model and its parameters are known. If the procedure works well, it should consistently identify the true model as the best fit during model comparison (i.e., model recovery analysis), and the estimated parameters should align with the true parameter values (i.e., parameter recovery analysis).

Model recovery analysis. Model recovery analysis allows researchers to confirm the identifiability of models under current experimental settings by creating a "confusion matrix" and "inversion matrix" from simulated data (Wilson and Collins 2019). In this process, data are simulated with the same experimental design (e.g., the decision-making task, the number of participants, and the number of trials) as in the actual experiment. This simulation is performed multiple times for each of the competing models to generate the confusion and inversion matrices. In the confusion matrix, each row indicates the probability that a particular model provides the best fit—determined by metrics such as exceedance probability, model evidence, or WAIC—for data generated by the corresponding model (see Fig. 14.3a for details). Conversely, in the inversion matrix, each row indicates the probability that data best fitted by a particular model originated from the corresponding model (see Fig. 14.3b for details). Note that the inversion matrix can be derived from the confusion matrix by Bayes rule (Wilson and Collins 2019). If the models were perfectly identifiable, given a set of competing models and experimental settings, the confusion and inversion matrices should be the identity matrix, where data generated by a specific model are best explained by that same model. By evaluating the similarity of these matrices to the identity matrix, we can gauge how accurately the model comparison procedure can identify the true generative model, at least in ideal scenarios where the true model is known.

Parameter recovery analysis. Using simulated data, we can also assess the reliability of parameter estimation procedures, specifically how accurately they estimate the true parameter values given the experimental design. In parameter recovery analysis, we test to what extent the true parameters used to generate the simulated data are accurately recovered by the estimation procedure (Fig. 14.3c). Successful recovery indicates that the procedurse can reliably estimate parameters when the true generative model and its parameter values are known. Through this analysis, we can evaluate the effectiveness of different parameter estimation methods, such as individual and hierarchical modeling approaches, within the context of the given experimental settings and model.

These analyses of simulated data are critical for verifying the model-fitting procedures employed in computational psychiatry studies (Safra et al. 2019; Suzuki et al. 2023). It is advisable for researchers to perform model and parameter recovery analyses during the research design phase, including the setting of the experimental tasks, the construction of competing models, and the selection of methods for model fitting. If not feasible during the initial stages, these analyses should be conducted a posteriori. In any case, we believe that it is imperative to report the capabilities and limitations of the model-fitting procedures transparently on the basis of simulation data analyses.

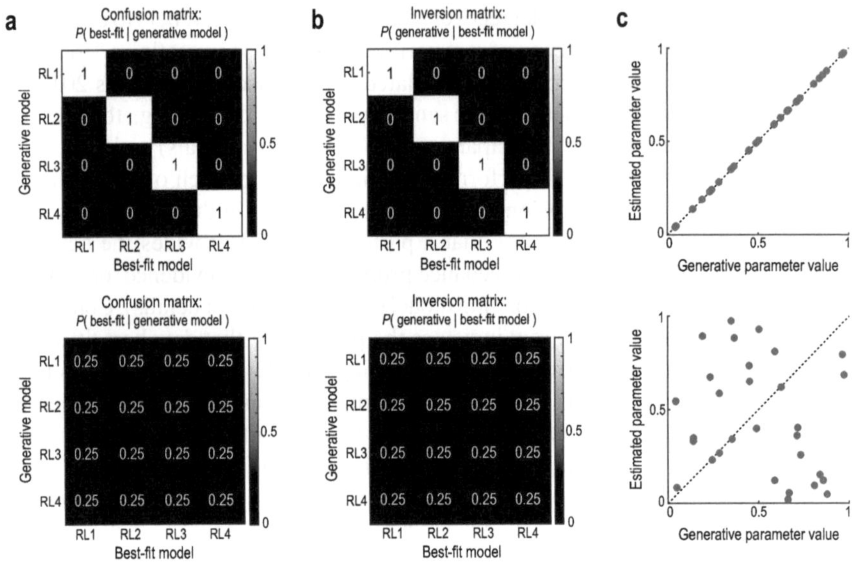

Fig. 14.3 Examples of model and parameter recovery analyses. **a** Examples of the confusion matrix in model recovery analysis. Each row denotes the probability that each of the four competing models (RL1, RL2, RL3, and RL4) provides the best fit on the data generated by the corresponding model, given a model comparison procedure. For instance, in the 1st row, the 2nd column indicates the probability that RL2 outperforms the other three models when the simulation data are generated by RL1. *Top*, the theoretically best case (i.e., close to the identity matrix). *Bottom*, an example of an underperforming case (i.e., far from the identity matrix). **b** Inversion matrix in model recovery analysis. Each row denotes the probability that the data best fitted by the corresponding model are actually generated from each of the four competing models. For instance, in the 1st row, the 2nd column shows the probability that the simulation data are generated by RL1 when RL2 is specified as the best-fit. *Top*, the theoretically best case (i.e., close to the identity matrix). *Bottom*, an example of an underperforming case (i.e., far from the identity matrix). **c** Parameter recovery analysis. Estimated parameter values are plotted against the true generative parameters. Each data point denotes the estimated and true generative parameter values for one participant. *Top*, the theoretically best case in which the estimated parameters are perfectly aligned with the true generative values (i.e., all the points are on the 45° line). *Bottom*, an example of an underperforming case

14.5 Conclusions

In this chapter, we reviewed the application of the formal RL framework to understanding the cognitive and neural mechanisms of mental disorders, drawing on computational psychiatry studies of OCD and GD as illustrative examples. Our review underscores the value of computational model fitting in uncovering the latent decision-making processes that are altered in mental disorders, emphasizing a critical need for the careful design of model-fitting procedures. This involves leveraging parameter and model recovery analyses of synthesized data to verify the procedures within a given experimental context.

References

Akaike H (1992) Breakthroughs in statistics, foundations and basic theory. Springer Series in Statistics, pp 610–624. https://doi.org/10.1007/978-1-4612-0919-5_38

Aylward J, Valton V, Ahn W-Y, Bond RL, Dayan P, Roiser JP, Robinson OJ (2019) Altered learning under uncertainty in unmedicated mood and anxiety disorders. Nat Hum Behav, pp 1–8. https://doi.org/10.1038/s41562-019-0628-0

Bishop CM (2006) Pattern recognition and machine learning. Springer

Clark L, Studer B, Bruss J, Tranel D, Bechara A (2014) Damage to insula abolishes cognitive distortions during simulated gambling. Proc Natl Acad Sci 111(16):6098–6103. https://doi.org/10.1073/pnas.1322295111

Corlett PR, Mollick JA, Kober H (2022) Meta-analysis of human prediction error for incentives, perception, cognition, and action. Neuropsychopharmacology, pp 1–11. https://doi.org/10.1038/s41386-021-01264-3

Daw ND (2011) Trial-by-trial data analysis using computational models, Delgado MR, Phelps EA, Robbins TW (Eds.). Oxford Scholarship Online

Dayan P, Daw ND (2008) Decision theory, reinforcement learning, and the brain. Cogn Affect Behav Neurosci 8(4):429–453. https://doi.org/10.3758/cabn.8.4.429

Figee M, Pattij T, Willuhn I, Luigjes J, van den Brink W, Goudriaan A, Potenza MN, Robbins TW, Denys D (2016) Compulsivity in obsessive–compulsive disorder and addictions. Eur Neuropsychopharmacol 26(5):856–868. https://doi.org/10.1016/j.euroneuro.2015.12.003

Fontenelle LF, Oostermeijer S, Harrison BJ, Pantelis C, Yücel M (2011) Obsessive-compulsive disorder, impulse control disorders and drug addiction. Drugs 71(7):827–840. https://doi.org/10.2165/11591790-000000000-00000

Gillan CM, Robbins TW (2014) Goal-directed learning and obsessive-compulsive disorder. Philos Trans R Soc B: Biol Sci 369(1655):20130475–20130475. https://doi.org/10.1098/rstb.2013.0475

Gillan CM, Papmeyer M, Morein-Zamir S, Sahakian BJ, Fineberg NA, Robbins TW, de Wit S (2011) Disruption in the Balance Between Goal-Directed Behavior and Habit Learning in Obsessive-Compulsive Disorder. Am J Psychiatry 168(7):718–726. https://doi.org/10.1176/appi.ajp.2011.10071062

Glimcher PW (2011) Understanding dopamine and reinforcement learning: the dopamine reward prediction error hypothesis. Proc Natl Acad Sci 108(Supplement 3):15647–15654. https://doi.org/10.1073/pnas.1014269108

Hauser TU, Iannaccone R, Dolan RJ, Ball J, Hättenschwiler J, Drechsler R, Rufer M, Brandeis D, Walitza S, Brem S (2017) Increased fronto-striatal reward prediction errors moderate decision making in obsessive–compulsive disorder. Psychol Med 47(7):1246–1258. https://doi.org/10.1017/s0033291716003305

Huys QJ, Pizzagalli DA, Bogdan R, Dayan P (2013) Mapping anhedonia onto reinforcement learning: a behavioural meta-analysis. Biol Mood Anxiety Disord 3(1):12. https://doi.org/10.1186/2045-5380-3-12

Huys QJM, Maia TV, Frank MJ (2016) Computational psychiatry as a bridge from neuroscience to clinical applications. Nat Neurosci 19(3):404–413. https://doi.org/10.1038/nn.4238

Kanen JW, Ersche KD, Fineberg NA, Robbins TW, Cardinal RN (2019) Computational modelling reveals contrasting effects on reinforcement learning and cognitive flexibility in stimulant use disorder and obsessive-compulsive disorder: remediating effects of dopaminergic D2/3 receptor agents. Psychopharmacology 236(8):2337–2358. https://doi.org/10.1007/s00213-019-05325-w

Katahira K (2016) How hierarchical models improve point estimates of model parameters at the individual level. J Math Psychol 73:37–58. https://doi.org/10.1016/j.jmp.2016.03.007

Katahira K (2018) The statistical structures of reinforcement learning with asymmetric value updates. J Math Psychol 87(Computational Psychiatry 1 2017):31–45. https://doi.org/10.1016/j.jmp.2018.09.002

Kunisato Y, Okamoto Y, Ueda K, Onoda K, Okada G, Yoshimura S, Suzuki S, Samejima K, Yamawaki S (2012) Effects of depression on reward-based decision making and variability of action in probabilistic learning. J Behav Ther Exp Psychiatry 43(4):1088–1094. https://doi.org/10.1016/j.jbtep.2012.05.007

Li J, Schiller D, Schoenbaum G, Phelps EA, Daw ND (2011) Differential roles of human striatum and amygdala in associative learning. Nat Neurosci 14(10):1250–1252. https://doi.org/10.1038/nn.2904

Limbrick-Oldfield EH, Mick I, Cocks RE, McGonigle J, Sharman SP, Goldstone AP, Stokes PRA, Waldman A, Erritzoe D, Bowden-Jones H, Nutt D, Lingford-Hughes A, Clark L (2017) Neural substrates of cue reactivity and craving in gambling disorder. Transl Psychiatry 7(1):e992–e992. https://doi.org/10.1038/tp.2016.256

Marzuki AA, Tomić I, Ip SHY, Gottwald J, Kanen JW, Kaser M, Sule A, Conway-Morris A, Sahakian BJ, Robbins TW (2021) Association of environmental uncertainty with altered decision-making and learning mechanisms in youths with obsessive-compulsive disorder. JAMA Netw Open 4(11):e2136195. https://doi.org/10.1001/jamanetworkopen.2021.36195

Mkrtchian A, Aylward J, Dayan P, Roiser JP, Robinson OJ (2017) Modeling avoidance in mood and anxiety disorders using reinforcement learning. Biol Psychiat 82(7):532–539. https://doi.org/10.1016/j.biopsych.2017.01.017

Montague PR, Dolan RJ, Friston KJ, Dayan P (2012) Computational psychiatry. Trends Cogn Sci 16(1):72–80. https://doi.org/10.1016/j.tics.2011.11.018

O'Doherty J, Dayan P, Schultz J, Deichmann R, Friston K, Dolan RJ (2004) Dissociable roles of ventral and dorsal striatum in instrumental conditioning. Science 304(5669):452–454. https://doi.org/10.1126/science.1094285

Oba T, Katahira K, Ohira H (2019) The effect of reduced learning ability on avoidance in psychopathy: a computational approach. Front Psychol 10:2432. https://doi.org/10.3389/fpsyg.2019.02432

Pike AC, Robinson OJ (2022) Reinforcement learning in patients with mood and anxiety disorders versus control individuals. JAMA Psychiat 79(4):313–322. https://doi.org/10.1001/jamapsychiatry.2022.0051

Piray P, Daw ND (2020) A simple model for learning in volatile environments. PLoS Comput Biol 16(7):e1007963. https://doi.org/10.1371/journal.pcbi.1007963

Rangel A, Camerer C, Montague PR (2008) A framework for studying the neurobiology of value-based decision making. Nat Rev Neurosci 9(7):545–556. https://doi.org/10.1038/nrn2357

Robbins TW, Vaghi MM, Banca P (2019) Obsessive-compulsive disorder: puzzles and prospects. Neuron 102(1):27–47. https://doi.org/10.1016/j.neuron.2019.01.046

Safra L, Chevallier C, Palminteri S (2019) Depressive symptoms are associated with blunted reward learning in social contexts. PLoS Comput Biol 15(7):e1007224. https://doi.org/10.1371/journal.pcbi.1007224

Sakai Y, Sakai Y, Abe Y, Narumoto J, Tanaka SC (2022) Memory trace imbalance in reinforcement and punishment systems can reinforce implicit choices leading to obsessive-compulsive behavior. Cell Rep 40(9):111275. https://doi.org/10.1016/j.celrep.2022.111275

Sato R, Shimomura K, Morita K (2023) Opponent learning with different representations in the cortico-basal ganglia pathways can develop obsession-compulsion cycle. PLoS Comput Biol 19(6):e1011206. https://doi.org/10.1371/journal.pcbi.1011206

Schultz W, Dayan P, Montague PR (1997) A neural substrate of prediction and reward. Science 275(5306):1593–1599. https://doi.org/10.1126/science.275.5306.1593

Schwarz G (1978) Estimating the dimension of a model. Ann Stat 6(2). https://doi.org/10.1214/aos/1176344136

Stephan KE, Penny WD, Daunizeau J, Moran RJ, Friston KJ (2009) Bayesian model selection for group studies. Neuroimage 46(4):1004–1017. https://doi.org/10.1016/j.neuroimage.2009.03.025

Sutton RS, Barto AG (1998) Reinforcement learning: an introduction. MIT Press

Suzuki S, Harasawa N, Ueno K, Gardner JL, Ichinohe N, Haruno M, Cheng K, Nakahara H (2012) Learning to simulate others' decisions. Neuron 74(6):1125–1137. https://doi.org/10.1016/j.neuron.2012.04.030

Suzuki S, Yamashita Y, Katahira K (2021) Psychiatric symptoms influence reward-seeking and loss-avoidance decision-making through common and distinct computational processes. Psychiatry Clin Neurosci. https://doi.org/10.1111/pcn.13279

Suzuki S, Zhang X, Dezfouli A, Braganza L, Fulcher BD, Parkes L, Fontenelle LF, Harrison BJ, Murawski C, Yücel M, Suo C (2023) Individuals with problem gambling and obsessive-compulsive disorder learn through distinct reinforcement mechanisms. PLoS Biol 21(3):e3002031. https://doi.org/10.1371/journal.pbio.3002031

van Timmeren T, Daams JG, van Holst RJ, Goudriaan AE (2018) Compulsivity-related neurocognitive performance deficits in gambling disorder: a systematic review and meta-analysis. Neurosci Biobehav Rev 84:204–217. https://doi.org/10.1016/j.neubiorev.2017.11.022

Vehtari A, Gelman A, Gabry J (2017) Practical Bayesian model evaluation using leave-one-out cross-validation and WAIC. Stat Comput 27(5):1413–1432. https://doi.org/10.1007/s11222-016-9696-4

Watanabe S (2010) Asymptotic equivalence of bayes cross validation and widely applicable information criterion in singular learning theory. ArXiv

Watanabe S (2012) A widely applicable Bayesian information criterion. ArXiv. https://doi.org/10.48550/arxiv.1208.6338

Wiehler A, Chakroun K, Peters J (2021) Attenuated directed exploration during reinforcement learning in gambling disorder. J Neurosci 41(11):2512–2522. https://doi.org/10.1523/jneurosci.1607-20.2021

Wilson RC, Collins AG (2019) Ten simple rules for the computational modeling of behavioral data. Elife 8:e49547. https://doi.org/10.7554/elife.49547

Wyckmans F, Otto AR, Sebold M, Daw N, Bechara A, Saeremans M, Kornreich C, Chatard A, Jaafari N, Noël X (2019) Reduced model-based decision-making in gambling disorder. Sci Rep 9(1):19625. https://doi.org/10.1038/s41598-019-56161-z

Chapter 15
Decision-Making in Attention-Deficit/Hyperactivity Disorder (ADHD)

Eva Halbe, Marcel Schulze, Silke Lux, Dave Coghill, and Alexandra Philipsen

Abstract ADHD is characterized by a wide range of cognitive impairments that not only vary from childhood to adulthood but also exhibit significant inter-individual variability. Specifically, deficits in various executive functions like selective attention, cognitive flexibility, inhibition, and working memory as well as cognitive functions with low executive demand like immediate long-term and short-term memory, and sustained attention can lead to daily life challenges, which are often accompanied by impairments in decision-making. These challenges are frequently linked to neurobiological developmental aspects of ADHD, such as the effects of deficient dopamine and noradrenaline signalling pathways on affective and cognitive functions. Given the heterogeneity of the cognitive associations seen in ADHD, gaining a deeper understanding of the underlying mechanisms contributing to poor decision-making behaviour is crucial. This is particularly important because the impact of impulsivity, and diminished cognitive flexibility, on the processing of feedback on decision-making can have profound consequences for individuals with ADHD and those around them. The following chapter offers insights into the cognitive and affective functions underlying decision-making impairments in ADHD as well as the neurobiological principles associated with these challenges.

Keywords ADHD · Cognitive functions · Affective functions · Impulsivity · Default-mode network · Connectivity

---

E. Halbe (✉) · M. Schulze (✉) · S. Lux · A. Philipsen
Department of Psychiatry and Psychotherapy, University of Bonn, Bonn, Germany
e-mail: eva_halbe@web.de

M. Schulze
e-mail: marcel.schulze@ukbonn.de

D. Coghill
Department of Paediatrics, Faculty of Medicine, Dentistry and Health Sciences, The University of Melbourne, Melbourne, VIC, Australia

Department of Mental Health, The Royal Children's Hospital, Parkville, VIC, Australia

Neurodevelopment and Disability Research, Murdoch Children's Research Institute, The Royal Children's Hospital, Parkville, VIC, Australia

U. Ettinger et al. (eds.), *Decision Making*, Studies in Neuroscience, Psychology and Behavioral Economics, https://doi.org/10.1007/978-3-032-00880-0_15

15.1 Introduction

Attention-Deficit/Hyperactivity Disorder (ADHD) is a common neurodevelopmental condition characterized by core symptoms such as inattention, hyperactivity, and impulsivity (American Psychiatric Association in 2013). These core symptoms can manifest differently, leading to three presentations of ADHD: predominantly hyperactive, predominantly inattentive, and most commonly, combined (Willcutt 2012). With an estimated prevalence of around 5–7%, ADHD is one of the most common psychiatric disorders in childhood. It is estimated that 40–50% of affected youth continue to experience symptoms into adulthood, resulting in a global prevalence of adult ADHD of approximately 2.5% (Uchida et al. 2018). This persistence of ADHD into adulthood is interesting, as some symptoms appear to change across the lifespan. While children with ADHD often exhibit obvious hyperactivity during adulthood people with ADHD increasingly describe an inner restlessness and frequently complain about heightened emotional dysregulation or instability, mood swings or depressive symptoms (Francx et al. 2015; Gibbins et al. 2010). In this regard, it is also important to consider that female adolescents and adults with ADHD are more frequently affected by these deficits in affective functions (Quinn and Madhoo 2014). As women with ADHD are less likely to present with hyperactivity and associated oppositional behaviours, their ADHD symptoms are often overlooked (Quinn and Madhoo 2014). Taken together these findings highlight that there is considerable heterogeneity in the clinical profile of ADHD as well as important sex differences interindividual variability across the lifespan.

The associations and impact of ADHD extends beyond the core symptoms of inattention, hyperactivity, and impulsivity. ADHD is associated with a wide range of cognitive functions and results in significant impairments in daily life (Schulze et al. 2021). Individuals with ADHD face significant challenges that can be specifically assigned to six key cognitive domains. Firstly, deficits in *inhibition,* reflecting the reduced ability to regulate one's own behaviour, can lead to difficulties in controlling impulsive actions. *Memory* problems, including working memory, may affect the patients' ability to hold, manipulate, and recall information which impacts task execution particularly. Furthermore, *temporal discounting*, another domain being impaired in patients with ADHD, describes the tendency to favour immediate rewards over future benefits, undermining long-term planning and consideration. *Timing problems*, which reflect difficulties in regulating responses within appropriate time frames, are also common in ADHD. This affects not only the individual's ability to complete tasks in a timely manner but also disrupts daily time management and self-organisation. Additionally, intraindividual *variability of reaction times* presents a significant challenge. Individuals with ADHD often exhibit a wide range of response times, fluctuating between unusually rapid and exceedingly slow reactions. This variability can severely impact various aspects of life, from academic performance to social interactions. Lastly, the intricate interplay and dynamic of neural processes in patients with ADHD is also reflected in impaired *decision-making*, which draws on a set of cognitive processes that rely on many different cognitive resources (Coghill et al.

2018, 2014). Together, these multifaceted challenges contribute to significant impairments in daily life, underscoring the profound impact ADHD has on individuals and those around them (Diamond 2013). However, it is often reported that across patients with ADHD, a consistent clinical or cognitive profile cannot necessarily be observed, reflecting the heterogeneous and complex nature of the disorder (Coghill et al. 2014).

In adulthood, the behavioural problems associated with ADHD can often be traced back to disadvantageous decision-making abilities, which are often seen in decisions that lead to risky or inappropriate behaviours (Matthies et al. 2012). The ability to appropriately estimate risk propensity or personal benefit from choices has been found to be altered in adults with ADHD compared to healthy controls (Mäntylä et al. 2012). These include everyday behaviours such as butting in to conversations, skipping questions, or avoiding tasks. Moreover, several studies indicate an increased tendency towards risky behaviour in various life domains, such as reckless driving, unsafe sexual behaviour, criminal behaviour, and pathological gambling (Dekkers et al. 2022; Faregh and Derevensky 2011; Flory et al. 2006; Thompson et al. 2007). As defined in this chapter, decision-making represents a cognitive competence that encompasses a wide range of functions, such as affective capacities, which are also crucial for effective decision-making. Deficiencies in these functions are particularly evident in individuals with ADHD. In the last decade, numerous studies have investigated different aspects of decision-making in ADHD, with a focus on the underlying cognitive functions and their interplay. Studies have often focussed on the cognitive-analytic aspects of behavioural impairments during decision-making (Dekkers et al. 2016, 2021; Mäntylä et al. 2012; Mowinckel et al. 2015). However, recent studies have expanded the understanding of decision-making behaviours, emphasising emotional-motivational aspects that are primarily unconsciously processed, being a crucial component of decision-making (Matthies et al. 2012; Mäntylä et al. 2012; Ibanez et al. 2012; Halbe et al. 2023, 2024, 2025).

The following sections aim to elucidate decision-making qualities in ADHD. We specifically focus on the impulsive and affective components, as well as the underlying neural mechanisms of decision-making in ADHD, representing critical elements in understanding the variability and complexity of decision-making in individuals with ADHD. We address not only the interindividual differences influenced by these components, but also the challenges in capturing them, underscoring the incomplete nature of our understanding of decision-making behaviour in ADHD and emphasizing the need for further research.

15.2 Impulsivity and Risky Behaviour

Impulsivity, one of the core components of ADHD, is considered a crucial factor in decision-making difficulties among patients. In individuals with ADHD, impulsivity is characterized by actions that are poorly conceived, prematurely expressed, unduly risky, or inappropriate to the situation, thereby often resulting in undesirable outcomes (Halbe et al. 2023). This can manifest in various decision-making contexts, from everyday choices to significant life decisions. Research suggests that impulsivity in ADHD is associated with dysregulation in neurotransmitter systems, particularly dopamine. Positron emission tomography (PET) scans have shown that individuals with ADHD have lower dopamine receptor and transporter levels, and that these correlate with measures of inattention and impulsivity (Volkow et al. 2009). Impulsivity in ADHD is also linked to deficits in executive functioning, including planning, inhibition, and working memory. These cognitive processes are crucial for effective decision-making, as they enable individuals to consider the consequences of their actions, generate alternative solutions, and hold information in mind while making decisions (Wilcutt et al. 2005). Regarding the significant impact of impulsivity and impairments in daily life, research indicates that interventions including behavioural strategies, medication, or cognitive training can lead to measurable improvements in impulsivity and decision-making capabilities in individuals with ADHD (Lee et al. 2022). For example, stimulant medications, such as methylphenidate, have been shown to reduce impulsivity and enhance executive functioning in individuals with ADHD, thereby improving decision-making capacity (Faraone and Biederman 2002). However, it is critical to note that while these interventions show promise in enhancing specific cognitive measures and, in some cases, symptom reduction, a substantial challenge within the field remains the lack of a robust correlation between changes in symptoms and cognition, and between these and daily functioning or overall cognition (Lee et al. 2022; O'Neil et al. 2018). This discrepancy highlights the complexity of translating cognitive improvements into clinical changes and then understanding how these are expressed as meaningful real-world outcomes.

Studies investigating impulsivity across a wider age range in ADHD have found a linear decrease in impulsive decision-making with age, indicating less impulsive behaviour in adolescents and adults compared to children (Scheres et al. 2014; Achterberg et al. 2016). A commonly used type of decision-making task assessing impulsive decision-making behaviour is the intertemporal choice paradigm, which often relies on monetary incentive delay tasks and assesses the degree to which participants trade off the reward size and delay to receiving the reward (Knutson et al. 2001). These tasks characterise the degree of delay discounting, thereby allowing to evaluate an individual's impulse control and tendency to devalue future rewards in favour of immediate gratification. Among children, studies have shown that those with an ADHD diagnosis have slower reaction times and greater omission errors (Demurie et al. 2012, 2016; Van Hulst et al. 2017). However, other findings report conflicting results with faster reaction times or no differences in accuracy, again reflecting current inconsistencies in literature (Kappel et al. 2015; Demurie et al. 2011; Scheres et al. 2007). Most studies of adults with ADHD did not find behavioural differences in performance on delay discounting tasks, suggesting that decision-making deficits related to reward processing ability are more pronounced in childhood ADHD (Edel et al. 2013; Hoofman et al. 2013; Wilbertz et al. 2013).

Another important aspect of decision-making, alongside the cognitive control of behavioural actions—such as retrieving choice options and analysing potential benefits and outcomes—is represented by affectively driven functions. These affective functions are responsible for emotional assessment, affective control, and also contribute to reward processing (Sonuga-Barke 2002). In the context of decision-making, affective functions play an important role by introducing an emotional aspect into the process of choosing between options. They enable individuals to weigh the emotional value of potential outcomes, significantly influencing decisions by prioritizing choices that promise greater emotional satisfaction or minimize emotional discomfort. In risky decision-making in particular, affective functions are critical to rational thinking as they guide individuals through the complexity of ambiguity and uncertainty (Damasio 1996). Thereby, most people now agree that the assessment of risk is not just based on pure logic, but is also guided by how a person feels about a situation, something that is often described as "gut feeling" (Mäntylä et al. 2012; Shoham et al. 2016; Sonuga-Barke 2002). In this context, we have recently focussed on describing the impact of affective functioning on decision making in ADHD (Halbe et al. 2023). The results shed light on how emotional processing and regulation contribute to the disorder's complexity and manifest in behavioural outcomes. We discussed the critical role of self-perception and awareness of affective states on decision-making abilities (Halbe et al. 2023). Based on this work, our findings suggest that patients with ADHD exhibit unfavourable decision-making behaviours, particularly in situations requiring a high level of affective functioning.

A further notable aspect regarding the relation between affective functioning and decision-making is the differences in symptomatology between men and women. For instance, adult women with ADHD experience emotional symptoms more frequently than their male counterparts and have a higher probability of developing a comorbid affective disorder (Biederman et al. 2008). Regarding these differences in clinical presentation, there are some indications of sex-specific variabilities in the context of decision-making behaviour that need to be considered when investigating decision-making in ADHD. However, most ADHD studies still exhibit a gender imbalance, relying predominantly on data from male participants and thereby indicating a gender publication bias.

Consequently, understanding deficits in decision-making in ADHD necessitates a more comprehensive approach that at a very minimum accounts for both cognitive and affective dimensions, acknowledging how impulsivity, emotional processing, and reward sensitivity collectively influence the unique decision-making difficulties that are observed in individuals with this condition. However, it remains difficult to distinguish between single cognitive functions and to what extent affective functions impact the ability of decision-making in patients with ADHD. Further research is needed to explore unconscious processes and interoceptive perception that might contribute to impulsive and risky decisions. In this context, it is also important to consider the diversity and complexity of ADHD, highlighting that, as is the case for the more general symptoms and cognitive deficits, there is no homogeneous pattern of difficulties in decision-making across all individuals with ADHD. Furthermore, it is crucial to recognize the limitations of laboratory studies in fully capturing the diverse pattern of real-life scenarios. Such constraints can result in inconsistent findings in behavioural research, underscoring the need for approaches that account for the wide-ranging manifestations of ADHD. Thus, study results have reported alterations in decision-making performance that can depend on age, sex, and individual clinical profiles, as well as on the chosen paradigm, method, and laboratory settings (Drechsler et al. 2010; Groen et al. 2013; Halbe et al. 2024). This highlights the importance of integrating diverse research methods and clinical observations to better understand and address the unique challenges faced by individuals with ADHD in their decision-making processes.

15.3 Underlying Theories and the Neurobiology of Decision Making in ADHD

Decision-making is a complex set of processes that involves several components, including delay discounting, reward valuation, cognitive control, and predictions of future rewards/costs (Noda et al. 2020). These functions are dependent on a complex array of neural networks. For example, reward evaluation is associated with activity in the prefrontal cortex (PFC) and the ventral striatum, which receives input from the orbitofrontal cortex (OFC) and cingulate cortex, areas that are thought to be involved in the adaption process and emotional evaluation of the reward (Protopopescu et al. 2005). Outcome estimation of the probability and expected value is related to activity in the medio prefrontal cortex (MPFC) and ventromedial prefrontal cortex (VMPFC). For cognitive strategy adaptation and conflict adaption, the network comprises DLPFC, DMPFC, inferior parietal cortex, cingulate cortex, and Precuneus (Noda et al. 2020). Future predictions and affective evaluation of a past decision are associated with neural activity in the medial temporal lobe, hippocampus, putamen, and amygdala (Peters et al. 2011). This complex interplay of distributed networks plays an important role, amongst others, in ensuring that we can make decisions that balance immediate pleasures with future benefits.

ADHD, as introduced earlier, is a neurodevelopmental disorder and associated with impairments in a spectrum of cognitive functions. While the relationship between cognition and symptoms is complex and not fully understood, it is at least clear that together they have a multi-faceted impact on an individual's day-to-day functioning. The neuropsychopathology of these cognitive domains is central to understanding the functional impairments associated with decision-making in ADHD. Working memory and attention, essential for holding and manipulating information to make informed decisions, are modulated in fronto-parietal networks commonly reported to be altered in ADHD (Cortese et al. 2012). Moreover, inhibitory control and delay aversion play a major role in a decision-making context. Inhibitory control involves the ability to suppress prepotent (automatic) responses to stimuli. The pathway underlying this ability closely overlaps with dysfunction in fronto-striatal circuits in ADHD, involving regions such as the prefrontal cortex (PFC) and the basal ganglia. The delay aversion pathway is based on the premise that individuals with ADHD have a heightened sensitivity to delay, leading to an aversive response to situations that involve waiting or delayed gratification. This pathway suggests that the motivational and reward systems are altered in ADHD, particularly in the glutaminergic pathways that connect the ventral striatum and the orbitofrontal cortex. Individuals with ADHD, according to this pathway, prefer immediate over delayed rewards, not merely because of an inability to delay gratification but also due to an intrinsic aversion to delay itself (Dekkers et al. 2015; Mowinckel et al. 2015). For people with ADHD, regions associated with inhibition and delay aversion are hypoactive when performing decision-making tasks, e.g., gambling tasks and delay discounting. More specifically, individuals with ADHD show less activation in the PFC, insula, anterior cingulate cortex, ventral striatum, insula, and OFC (Ernst et al. 2003; Norman et al. 2018; Shen et al. 2020; Wilbertz et al. 2012; Halbe et al. 2025). The lower activity of those brain regions responsible for cognitive-emotional processes is thought to reflect superior involvement in analysing the sensorimotor aspects of the stimuli rather than in the analysis of the decision itself (Ernst et al. 2003).

Besides the pathways that interfere with the affective evaluation and cognitive aspects in decision-making in ADHD, there also appears to be an imbalance between "task-positive" and "task-negative" functional networks (Konrad and Eickhoff 2010). Task-positive networks are involved in executing cognitive tasks while task-negative networks are associated with resting-state and more internally oriented processes, e.g., self-referential thoughts (Andrews-Hanna et al. 2014; Buckner et al. 2008; Sidlauskaite et al. 2016; Sonuga-Barke and Castellanos 2007). The two networks are anti-correlated, i.e., the task-negative network is suppressed when an external cognitive task needs to be processed. The task-negative network, also called default-mode network (DMN), involves the posterior cingulate cortex, precuneus, and medial frontal gyrus (Raichle 2015; Uddin et al. 2008). In ADHD, the default mode network interference hypothesis posits an overactive task-negative network for goal-directed tasks (Pereira-Sanchez and Castellanos 2021; Sonuga-Barke and Castellanos 2007). In other words, the switch between task-negative and task-positive networks may be altered in ADHD due to a reduced suppression of the default mode network. The insufficient down-regulation of the DMN leads to higher trial-by-trial variability in cognitive tasks associated with attentional lapses in patients with ADHD (Weissman et al. 2006). Mowinckel and colleagues (Mowinckel et al. 2017) investigated the role of DMN activity in a reward-based decision task and its modulation by ADHD medication. Unmedicated patients with ADHD, compared to neurotypical controls, exhibited higher DMN activity during decision phases, which was disadvantageous for task performance. In contrast, DMN activity in patients that received medication was similar to neurotypical controls (Mowinckel et al. 2017).

In conclusion, several of the neural underpinnings of ADHD, characterized by disrupted network connectivity and altered brain function across multiple neural pathways, have significant indirect consequences for decision-making processes characterized by hastiness, lack of foresight, and a failure to consider the long-term consequences of actions. These neurobiological disruptions contribute to a range of decision-making difficulties, including impaired attention to relevant cues, increased susceptibility to distractions, and a diminished capacity for delay of gratification.

Understanding these neural mechanisms is crucial for developing targeted interventions that aim to improve decision-making abilities in individuals with ADHD, thereby enhancing their overall quality of life and ability to achieve long-term goals.

15.4 Conclusions

ADHD not only impacts the cognitive functions critical for daily decision-making but also predisposes individuals to a range of risky behaviours. Together these difficulties underscore the importance of understanding the cognitive and affective dimensions of decision-making in ADHD. Recent findings elucidate neurobiological deficits influencing impulsivity, with a specific focus on neurotransmitter systems such as dopamine. These studies also examine the complex interplay between cognitive functioning and altered impulsive and risky decision-making. These impairments affect the capacity to engage in thoughtful decision-making, reflecting the disorder's profound impact on individuals' lives. Moreover, recent insights into the neurobiological underpinnings of ADHD emphasize the role of disrupted network connectivity and altered brain function in shaping decision-making processes. Specifically, deficiencies in both the affective and cognitive control systems, alongside alterations in the default mode network, contribute to the challenges in decision-making faced by individuals with ADHD. However, it is important to recognise that many of the available studies have been based on small sample sizes, have had a limited focus and have used a wide range of paradigms to assess cognition and behaviour. Furthermore, several confounding factors such as age, medication, comorbidity, and sex should be considered further as influential factors.

References

Achterberg M, Peper JS, van Duijvenvoorde AC, Mandl RC, Crone EA (2016) Frontostriatal white matter integrity predicts development of delay of gratification: a longitudinal study. J Neurosci 10;36(6):1954–61. https://doi.org/10.1523/JNEUROSCI.3459-15.2016

American Psychiatric Association (2013) Diagnostic and statistical manual of mental disorders. American Psychiatric Publishing, Washington, DC, 991 p. ISBN: 978-0890425558. Svenaeus F (2014) Diagnosing mental disorders and saving the normal. Med Health Care and Philos 17, 241–244

Andrews-Hanna JR, Smallwood J, Spreng RN (2014) The default network and self-generated thought: component processes, dynamic control, and clinical relevance. Ann N Y Acad Sci 1316(1):29–52. https://doi.org/10.1111/NYAS.12360

Biederman J, Ball SW, Monuteaux MC, Mick E, Spencer TJ, McCreary M (2008) New insights into the comorbidity between ADHD and major depression in adolescent and young adult females. J Am Acad Child Adolesc Psychiatry 47(4):426–434

Buckner RL, Andrews-Hanna JR, Schacter DL (2008) The brain's default network. Ann N Y Acad Sci 1124(1):1–38. https://doi.org/10.1196/ANNALS.1440.011

Coghill DR, Seth S, Matthews K (2014) A comprehensive assessment of memory, delay aversion, timing, inhibition, decision making and variability in attention deficit hyperactivity disorder: advancing beyond the three-pathway models. Psychol Med 44(9):1989–2001. https://doi.org/10.1017/S0033291713002547

Coghill D, Toplak M, Rhodes S, Adamo N (2018) Cognitive functioning in ADHD: inhibition, memory, temporal discounting, decision-making, timing and reaction time variability. Case studies in clinical psychological science: bridging the gap from science to practice. https://doi.org/10.1093/MED/9780198739258.003.0010

Cortese S, Kelly C, Chabernaud C, Proal E, Di Martino A, Milham MP, Castellanos FX (2012) Toward systems neuroscience of ADHD: a meta-analysis of 55 fMRI sudies. Am J Psychiatry 169(10):1038–1055. https://doi.org/10.1176/APPI.AJP.2012.11101521/ASSET/IMAGES/LARGE/1038F3.JPEG

Damasio AR (1996) The somatic marker hypothesis and the possible functions of the prefrontal cortex. Philos Trans R Soc London Ser B, Biol Sci 351(1346):1413–1420. https://doi.org/10.1098/RSTB.1996.0125

Dekkers TJ, Popma A, van Rentergem JA, Huizenga HM (2015) Risky decision making in ADHD: a meta-regression analysis. Eur Child Adolesc Psychiatry 24:S25–S26

Dekkers TJ, Popma A, Agelink van Rentergem JA, Bexkens A, Huizenga HM (2016) Risky decision making in attention-deficit/hyperactivity disorder: a meta-regression analysis. Clin Psychol Rev 45:1–16. https://doi.org/10.1016/j.cpr.2016.03.001

Dekkers TJ, Agelink van Rentergem JA, Huizenga HM, Raber H, Shoham R, Popma A, Pollak Y (2021) Decision-making deficits in ADHD are not related to risk seeking but to suboptimal decision-making: meta-analytical and novel experimental evidence. J Atten Disord 25(4):486–501. https://doi.org/10.1177/1087054718815572/ASSET/IMAGES/LARGE/10.1177_1087054718815572-FIG4.JPEG

Dekkers TJ, de Water E, Scheres A (2022) Impulsive and risky decision-making in adolescents with attention-deficit/hyperactivity disorder (ADHD): the need for a developmental perspective. Curr Opin Psychol 44:330–336. https://doi.org/10.1016/J.COPSYC.2021.11.002

Demurie E, Roeyers H, Baeyes D, Sonuga-Barke E (2011) Common alterations in sensitivity to type but not amount of reward in ADHD and autism spectrum disorders. J Child Psychol Psychiatry 52:1164–1173

Demurie E, Roeyers H, Baeyens D, Sonuga-Barke E (2012) Temporal discounting of monetary rewards in children and adolescents with ADHD and autism spectrum disorders. Dev Sci 15:791–800

Demurie E, Roeyers H, Wiersema JR, Sonuga-Barke E (2016) No evidence for inhibitory deficits or altered reward processing in ADHD: data from a new integrated monetary incentive delay go/no-go task. J Atten Disord 20:353–367

Diamond A (2013) Executive functions 64:135–168. https://doi.org/10.1146/Annurev-Psych-113011-143750

Drechsler R, Rizzo P, Steinhausen HC (2010) Decision making with uncertain reinforcement in children with attention deficit/hyperactivity disorder (ADHD). Child Neuropsychol. https://doi.org/10.1080/09297040903190774

Edel JA, Enzi B, Witthaus H, Tegenthoff M, Peters S, Lissek S (2013) Differential reward processing in subtypes of adult attention deficit hyperactivity disorder. J Psychiatr Res 47:350–356

Ernst M, Kimes AS, London ED, Matochik JA, Eldreth D, Tata S, Contoreggi C, Leff M, Bolla K (2003) Neural substrates of decision making in adults with attention deficit hyperactivity disorder. Am J Psychiatry 160(6):1061–1070. https://doi.org/10.1176/APPI.AJP.160.6.1061

Faraone SV, Biederman J (2002) Efficacy of Adderall® for attention-deficit/hyperactivity disorder: a metaanalysis. J Attention Disord 6(2):69–75

Faregh N, Derevensky J (2011) Gambling behavior among adolescents with attention deficit/hyperactivity disorder. J Gambl Stud 27(2):243–256. https://doi.org/10.1007/s10899-010-9211-3

Flory K, Molina BSG, Pelham WE, Gnagy E, Smith B (2006) Childhood ADHD predicts risky sexual behavior in young adulthood. J Clin Child Adolesc Psychol 35(4):571–577. https://doi.org/10.1207/s15374424jccp3504_8

Francx W, Zwiers MP, Mennes M, Oosterlaan J, Heslenfeld D, Hoekstra PJ, Hartman CA, Franke B, Faraone SV, O'Dwyer L, Buitelaar JK (2015) White matter microstructure and developmental improvement of hyperactive/impulsive symptoms in attention-deficit/hyperactivity disorder. J Child Psychol Psychiatry 56(12):1289–1297. https://doi.org/10.1111/JCPP.12379

Gibbins C, Weiss MD, Goodman DW, Hodgkins PS, Landgraf JM, Faraone SV (2010) ADHD-hyperactive/impulsive subtype in adults. Ment Illn 2(1):41–45. https://doi.org/10.4081/MI.2010.E9

Groen Y, Gaastra GF, Lewis-Evans B, Tucha O (2013) Risky behavior in gambling tasks in individuals with ADHD—A systematic literature review. PLoS ONE 8(9):e74909. https://doi.org/10.1371/journal.pone.0074909

Halbe E, Kolf F, Heger AS, Hüpen P, Bergmann M, Aslan B, Harrison BJ, Davey CG, Philipsen A, Lux S (2023) Altered interaction of physiological activity and behavior affects risky decision-making in ADHD. Front Hum Neurosci 17:1147329. https://doi.org/10.3389/FNHUM.2023.1147329/BIBTEX

Halbe E, Heger AS, Kolf F, Bergmann M, Harrison BJ, Davey CG, Philipsen A, Lux S (2024) Sex differences in physiological correlates of affectively driven decision-making behavior in adult ADHD. BMC Psychiatry 24, 602. https://doi.org/10.1186/s12888-024-06040-3

Halbe E, Jamieson A, Bergmann M, Mehren A, Harrison BJ, Davey CG, Stöcker T, Philipsen A, Lux S (2025) Neural Correlates and Sex-Specific Effects of Affectively Driven Processes Underlying Decision-Making in Adult ADHD. Brain Behav 15:e70215. https://doi.org/10.1002/brb3.70215

Hoofman M, Onnink M, Cools R, Aarts E, Kan C, Franke B (2013) The dopamine transporter hyplotype and reward-related striatal responses in adult ADHD. Eur Neuropsychopharmacol 23:469–478

Ibanez A, Cetkovich M, Petroni A, Urquina H, Baez S, Gonzalez-Gadea ML, Manes F (2012) The neural basis of decision-making and reward processing in adults with euthymic bipolar disorder or attention-deficit/hyperactivity disorder (ADHD). PLoS ONE 7(5). https://doi.org/10.1371/journal.pone.0037306

Kappel V, Lorenz RC, Streifling M, Renneberg B, Lehmkuhl U, Ströhle A, Salbach-Andrae H, Beck A (2015) Effect of brain structure and function on reward anticipation in children and adults with attention deficit hyperactivity disorder combined subtype. Soc Cogn Affect Neurosci 10(7):945–951. https://doi.org/10.1093/scan/nsu135

Knutson B, Adams CM, Fong GW, Hammer D (2001) Anticipation of increasing monetary reward selectively recruits nucleus accumbens. J Neurosci Res 21(1–5):RC159

Konrad K, Eickhoff SB (2010) Is the ADHD brain wired differently? A review on structural and functional connectivity in attention deficit hyperactivity disorder. Hum Brain Mapp 31(6):904–916. https://doi.org/10.1002/HBM.21058

Lee S, Hill TR, Johnson B, Testa R, Priya V, Spencer-Smith M, Coghill D (2022) Can neurocognitive outcomes assist measurement-based care for children with attention-deficit/hyperactivity disorder? A systematic review and meta-analyses of the relationships among the changes in neurocognitive functions and clinical outcomes of attention-deficit/hyperactivity disorder in pharmacological and cognitive training interventions. J Child Adolesc Psychopharmacol 32(5):250–277. https://doi.org/10.1089/CAP.2022.0028

Mäntylä T, Still J, Gullberg S, Del Missier F (2012) Decision making in adults with ADHD. J Atten Disord 16(2):164–173. https://doi.org/10.1177/1087054709360494

Matthies S, Philipsen A, Svaldi J (2012) Risky decision making in adults with ADHD. J Behav Ther Exp Psychiatry. https://doi.org/10.1016/j.jbtep.2012.02.002

Mowinckel AM, Pedersen ML, Eilertsen E, Biele G (2015) A meta-analysis of decision-making and attention in adults With ADHD. J Atten Disord. https://doi.org/10.1177/1087054714558872

Mowinckel AM, Alnæs D, Pedersen ML, Ziegler S, Fredriksen M, Kaufmann T, Sonuga-Barke E, Endestad T, Westlye LT, Biele G (2017) Increased default-mode variability is related to reduced task-performance and is evident in adults with ADHD. NeuroImage Clin 16:369–382. https://doi.org/10.1016/J.NICL.2017.03.008

Noda Y, Barr MS, ElSalhy M, Masuda F, Tarumi R, Ogyu K, Wada M, Tsugawa S, Miyazaki T, Nakajima S, Mimura M (2020) Neural correlates of delay discount alterations in addiction and psychiatric disorders: a systematic review of magnetic resonance imaging studies. Prog Neuropsychopharmacol Biol Psychiatry 99:109822. https://doi.org/10.1016/J.PNPBP.2019.109822

Norman LJ, Carlisi CO, Christakou A, Murphy CM, Chantiluke K, Giampietro V, Simmons A, Brammer M, Mataix-Cols D, Rubia K (2018) Frontostriatal dysfunction during decision making in attention-deficit/hyperactivity disorder and obsessive-compulsive disorder. Biol Psychiatry: Cogn Neurosci Neuroimaging 3(8):694–703. https://doi.org/10.1016/J.BPSC.2018.03.009

O'Neil S, Halperin JM, Coghill D (2018) Neuropsychological functioning and ADHD: a developmental perspective. In Banaschewski T, Coghill D, Zuddas A (Eds.), Oxford textbook of attention deficit hyperactivity disorder. Oxford University Press, pp 118–130

Pereira-Sanchez V, Castellanos FX (2021) Neuroimaging in attention-deficit/hyperactivity disorder. Curr Opin Psychiatry 34(2):105–111. https://doi.org/10.1097/YCO.0000000000000669

Peters J, Bromberg U, Schneider S, Brassen S, Menz M, Banaschewski T, Conrod PJ, Flor H, Gallinat J, Garavan H, Heinz A, Itterman B, Lathrop M, Martinot JL, Paus T, Poline JB, Robbins TW, Rietschel M, Smolka M, Büchel C (2011) Lower ventral striatal activation during reward anticipation in adolescent smokers. Am J Psychiatry 168(5):540–549. https://doi.org/10.1176/APPI.AJP.2010.10071024/ASSET/IMAGES/LARGE/APPI.AJP.2011.540C.JPEG

Protopopescu X, Pan H, Altemus M, Tuescher O, Polanecsky M, McEwen B, Silbersweig D, Stern E (2005) Orbitofrontal cortex activity related to emotional processing changes across the menstrual cycle. Proc Natl Acad Sci USA 102(44):16060–16065. https://doi.org/10.1073/PNAS.0502818102/SUPPL_FILE/02818FIG5.PDF

Quinn PO, Madhoo M (2014) A review of attention-deficit/hyperactivity disorder in women and girls: uncovering this hidden diagnosis. Prim Care Companion CNS Disord 16(3). https://doi.org/10.4088/PCC.13R01596

Raichle ME (2015) The brain's default mode network. Annu Rev Neurosci. https://doi.org/10.1146/annurev-neuro-071013-014030

Scheres A, Milham MP, Knutson B, Castellanos FX (2007) Ventral striatal hyporesponsiveness during reward anticipation in attention-deficit/hyperactivity disorder. Biol Psychiat 61(5):720–724

Scheres A, Tontsch C, Thoeny AL, Sumiya M (2014) Temporal reward discounting in children, adolescents, and emerging adults during an experiential task. Front Psychol 5:711. https://doi.org/10.3389/fpsyg.2014.00711

Schulze M, Coghill D, Lux S, Philipsen A (2021) Disentangling ADHD's presentation-related decision-making-a meta-analytic approach on predominant presentations. Front Psychiatry 12. https://doi.org/10.3389/FPSYT.2021.519840

Shen C, Luo Q, Jia T, Zhao Q, Desrivières S, Quinlan EB, Banaschewski T, Millenet S, Bokde ALW, Büchel C, Flor H, Frouin V, Garavan H, Gowland P, Heinz A, Ittermann B, Martinot JL, Artiges E, Paillère-Martinot ML, Sahakian BJ (2020) Neural correlates of the dual-pathway model for ADHD in adolescents. Am J Psychiatry 177(9):844–854. https://doi.org/10.1176/APPI.AJP.2020.19020183

Shoham R, Sonuga-Barke EJS, Aloni H, Yaniv I, Pollak Y (2016) ADHD-associated risk taking is linked to exaggerated views of the benefits of positive outcomes. Sci Rep 6. https://doi.org/10.1038/SREP34833

Sidlauskaite J, Sonuga-Barke E, Roeyers H, Wiersema JR (2016) Default mode network abnormalities during state switching in attention deficit hyperactivity disorder. Psychol Med 46(3):519–528. https://doi.org/10.1017/S0033291715002019

Sonuga-Barke EJS (2002) Psychological heterogeneity in AD/HD—A dual pathway model of behaviour and cognition. Behav Brain Res 130(1–2):29–36

Sonuga-Barke EJS, Castellanos FX (2007) Spontaneous attentional fluctuations in impaired states and pathological conditions: a neurobiological hypothesis. Neurosci Biobehav Rev 31(7):977–986. https://doi.org/10.1016/J.NEUBIOREV.2007.02.005

Thompson AL, Molina BSG, Pelham W, Gnagy EM (2007) Risky driving in adolescents and young adults with childhood ADHD. J Pediatr Psychol 32(7):745–759. https://doi.org/10.1093/jpepsy/jsm002

Uchida M, Spencer TJ, Faraone SV, Biederman J (2018) Adult outcome of ADHD: an overview of results from the MGH longitudinal family studies of pediatrically and psychiatrically referred youth with and without ADHD of both sexes. J Atten Disord 22(6):523–534

Uddin LQ, Kelly AMC, Biswal BB, Margulies DS, Shehzad Z, Shaw D, Ghaffari M, Rotrosen J, Adler LA, Castellanos FX, Milham MP (2008) Network homogeneity reveals decreased integrity of default-mode network in ADHD. J Neurosci Methods 169(1):249–254. https://doi.org/10.1016/J.JNEUMETH.2007.11.031

Van Hulst BM, de Zeeuw P, Bos DJ, Rijks Y, Neggers SFW, Durston S (2017) Defective somatic markers in subclinical psychopathy. NeuroReport 13:1025–1027

Volkow ND, Wang GJ, Newcorn JH, Telang F, Solanto MV, Fowler JS, Swanson JM (2009) Depressed dopamine activity in caudate and preliminary evidence of limbic involvement in adults with attention-deficit/hyperactivity disorder. Arch Gen Psychiatry 66(8):838–846

Weissman DH, Roberts KC, Visscher KM, Woldorff MG (2006) The neural bases of momentary lapses in attention. Nature Neurosci 9(7):971–978. https://doi.org/10.1038/nn1727

Wilbertz G, Trueg A, Sonuga-Barke EJ, Blechert J, Philipsen A, Tebartz van Elst L (2013) Neural and psychophysiological markers of delay aversion in attention-deficit hyperactivity disorder. J Abnorm Psychol 122(2):566

Wilbertz G, van Elst LT, Delgado MR, Maier S, Feige B, Philipsen A, Blechert J (2012) Orbitofrontal reward sensitivity and impulsivity in adult attention deficit hyperactivity disorder. Neuroimage 60(1):353–361. https://doi.org/10.1016/j.neuroimage.2011.12.011

Willcutt EG (2012) The prevalence of DSM-IV attention-deficit/hyperactivity disorder: a meta-analytic review. Neurother 9(3):490–499. https://doi.org/10.1007/s13311-012-0135-8

Willcutt EG, Doyle AE, Nigg JT, Faraone SV, Pennington BF (2005). Validity of the executive function theory of attention-deficit/hyperactivity disorder: a meta-analytic review. Biol Psychiatry 57(11):1336–1346

Chapter 16
Insights into Effort-Based Decision Making in Schizophrenia

Chao Yan and Lei Shen

Abstract Apathy, characterized by impaired motivation and diminished goal-directed behavior, is a prominent negative symptom in schizophrenia, significantly impacting psychological well-being, social functioning, and quality of life. Unfortunately, it often remains overlooked, leading to limited understanding of psychopathology and ineffective interventions. This chapter provide a comprehensive characterisation of apathy in schizophrenia from the perspective of effort-based decision making (EBDM). We begin by presenting an updated conceptual framework of EBDM and summarizing laboratory paradigms employed in human subjects. Subsequently, we summarize the behavioral manifestations of EBDM impairment and the dysfunctional neural circuits observed in patients with schizophrenia across various stages of the illness. Finally, we address the limitations of existing evidence and discuss future research directions in this domain.

Keywords Schizophrenia · Apathy · Motivation · Effort-based decision making · Computational modeling

16.1 Overview of Apathy

Apathy is a prevalent neuropsychiatric syndrome characterized by a decrease in motivation and goal-directed activity compared to an individual's previous level of functioning (Marin 1990; Robert et al. 2018). This symptom is evident across a wide spectrum of neurological and neuropsychiatric disorders, encompassing conditions such as Parkinson's disease, Alzheimer's disease, schizophrenia, and depression. In recent decades, increasing attention has been devoted to understanding the underlying mechanisms of apathy across these diverse disorders, aiming to identify robust psychological and biological markers for future precision medication. This chapter, however, primarily concentrates on the examination of apathy in one severe psychiatric disorder: schizophrenia.

C. Yan (✉) · L. Shen
School of Psychology and Cognitive Science, East China Normal University, Shanghai, China
e-mail: cyan@psy.ecnu.edu.cn

U. Ettinger et al. (eds.), *Decision Making*, Studies in Neuroscience, Psychology and Behavioral Economics, https://doi.org/10.1007/978-3-032-00880-0_16

In schizophrenia, apathy has been regarded as a core negative symptom (Andreasen 1989), with around 30–50% of patients displaying apathy symptoms across various stages of this disorder (Evensen et al. 2012). Approximately 30% of patients who have had their first episode of psychosis still showed apathy symptoms at the 10-year follow-up, contributing to impairments in cognitive/behavioral processes (e.g., persistence of activities, interest in external issues), emotional responsiveness (e.g., emotional reactions to the environment, empathy), and social interactions (e.g., relationship with family members) (Evensen et al. 2012; Robert et al. 2018). In the subsequent sections, we present the conceptual framework of effort-based decision-making (EBDM), discuss the measurements used to assess EBDM, summarize the behavioral patterns and neural responses associated with EBDM in individuals with schizophrenia, and identify research gaps and future directions in this area.

16.2 Apathy, Motivation and Goal-Directed Behavior

16.2.1 Conceptualization of Apathy

The initial conceptualization of apathy was based on the description of 'absence or lack of feeling, emotion, interest or concern.' However, despite this early understanding, there remains a lack of consensus regarding the precise definition of apathy, and psychiatric classification systems have struggled to provide clear diagnostic criteria for the symptom (Starkstein and Leentjens 2008). Drawing on clinical theories and empirical research on neurological disorders, Marin (1990) has highlighted that diminished motivation lies at the core of apathy. This is primarily characterized by a reduction in goal-directed behaviors, often accompanied by concurrent declines in cognitive and emotional function. Building upon this theoretical framework, apathy has evolved from a singular concept into a multidimensional structure, subsequently divided into cognitive, emotional, and behavioral domains. This evolution has paved the way for the development of standardized diagnostic assessments (Starkstein 2000). Beyond the psychological interpretations of internal states for apathy, Levy and Dubois (2006) argued for exploring the pathological mechanisms underlying apathy through more observable and objectively measurable means. They defined apathy as 'the quantitative reduction of self-generated voluntary and purposeful behaviors.' This conceptual shift has provided a clearer framework for understanding and assessing apathy.

As research into the clinical features and cognitive-biological processes of apathy progresses, the latest diagnostic criteria explicitly use goal-directed behavior to describe motivation (Robert et al. 2018). It is evident that apathy is not merely a singular symptom but rather a complex motivational disorder involving impairments in multiple domains such as cognition, emotion, and social interaction (Chong 2018). The core feature is the reduction of spontaneous goal-directed behavior (Levy 2012;

Levy and Dubois 2006; Marin 1990). This conceptual interpretation links the understanding of potential cognitive mechanisms underlying apathy to the processes of planning, formulation, and execution of goal-directed behaviors.

Although the conceptual understanding and theoretical frameworks for apathy were initially developed within the context of neurological disorders (Costello et al. 2024), these foundational perspectives have proven applicable to research of apathy within psychiatric disorders, particularly schizophrenia. In the early exploration and treatment of schizophrenia, research predominantly focused on positive symptoms such as hallucinations, delusions, and thought disorder (McCutcheon et al. 2020). This emphasis was partly due to the conspicuous nature of these symptoms and the notable efficacy of antipsychotic medications in improving them (Strauss et al. 2021). However, as research into schizophrenia progressed, a growing recognition of the profound and enduring adverse impact of negative symptoms on patients' social functioning, quality of life, and overall recovery trajectory emerged (Crespo-Facorro et al. 2021; Ristic and Batinic 2022). Research in recent decades based on the development of clinical assessments has revealed a two-factor model that includes the most stable dimensions of negative symptoms: amotivation and diminished expression (Marder and Galderisi 2017). Among these dimensions, apathy emerges as a central feature in schizophrenia's negative symptoms (Strauss et al. 2020). Early studies on clinical symptoms of schizophrenia also emphasized the term 'volitional impairment' to describe deficits in the motivation aspect, reflecting a decreased level of motivation for engaging in goal-directed activities (Foussias et al. 2014). While significant headway has been achieved in comprehending apathy, a conspicuous gap persists in our examination of the cognitive mechanisms that underlie this symptom. Understanding these mechanisms is crucial for elucidating the nature of apathy and guiding the formulation of effective prevention and intervention strategies in schizophrenia.

16.2.2 Theoretical Framework of EBDM

Conceptualizing the motivation process through goal-directed behavior involves a series of cognitive stages. These stages, illustrated in Fig. 16.1 by Husain and Roiser (2018), encompass option generation, cost–benefit decision, anticipatory phase, initiation of action and sustaining effort, interaction with the behavioral goal, and learning from outcome. Option generation marks the initial phase, where individuals generate various options, prompted by either self-generated or environmental cues. Following this, individuals assess the potential rewards against the effort required for their pursuit in the cost–benefit decision stage. Motivation is intricately linked to the reward processing system, as highlighted by neuroeconomic theories, which emphasize that a key aspect of cognitive motivation involves assessing whether potential rewards are worth the effort required to obtain them (Le Heron et al. 2018; Saleh et al. 2021). In essence, goal-directed behavior hinges on striking a balance between the reward value associated with the target behavior and the effort cost, commonly known as cost–benefit analysis. This evaluation significantly influences motivation toward a specific goal and impairments in this function are considered a core component of apathy (Costello et al. 2024).

Previous research on EBDM in rodents has laid a robust foundation for understanding the complexities of motivational processes in decision-making. Through training animals to perform tasks requiring varying levels of effort to obtain rewards, these studies have effectively delved into the delicate balance between rewards and effort, shedding light on the underlying motivational processes driving animal decision-making (Salamone et al. 1994a, b; Salamone 1996). Significantly, the findings from rodent studies have profound implications for understanding human motivational behavior.

In EBDM, motivational behavior is defined as the level of effort individuals are willing to exert for different levels of reward (Green et al. 2015). The translation of rewards into action depends on the cost–benefit calculation process, which generates and executes goal-directed actions necessary to achieve valuable outcomes (Chong et al. 2016). During this stage, the decision-making process involves calculating and evaluating the cost of effort (e.g., the effort required to obtain a specific reward or the risks involved), as well as assessing the value of rewards (e.g., reward probability or reward amount). Additionally, it encompasses evaluating how decisions influence subsequent behavior (Husain and Roiser 2018). In recent years, the EBDM framework has bridged preclinical cost–benefit analysis models with motivation deficit of schizophrenia, providing a theoretical framework for objectively measuring apathy in schizophrenia patients. By elucidating these cognitive processes, researchers may gain insights into the underlying mechanisms of goal-directed behavior and the development of apathy symptoms.

16.2.3 Effort and Reward in Decision-Making Paradigms

Two main types of EBDM paradigms can be distinguished: progressive paradigms and non-progressive paradigms (see Table 16.1). Effort and reward dimensions are central components within this framework, where decision-related costs are manipulated, encompassing both physical and cognitive efforts. Non-progressive paradigms represent typical binary decision-making scenarios, where individuals must choose between high-effort/high-reward (i.e., hard tasks) and low-effort/low-reward (i.e., easy tasks) options. In classical EBDM paradigms, the research primarily revolves around manipulating physical effort, such as lever pressing amount or obstacle height for animals, and key pressing or grip strength for humans. For instance, in early animal studies like the T-maze barrier task by Cousins et al. (1996), varying quantities of food were placed on each arm of the maze to manipulate the reinforcement level of reward stimuli, with barriers added to adjust task difficulty.

The Effort-Expenditure for Rewards Task (EEfRT), among the most commonly utilized tasks, has effectively translated preclinical research insights into an in-depth exploration of cognitive mechanisms underpinning human motivational processes (Treadway et al. 2009). The EEfRT paradigm and its variants, such as the Effort

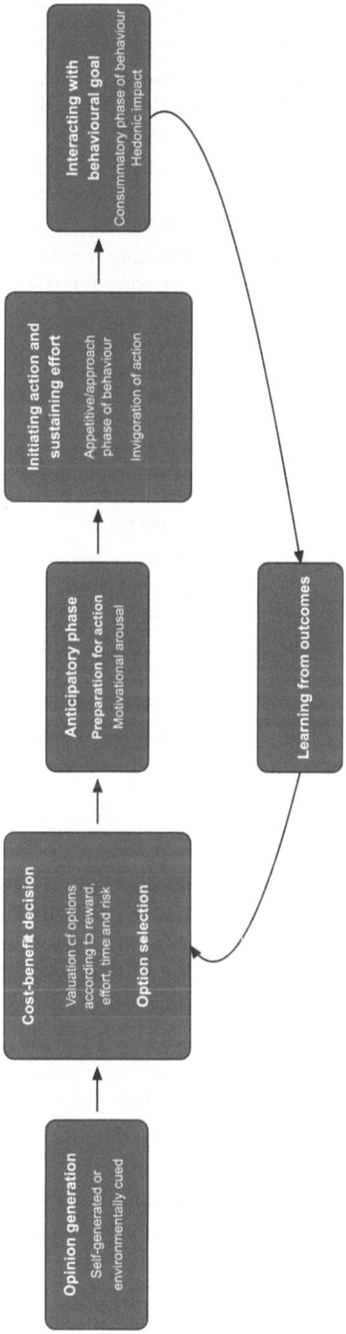

Fig. 16.1 “Mechanisms underlying EBDM” was adapted from Husain and Roiser (2018), Springer Nature (*Rights obtained via Copyright Clearance Center*)

Table 16.1 Paradigms assessing EBDM

References	Paradigms	Reward/effort types	Outcome variables
Non-progressive effort manipulation: High effort/high reward and low-effort/low reward options			
Treadway et al. 2009	Effort-expenditure for rewards task (EEfRT)	Monetary reward/ physical effort (Button pressing)	Hard trial choice %
Gold et al. 2013	Effort-cost computation task	Monetary reward/ Physical effort (Button pressing)	Hard trial choice %
Wang et al. 2015	Effort-based Pleasure experience task (E-pet)	Monetary reward/ Physical effort (Hand grip)	Hard trial choice %
Reddy et al. 2015	Deck choice effort task	Monetary reward/ Cognitive effort (Cognitive switching)	Hard trial choice %
Progressive effort manipulation: increasing effort at a given reward level			
Hartmann et al. 2013	Effort discounting task (EDT)	Monetary reward/ physical effort (Hand grip)	Breakpoint
Strauss et al. 2016	Progressive ratio task (PRT)	Monetary reward/ physical effort (Button pressing)	Breakpoint
Wolf et al. 2014	Progressive ratio task (PRT)	Monetary reward/ cognitive Effort (Bigger number task)	Breakpoint
Culbreth et al. 2016a	Cognitive effort discounting (COG-ED)	Monetary reward/ cognitive effort (n-back)	Breakpoint

Cost Computation Task (Gold et al. 2013) and the Effort-Based Pleasantness Experience Task (Wang et al. 2015), have been widely applied in research involving both healthy populations and patients with psychiatric disorders. This paradigm was primarily utilized to investigate the relationship between motivation deficits and EBDM concerning factors such as reward value, reward probability, and anticipated rewards. Participants are presented with choices between options that involve fewer key presses or lower monetary rewards and options that require more key presses or offer higher monetary rewards. After expending effort within a designated time frame, participants receive monetary rewards based on the likelihood of winning for that round.

So far, researchers have explored the impact of effort on reward decision-making by manipulating various parameters. The primary behavioral metric employed for assessment is the percentage of difficult task choices for the non-progressive paradigms (Luther et al. 2019). For instance, in the Effort-Cost Computation Task, participants are offered choices between easy and hard trials. To obtain a certain monetary reward (e.g., $0.50), participants need to complete a number of button presses (e.g., 100) to inflate a balloon until it bursts (Gold et al. 2013). In contrast, progressive paradigms involve determining an 'effort breakpoint,' indicating the maximum level of effort an individual is willing to exert to attain a particular reward value (Saleh et al. 2021). A higher effort breakpoint reflects a greater willingness to exert effort for rewards. In experiments like the ratio progressive task, participants engage in seven sets of trials across three reward levels ($0.10, $0.25, $0.50). The effort required to earn each reward (e.g., $0.10) parametrically increases within each set, revealing participants' maximum willingness to expend physical effort for varying reward values (Strauss et al. 2016).

The investigation of decision-making processes in daily life has advanced beyond simply considering physical exertion, extending its scope to encompass cognitive effort as well. Researchers are currently investigating decision-making based on cognitive effort, where tasks involve both simple (e.g., judging number sizes) and complex (e.g., alternating between judging number sizes or odd/even) cognitive activities (Reddy et al. 2016). Another prevalent paradigm for assessing cognitive effort in decision-making is the n-back task, which manipulates the cognitive load of working memory. Participants are presented with choices between low cognitive effort (i.e., 1-back) and high cognitive effort (i.e., 2–6 back), with the reward value gradually increasing if they opt for the challenging task until the point where individuals are unwilling to exert further effort (Culbreth et al. 2020).

Moreover, recent research has extended EBDM into the realm of social interactions, shedding light on the role of social factors in decision-making processes. In studies examining effort-based social reward decision-making, individuals with autistic traits demonstrated a greater propensity to expend effort for non-social rewards (movies without social information) (Dubey et al. 2015). Other researchers explored the level of effort individuals were willing to invest to obtain rewards in social incentive contexts, finding that social incentives heighten individuals' willingness to exert effort (Fulford et al. 2018). Although research on effort-based social decision-making is still in its early stages of exploration, further exploration in this field holds promise for advancing our understanding of the influence of social factors on decision-making processes.

16.3 Aberrant EBDM in Schizophrenia Spectrum Disorders

16.3.1 Cognitive Mechanisms of EBDM in Schizophrenia

It has been suggested that individuals diagnosed with schizophrenia typically exhibited attenuated pleasure in anticipation of future reward that may lead to decreased motivation for pleasure but comparable consummatory pleasurable experiences compared to healthy controls (HC) when receiving positive outcomes. Such impairment in anticipatory pleasure may hinder goal-directed behavior, making it challenging for individuals to utilize previous rewarding experiences to motivate their actions (Gold et al. 2013; Saleh et al. 2021).

Evidence from numerous studies in EBDM consistently indicates abnormalities in cost–benefit calculations among patients with schizophrenia. These individuals tend to overestimate the costs associated with effort and undervalue subjective rewards. Compared to HC, they exhibit a diminished inclination to exert effort for rewards, as evidenced by a lower proportion of selecting high-effort options under conditions of increased reward magnitude and probability, as well as lower breakpoint scores (Gold et al. 2013; Reddy et al. 2015; Strauss et al. 2016; Wang et al. 2015; Whitton et al. 2020). Such impairment is particularly evident in patients with more severe negative symptoms (Barch et al. 2014; Hartmann et al. 2013; Treadway et al. 2015). Similarly, in decision-making tasks involving cognitive effort, schizophrenia patients exhibit a reluctance to select difficult options and this decreased willingness to invest cognitive effort correlates negatively with the severity of negative symptoms (Culbreth et al. 2016b, 2023; Wolf et al. 2014). However, some researchers found that as the level of physical effort increases, both patients with schizophrenia and HC exhibited a decrease in the subjective value of rewards, with effort discounting showing no significant correlation with negative symptoms (Docx et al. 2015). These inconsistent findings might reflect the distinct psychological processes behind decisions based on physical and cognitive effort. Despite the heterogeneity of findings across studies, a recent meta-analysis investigating effort-based decision-making in schizophrenia, encompassing 20 studies involving a total of 666 HC and 837 patients, revealed moderate deficits in patients' decision-making processes. Particularly noteworthy was the significantly greater deficit magnitude observed in patients during high-reward and high-probability trials. Additionally, their decision-making performance exhibited a moderate negative correlation with the severity of negative symptoms (Blouzard et al. 2023).

Furthermore, a recent study examined the performance of patients with schizophrenia in EBDM involving social rewards. Participants were asked to rate their anticipated pleasure from online interactions with social partners and to gauge their level of trust toward these partners by choosing to send points. Subsequently, participants were asked to use key presses to determine whether to increase or decrease effort, reflecting their willingness to cooperate again (Campellone et al. 2018). The findings revealed that patients exhibited lower levels of trust in their partners and were less inclined to exert effort to attain positive social interaction outcomes. This discovery implied that patients with schizophrenia also displayed abnormalities in their willingness to invest effort when pursuing social rewards. The evidence from these studies suggested that deficits in EBDM were prevalent among schizophrenia patients across various domains and were closely linked with negative symptoms.

It is worth noting that EBDM paradigms encompass the simultaneous manipulation of rewards and effort, posing challenges in isolating the underlying processing mechanisms of these information. Assuming that both reward and effort information guide decision-making, the integration of these signals to guide goal-directed behavior remains elusive. In recent years, computational models have emerged as invaluable tools within the EBDM framework, offering insights into the cognitive mechanisms underlying apathy in schizophrenia. By employing precise mathematical algorithms to capture underlying psychological computational processes and integrating cognitive computational modeling of decision-making, these models quantify individuals' sensitivity to reward and effort information during decision-making.

For instance, a recent study employing a drift–diffusion model investigated decision-making processes, revealing that treatment-resistant schizophrenia patients exhibited diminished responses to changes in reward and effort compared to HC. Patients displayed reduced sensitivity to both monetary rewards and effort, and reduced effort aversion correlated with the severity of negative symptoms (Saleh et al. 2023). Furthermore, abnormalities in the cost–benefit calculation process may also be influenced by cognitive impairments in patients. In exploring alterations in EBDM, some researchers have turned to computational modeling to investigate whether these changes stem from deficits in motivation, cognitive impairment, or a combination of both. Model findings indicate that schizophrenia patients who struggle to integrate information regarding reward magnitude and probability into the effort allocation process demonstrated cognitive functional deficits (Cooper et al. 2019). This might suggest that effort-based reward decision-making was intricately linked with both cognitive function and motivational deficits in schizophrenia.

16.3.2 EBDM Deficits Across Stages of Schizophrenia

Previous research on EBDM has predominantly focused on patients with chronic schizophrenia. However, there remains a relatively limited understanding of the developmental psychopathological mechanisms driving apathy across various stages of schizophrenia. Here, we summarize EBDM studies in clinical patients with first-episode psychosis (FEP) and non-clinical individuals at high risk for psychosis (see Table 16.2).

Reports from several studies have highlighted the prevalence of apathy in individuals with FEP (Fervaha et al. 2015). Therefore, it is imperative to direct attention towards this early stage of schizophrenia to acquire a more holistic comprehension of the pathological trajectory of apathy and the underlying mechanisms contributing to decision-making abnormalities. Recent research found that FEP patients demonstrated diminished willingness to exert effort for rewards and that this reduction primarily manifested as decreased sensitivity to rewards rather than effort itself, particularly evident in those with higher levels of apathy (Chang et al. 2019, 2020). Moreover, there was a noticeable increase in the frequency of skipping difficult trials among this population (Fortgang et al. 2020). As indicated by current research findings, FEP patients exhibited comparable abnormal decision-making patterns to those

Table 16.2 Summary of studies on EBDM in first episode psychosis and individuals with clinical and non-clinical high risk for psychosis

Reference	Sample	Paradigm	Reward	Effort	Performance outcome	Major findings
Chang et al. 2019	45 FEP versus 45 HC	EEfRT	Money	Physical	Hard trial chosen %	FEP < HC; high-value/ high-probability reward
Chang et al. 2020	40 FEP versus 44 HC	CO-GED	Money	Cognitive	Subjective value by effort level	FEP < HC;
Fortgang et al. 2020	32 FEP versus 27 HC	Cognitive effort and DisEngagement (CEDE)	Token	Cognitive	Skips	FEP > HC; difficult trials
Strauss et al. 2023	44 CHR versus 32 HC	EEfRT	Money	Physical	Hard trial chosen %	CHR < HC; high-value/ high-probability reward
Chu et al. 2023	40 Schizotypy versus 40 HC	EEfRT	Money	Physical	Hard trial chosen %	High-amotivation schizotypy < HC
Terenzi et al. 2019	19 High-PS/ 21 medium-PS/ 20 Low-PS	Concurrent schedule task (CST)	Food	Physical	Number of presses	High-PS < Low-PS; high effort
McCarthy et al. 2015	30 SA versus 40 HC	EEfRT	Money	Physical	Hard trial chosen %	SA > HC; medium and low reward
Wang et al. 2018	26 SA versus 28 HC	E-Pet	Money	Physical	Hard trial chosen %	SA = HC

Notes HC: healthy controls; FEP: first-episode psychosis; CHR: clinical high-risk; PS: psychotic symptom; SA: social anhedonia; EEfRT: Effort-Expenditure for Rewards Task; CO-GED: Cognitive Effort Discounting; E-pet: Effort-based Pleasure Experience Task

observed in chronic patients. These collective observations suggested that deficits in EBDM might manifest early in the course of psychosis.

Theories in psychiatry emphasize a dimensional understanding of psychiatric disorders, moving beyond traditional binary classifications (Eaton et al. 2023). Notably, research has revealed that around 50% of individuals at high risk for mental illness, who have not yet reached clinical levels, experience symptoms of apathy (Sauvé et al. 2019). However, there is limited research on EBDM in individuals at high risk for schizophrenia. A recent study indicated that clinical high-risk (CHR) individuals for psychosis exhibited reduced willingness to exert effort when faced with high reward probabilities and magnitudes, and such reduced willingness was significantly negatively correlated with the severity of negative symptoms (Strauss et al. 2023).

Characterized by experiences resembling symptoms of schizophrenia, schizotypy is recognized as a non-clinical population at high risk for psychosis within the schizophrenia-spectrum psychopathology (Ettinger et al. 2015; Kwapil and Barrantes-Vidal 2015). Among schizotypy individuals with high levels of amotivation, researchers noted a significant decrease in choosing options with high reward values and probabilities (Chu et al. 2023). Furthermore, subclinical psychotic symptom (PS) individuals with high levels of negative symptoms also exhibited a diminished inclination to complete tasks requiring higher effort levels (Terenzi et al. 2019). These findings were consistent with decision-making impairment patterns observed in clinical populations.

Additionally, individuals experiencing social anhedonia (SA) are also considered a risk population within the schizophrenia spectrum. Research on this population has revealed contrasting results regarding the impact of reward probability and amount on task choices among individuals with SA. One study found that when the probability of reward is low or moderate, there is an increase in the proportion of hard task choices among individuals with SA (McCarthy et al. 2015). However, contradictory findings emerged from another study, where researchers observed that despite increases in reward probability and amount, the proportion of hard task choices did not significantly increase compared to HC among individuals with SA (J. Wang et al. 2018). Given that these disparities might arise from variances in the experimental tasks employed (i.e., EEfRT/E-Pet) and the heterogeneous nature of the population with SA, it emphasizes the necessity for further investigation to elucidate the underlying mechanisms of decision-making. This may offer insights towards a more comprehensive understanding of the manifestations and underlying mechanisms of apathy within the schizophrenia spectrum.

16.4 Neural Correlates of Apathy in Schizophrenia

The neural circuits underlying reward processing encompass a network of brain regions, including the ventral tegmental area (VTA), ventral pallidum (VP), paraventricular nucleus of thalamus (PVT), nucleus accumbens (NAcc), ventral striatum (VS), and prefrontal cortex (PFC) (Birnie et al. 2022; Sesack and Grace 2010). Among these, the VTA is a midbrain structure with two major dopamine pathways: the mesolimbic pathway projecting from VTA to VS (comprising the NAcc and VP), and the mesocortical pathway projecting from striatal regions through the thalamus to prefrontal cortex, which includes ventromedial prefrontal cortex (vmPFC), dorsomedial prefrontal cortex (dmPFC), and anterior cingulate cortex (ACC) (Husain and Roiser 2018). The dopamine system assumes a pivotal role in modulating motivational behaviors, including the regulation of behavioral activation, effort expenditure, and decision-making processes pertaining to effort (Salamone et al. 2016). Accumulating evidence has suggested that impairment within the prefronto-striatal circuits

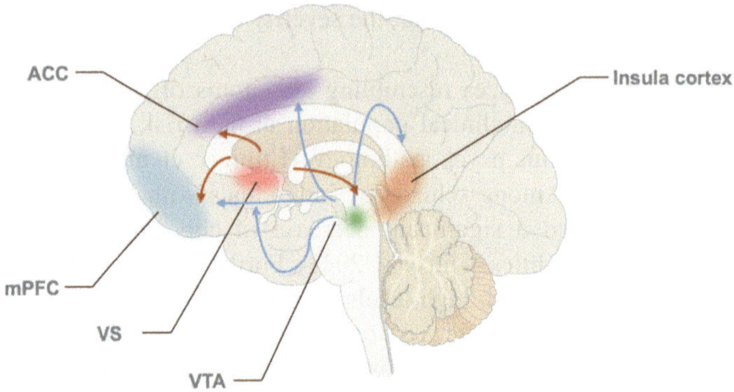

Fig. 16.2 "Neural basis of EBDM", was adapted from Le Heron et al. (2018), used under *CC BY 4.0*, and from Husain and Roiser (2018), Springer Nature (*Rights obtained via Copyright Clearance Center*). This adapted version was licensed under *CC BY 4.0* by Lei Shen and Chao Yan. ACC: anterior cingulate cortex; mPFC: medial prefrontal cortex; VS: ventral striatum; VTA: ventral tegmental area

may represent a central mechanism underlying apathy (Heron et al. 2019; Moran et al. 2019; Shukla et al. 2019). Recent meta-analyses utilizing human functional magnetic resonance imaging (fMRI) data have indicated the critical involvement of vmPFC, orbitofrontal cortex (OFC), and VS regions in the generation, maintenance, and updating of value representations, while ACC and insula are implicated in encoding effort-related information, serving as crucial neural substrates for anticipating the effort demands of tasks by integrating signals related to reward and effort to encode subjective value (Pessiglione et al. 2018) (Fig. 16.2).

16.4.1 Cost and Benefits Evaluation in the ACC and the VS

In the effortful pursuit of rewards, the VS and ACC regions play critical roles for reward valuation and effort allocation. Evidence from both animal and human studies in EBDM has indicated that the ACC regions encode the valuation of effort costs and integrate reward and effort signals (Hogan et al. 2019; Husain and Roiser 2018). Studies have shown that damage to the ACC leads to decision-making abnormalities, affecting the balance between effort expenditure and reward value, and closely correlates with motivational deficits (Vassena et al. 2017). Similarly, the VS is considered a core region for cost–benefit analysis, encoding not only the subjective value associated with effort but also being closely linked to effort activation and discounting (Dobryakova et al. 2017; Suzuki et al. 2021). Recent studies investigating PRT in schizophrenia have found a close association between the VS hypofunction and less willingness to exert effort (Wolf et al. 2014). During effortful decisions, consistent findings have revealed weaker activation in the VS region in schizophrenia patients, significantly correlating with the severity of clinical apathy symptom (Culbreth et al. 2020; Park et al. 2017; Prettyman et al. 2021; see Table 16.3).

Furthermore, NAcc as a fundamental sub-region of the VS, also plays a vital role in regulating reward processing and goal-directed behavior (Botvinick et al. 2009; Farrar et al. 2010; Mai et al. 2012; Salamone et al. 2007). Early studies in rodents underscored the significance of dopamine transmission in the NAcc, as its disruption significantly impacted cost–benefit calculations, thereby influencing behavioral choices (Salamone et al. 1994a, b; Salamone 1996). Specifically, researchers observed that injecting dopamine antagonists into the NAcc of animals led to profound changes in behavior, with animals opting for less effortful tasks (i.e., directly consuming lab chow) over tasks requiring greater effort (such as lever pressing to attain the more-preferred food) (Salamone et al. 1991). Similar findings have been observed in schizophrenia patients, particularly during the EEfRT, where patients exhibited reduced willingness to pursue rewards accompanied by attenuated neural activation in the NAcc compared to healthy individuals (Huang et al. 2016).

Table 16.3 Summary of fMRI studies on EBDM in schizophrenia

Reference	Sample	Paradigm	Reward	Effort	Major findings
Culbreth et al. 2020	28 schizophrenia versus 30 HC	CO-GED	Money	Cognitive	The bilateral VS ↓ was associated with negative symptoms
Huang et al. 2016	23 schizophrenia versus 23 HC	EEfRT	Money	Physical	The NAcc, vmPFC, the posterior cingulate gyrus and the left medial frontal gyrus ↓ in schizophrenia
Park et al. 2017	23 schizophrenia versus 31 HC	Effort-based Reinforcement Task	Money	Physical	Dysfunctional activation in meso/nigro-striatal-prefrontal circuitry in schizophrenia
Prettyman et al. 2021	21 schizophrenia versus 23 HC	EDT	Money	Physical	VS ↓ in schizophrenia and was correlated with amotivation severity
Wolf et al. 2014	41 schizophrenia versus 37 HC	PRT	Money	Physical	The VS hypofunction was associated with lower motivation in schizophrenia

Notes HC: healthy controls; CO-GED: Cognitive Effort Discounting; EEfRT: Effort-Expenditure for Rewards Task; EDT: Effort Discounting Task; PRT: Progressive Ratio Task; VS: ventral striatum; NAcc: nucleus accumbens; vmPFC: ventromedial prefrontal cortex

16.4.2 Representing Value in the OFC and the vmPFC

The OFC and vmPFC are pivotal in processing and integrating information associated with rewards (Walton et al. 2015). Specifically, the vmPFC is responsible for encoding value signals and subjective values of options during decision-making processes (Bartra et al. 2013). The apathy observed in various neurodegenerative conditions, such as Parkinson's disease, Alzheimer's disease, and frontotemporal dementia, has been associated with dysfunction in these specific brain regions (Le Heron et al. 2018).

Despite the critical role of the OFC and vmPFC in reward processing and decision-making, research investigating the neural involvement in EBDM abnormalities in schizophrenia is limited. Among the few existing fMRI studies in this area, Huang et al. (2016) found that schizophrenia patients exhibited reduced neural activity in the vmPFC compared to healthy individuals during effort-related decisions. Notably, this diminished neural response in the vmPFC was correlated with the magnitude of monetary rewards. This finding suggested a potential link between the neural underpinnings of apathy in schizophrenia and the dysfunction of prefrontal cortex regions involved in representing value during decision-making processes.

Collectively, the neural substrates underlying these abnormalities involve the coordinated interplay among various cortical and subcortical regions. Therefore, further extensive research is imperative to achieve a comprehensive understanding of the neural underpinnings of EBDM abnormalities in schizophrenia.

16.5 Research Gaps and Future Directions

16.5.1 Exploring Social Effort-Based Decision Making

Though existing evidence has shown that patients with schizophrenia have decreased willingness to expend physical and cognitive effort towards non-social reward (e.g., monetary reward), few studies have focused on social motivation in schizophrenia, and even less on social EBDM (Blouzard et al. 2023; Catalano and Green 2023). Understanding the decision-making process of individuals with schizophrenia in social scenarios is crucial due to their significant deficits in establishing and maintaining social relationships. These deficits can severely impact academic and work performance, as well as family intimacy. However, the methods for measuring social EBDM remain constrained. In one of the few studies, Abplanalp et al. (2022) have introduced the Social Effort and Conscientiousness Scale to assess social effort. This scale comprises two factors, delineating content related to general social effort and social effort in adherence to social norms. Scheffer et al. (2022) also developed a laboratory-based social EBDM task. In this task, participants were presented with a choice between an empathy task (e.g., observing and empathizing with a refugee child and writing a sentence to describe their emotional state) and a description task (e.g., focusing on the details of the refugee child and describing their characteristics) (Scheffer et al. 2022). Most studies have been conducted on non-clinical samples, with few directly measuring social EBDM in schizophrenia. In a recent study, Shao et al. (in press) proposed a novel paradigm called the Mock Job Interview Task to assess willingness to expend cognitive effort to obtain a job offer in an interview scenario. The findings revealed that patients with schizophrenia were less willing than HC to choose the hard option when faced with strict interviewers, particularly when the bonus points for the hard option were moderate or high. This impaired EBDM outcome was significantly associated with more severe symptoms of apathy and anhedonia in schizophrenia. However, research in this area is still in its infancy, and further studies are needed to fully explore the new framework of social motivation (e.g., Social Motivation Profile, see Catalano and Green 2023 for a review) to devise more robust empirical methods for a deeper comprehension of the diminished social drive observed in individuals with schizophrenia.

16.5.2 Leveraging Computational Modeling

Applying the EBDM paradigms (e.g., EEfRT) provides insights into pathological reductions to motivation in schizophrenia. However, such paradigms ignore an important part in theoretical accounts of motivation, which can help to elaborate the reason why patients with schizophrenia are less willing to exert effort in everyday life. Before making a decision, we must first gather sufficient information about the reward that can be received and the effort required to obtain it. Thus, how patients seek and perceive information might significantly affect their decision for reward. Specifically, patients with schizophrenia may exhibit insensitivity to rewarding information such that they show no tendency to the high reward options. Or, they may have heightened sensitivity to effort allocation so that they tend to choose more low-effort options. As mentioned earlier, computational modeling may serve as a value tool for disentangling between such alternative explanations of behavioral alterations (Husain and Roiser 2018; Rigoli and Pezzulo 2022). For instance, research into reward decision-making involving different types of effort (physical/cognitive) revealed that individuals displayed varying sensitivities to cognitive and physical effort (Chong et al. 2017). In EBDM tasks, choices were modeled using a softmax function and maximum likelihood estimation to fit distinct models. The equations for the best-fitting models of cognitive effort (using a hyperbolic function) and physical effort (using a parabolic function) can be expressed as follows:

$$\text{Hyperbolic}: \mathrm{SV(t)} = \mathrm{R(t)} \cdot \frac{1}{1 + \mathrm{k} \cdot \mathrm{E(t)}}$$

$$\text{Parabolic}: \mathrm{SV(t)} = \mathrm{R(t)} - \mathrm{k} \cdot \mathrm{E(t)}^2$$

where, SV(t) denotes subjective value, R represents the reward level, E represents the effort level, and k represents the discounting parameter. These models assumed how the discounting function reflects the impact of increasing effort levels on subjective value (Hartmann et al. 2013). Specifically, the hyperbolic model implies that changes at lower effort levels will yield a more pronounced impact on subjective value. Conversely, the parabolic model predicts the opposite trend, with the influence of effort on subjective value growing as effort level increases (see Fig. 16.3).

These models have been employed in EBDM tasks in degenerative neurological disorders like Parkinson's Disease and Alzheimer's disease (see Pessiglione et al. 2018 for a review). For example, a study utilizing computational modeling demonstrated that improvements in apathy scores among individuals with Parkinson's disease receiving dopaminergic medication were associated with increased reward sensitivity (i.e., the anticipated value of reward), rather than changes in effort sensitivity (Bouc et al. 2016). However, there has been a dearth of studies applying computational models to elucidate apathy as a manifestation of impairments in reward and effort evaluation in schizophrenia (Cooper et al. 2019; Saleh et al. 2023). This remains an important avenue for future research.

16.5.3 Applying Transdiagnostic Dimensional Approaches

As discussed previously, apathy symptoms manifest across various neuropsychiatric disorders, including schizophrenia, major depressive disorder (MDD), bipolar disorder (BD), eating disorders, and substance use disorders. Understanding the transdiagnostic mechanisms of EBDM is crucial, particularly within the context of the positive valence systems component of the Research Domain Criteria framework, to elucidate the heterogeneity observed in psychotic illnesses. However, there is a lack of studies investigating physical and cognitive EBDM across different neuropsychiatric disorders. Studies by Barch et al. (2023), Y. Wang et al. (2022), Yang et al. (2021), and Zou et al. (2020) revealed that individuals with schizophrenia and BD exhibited reduced willingness to expend physical and cognitive effort with increasing reward magnitude or probability. Conversely, Barch et al. (2023) and Yang et al., (2021) found that patients with current or remitted MDD did not differ from controls in their willingness to allocate effort toward higher rewards, suggesting a potential divergence in EBDM mechanisms between psychosis spectra and depression. Cognitive impairments commonly observed in psychosis spectrum disorders, but not necessarily in MDD, might play a role in the impaired allocation of cognitive effort among patients with schizophrenia and MDD. Nonetheless, additional research is required to thoroughly investigate these intricate matters. Such research endeavors have the potential to enhance the assessment of motivation deficits and refine the diagnosis of mental illnesses. Moreover, forthcoming studies should consider incorporating transdiagnostic dimensional approaches like the Hierarchical Taxonomy of Psychopathology (HiTOP). HiTOP is a contemporary conceptual framework crafted to classify and comprehend mental disorders grounded in empirical evidence.

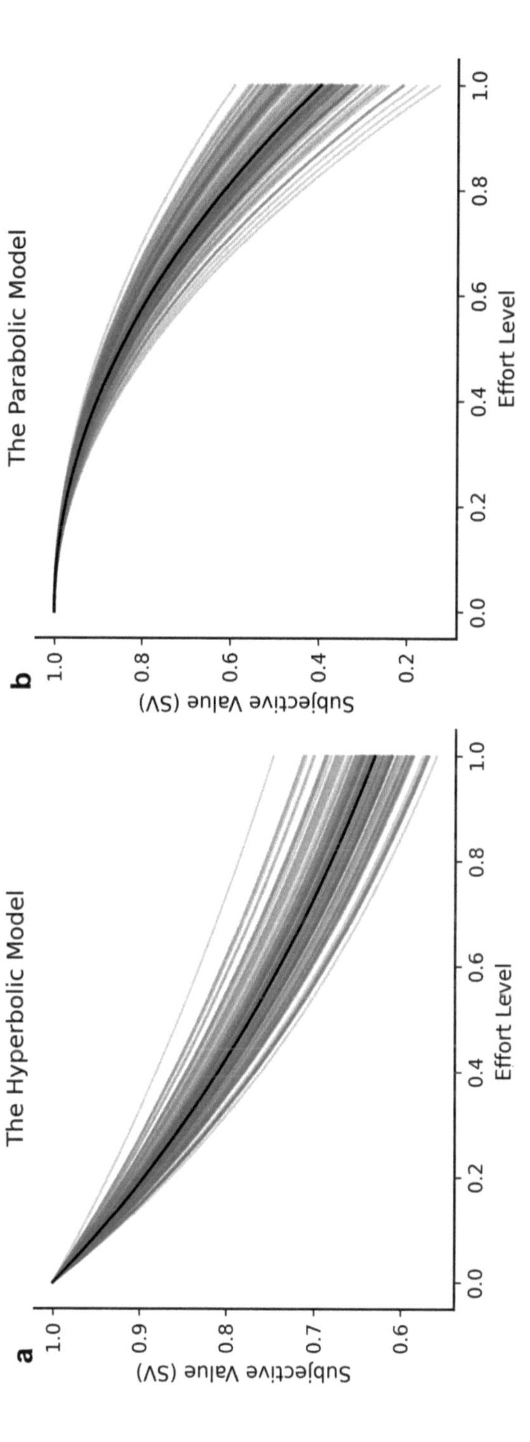

Fig. 16.3 Simulated subjective value across 100 participants at varying effort levels for the hyperbolic (**a**) and parabolic (**b**) models. Individual fits were shown as gray lines, and the group fits were shown as a black line

It comprises six spectra: internalizing (e.g., anxiety, depressive disorders, borderline disorder), thought disorder (e.g., psychotic symptoms), disinhibited externalizing (e.g., substance use disorders), antagonistic externalizing (e.g., narcissistic personality disorder), detachment (e.g., schizoid), and somatoform (e.g., somatoform symptom disorder) (Eaton et al. 2023; Kotov et al. 2017). Given that apathy is present across diverse mental disorders (e.g., thought disorder, internalizing disorder) and personality disorders (e.g., detachment), EBDM could serve as a pivotal factor in providing a more comprehensive and integrative understanding of mental disorders by capturing their shared and unique features.

In summary, the enduring impairment in EBDM and its associated prefrontal-mesolimbic abnormalities provide valuable insights into the developmental psychopathology of apathy in schizophrenia. This understanding can contribute to precise classification and the development of effective interventions across diverse neuropsychiatric disorders, including schizophrenia.

References

Abplanalp SJ, Mote J, Uhlman AC, Weizenbaum E, Alvi T, Tabak BA, Fulford D (2022) Parsing social motivation: development and validation of a self-report measure of social effort. J Ment Health 31(3):366–373. https://doi.org/10.1080/09638237.2021.1952948

Andreasen NC (1989) The scale for the assessment of negative symptoms (SANS): conceptual and theoretical foundations. Br J Psychiatry 155(S7):49–52. https://doi.org/10.1192/S0007125000291496

Barch DM, Treadway MT, Schoen N (2014) Effort, anhedonia, and function in schizophrenia: Reduced effort allocation predicts amotivation and functional impairment. J Abnorm Psychol 123:387–397. https://doi.org/10.1037/a0036299

Barch DM, Culbreth AJ, Ben Zeev D, Campbell A, Nepal S, Moran EK (2023) Dissociation of cognitive effort-based decision making and its associations with symptoms, cognition, and everyday life function across schizophrenia, bipolar disorder, and depression. Biol Psychiat 94(6):501–510. https://doi.org/10.1016/j.biopsych.2023.04.007

Bartra O, McGuire JT, Kable JW (2013) The valuation system: a coordinate-based meta-analysis of BOLD fMRI experiments examining neural correlates of subjective value. Neuroimage 76:412–427. https://doi.org/10.1016/j.neuroimage.2013.02.063

Birnie MT, Levis SC, Mahler SV, Baram TZ (2022) Developmental trajectories of anhedonia in preclinical models. In: Pizzagalli DA (Ed.), Anhedonia: preclinical, translational, and clinical integration, pp 23–41. Springer International Publishing. https://doi.org/10.1007/7854_2021_299

Blouzard E, Pouchon A, Polosan M, Bastin J, Dondé C (2023) Effort-cost decision-making among individuals with schizophrenia: a systematic review and meta-analysis. JAMA Psychiat 80(6):548–557. https://doi.org/10.1001/jamapsychiatry.2023.0553

Botvinick MM, Huffstetler S, McGuire JT (2009) Effort discounting in human nucleus accumbens. Cogn Affect Behav Neurosci 9(1):16–27. https://doi.org/10.3758/CABN.9.1.16

Bouc RL, Rigoux L, Schmidt L, Degos B, Welter M-L, Vidailhet M, Daunizeau J, Pessiglione M (2016) Computational dissection of dopamine motor and motivational functions in humans. J Neurosci 36(25):6623–6633. https://doi.org/10.1523/JNEUROSCI.3078-15.2016

Campellone TR, Truong B, Gard D, Schlosser DA (2018) Social motivation in people with recent-onset schizophrenia spectrum disorders. J Psychiatr Res 99:96–103. https://doi.org/10.1016/j.jpsychires.2018.01.006

Catalano LT, Green MF (2023) Social motivation in schizophrenia: what's effort got to do with it? Schizophr Bull 49(5):1127–1137. https://doi.org/10.1093/schbul/sbad090

Chang WC, Chu AOK, Treadway MT, Strauss GP, Chan SKW, Lee EHM, Hui CLM, Suen YN, Chen EYH (2019) Effort-based decision-making impairment in patients with clinically-stabilized first-episode psychosis and its relationship with amotivation and psychosocial functioning. Eur Neuropsychopharmacol 29(5):629–642. https://doi.org/10.1016/j.euroneuro.2019.03.006

Chang WC, Westbrook A, Strauss GP, Chu AOK, Chong CSY, Siu CMW, Chan SKW, Lee EHM, Hui CLM, Suen YM, Lo TL, Chen EYH (2020) Abnormal cognitive effort allocation and its association with amotivation in first-episode psychosis. Psychol Med 50(15):2599–2609. https://doi.org/10.1017/S0033291719002769

Chong TT-J (2018) Updating the role of dopamine in human motivation and apathy. Curr Opin Behav Sci 22:35–41. https://doi.org/10.1016/j.cobeha.2017.12.010

Chong TT-J, Apps M, Giehl K, Sillence A, Grima LL, Husain M (2017) Neurocomputational mechanisms underlying subjective valuation of effort costs. PLoS Biol 15(2):e1002598. https://doi.org/10.1371/journal.pbio.1002598

Chong TT-J, Bonnelle V, Husain M (2016) Chapter 4—Quantifying motivation with effort-based decision-making paradigms in health and disease. In: Studer B, Knecht S (Eds.), Progress in brain research, vol 229, pp 71–100. Elsevier. https://doi.org/10.1016/bs.pbr.2016.05.002

Chu RST, Tong CCHY, Wong CSM, Chang WC, Tang WCY, Chan CCL, Lui SSY, Hui LM, Suen YN, Chan KW, Lee HM, Chen EYH (2023) Effort-based decision making in schizotypy and its relationship with amotivation and psychosocial functioning. Front Psychiatry 14. https://doi.org/10.3389/fpsyt.2023.1123046

Cooper JA, Barch DM, Reddy LF, Horan WP, Green MF, Treadway MT (2019) Effortful goal-directed behavior in schizophrenia: computational subtypes and associations with cognition. J Abnorm Psychol 128:710–722. https://doi.org/10.1037/abn0000443

Costello H, Husain M, Roiser JP (2024) Apathy and motivation: biological basis and drug treatment. Annu Rev Pharmacol Toxicol 64(1):313-338. https://doi.org/10.1146/annurev-pharmtox-022423-014645

Cousins MS, Atherton A, Turner L, Salamone JD (1996) Nucleus accumbens dopamine depletions alter relative response allocation in a T-maze cost/benefit task. Behav Brain Res 74(1):189–197. https://doi.org/10.1016/0166-4328(95)00151-4

Crespo-Facorro B, Such P, Nylander A-G, Madera J, Resemann HK, Worthington E, O'Connor M, Drane E, Steeves S, Newton R (2021) The burden of disease in early schizophrenia—A systematic literature review. Curr Med Res Opin 37(1):109–121. https://doi.org/10.1080/03007995.2020.1841618

Culbreth A, Westbrook A, Barch D (2016a) Negative symptoms are associated with an increased subjective cost of cognitive effort. J Abnorm Psychol 125(4):528–536. https://doi.org/10.1037/abn0000153

Culbreth A, Westbrook A, Barch D (2016b) Negative symptoms are associated with an increased subjective cost of cognitive effort. J Abnorm Psychol 125:528–536. https://doi.org/10.1037/abn0000153

Culbreth AJ, Moran EK, Kandala S, Westbrook A, Barch DM (2020) Effort, avolition, and motivational experience in schizophrenia: analysis of behavioral and neuroimaging data with relationships to daily motivational experience. Clin Psychol Sci 8(3):555–568. https://doi.org/10.1177/2167702620901558

Culbreth AJ, Dershwitz SD, Barch DM, Moran EK (2023) Associations between cognitive and physical effort-based decision making in people with schizophrenia and healthy control subjects. Biol Psychiatry: Cogn Neurosci Neuroimaging 8(7):695–702. https://doi.org/10.1016/j.bpsc.2023.02.003

Dobryakova E, Jessup RK, Tricomi E (2017) Modulation of ventral striatal activity by cognitive effort. Neuroimage 147:330–338. https://doi.org/10.1016/j.neuroimage.2016.12.029

Docx L, de la Asuncion J, Sabbe B, Hoste L, Baeten R, Warnaerts N, Morrens M (2015) Effort discounting and its association with negative symptoms in schizophrenia. Cogn Neuropsychiatry 20(2):172–185. https://doi.org/10.1080/13546805.2014.993463

Dubey I, Ropar D, de C Hamilton AF (2015) Measuring the value of social engagement in adults with and without autism. Mol Autism 6(1):35. https://doi.org/10.1186/s13229-015-0031-2

Eaton NR, Bringmann LF, Elmer T, Fried EI, Forbes MK, Greene AL, Krueger RF, Kotov R, McGorry PD, Mei C, Waszczuk MA (2023) A review of approaches and models in psychopathology conceptualization research. Nat Rev Psychol, pp 1–15. https://doi.org/10.1038/s44159-023-00218-4

Ettinger U, Mohr C, Gooding DC, Cohen AS, Rapp A, Haenschel C, Park S (2015) Cognition and brain function in schizotypy: a selective review. Schizophr Bull 41(suppl_2):S417–S426. https://doi.org/10.1093/schbul/sbu190

Evensen J, Røssberg JI, Barder H, Haahr U, Hegelstad W ten V, Joa I, Johannessen JO, Larsen TK, Melle I, Opjordsmoen S, Rund BR, Simonsen E, Sundet K, Vaglum P, Friis S, McGlashan T (2012) Apathy in first episode psychosis patients: a ten year longitudinal follow-up study. Schizophr Res 136(1):19–24. https://doi.org/10.1016/j.schres.2011.12.019

Farrar AM, Segovia KN, Randall PA, Nunes EJ, Collins LE, Stopper CM, Port RG, Hockemeyer J, Müller CE, Correa M, Salamone JD (2010) Nucleus accumbens and effort-related functions: behavioral and neural markers of the interactions between adenosine A2A and dopamine D2 receptors. Neuroscience 166(4):1056–1067. https://doi.org/10.1016/j.neuroscience.2009.12.056

Fervaha G, Foussias G, Takeuchi H, Agid O, Remington G (2015) Measuring motivation in people with schizophrenia. Schizophr Res 169(1):423–426. https://doi.org/10.1016/j.schres.2015.09.012

Fortgang RG, Srihari V, Cannon TD (2020) Cognitive effort and amotivation in first-episode psychosis. J Abnorm Psychol 129(4):422–431. https://doi.org/10.1037/abn0000509

Foussias G, Agid O, Fervaha G, Remington G (2014) Negative symptoms of schizophrenia: clinical features, relevance to real world functioning and specificity versus other CNS disorders. Eur Neuropsychopharmacol 24(5):693–709. https://doi.org/10.1016/j.euroneuro.2013.10.017

Fulford D, Campellone T, Gard DE (2018) Social motivation in schizophrenia: how research on basic reward processes informs and limits our understanding. Clin Psychol Rev 63:12–24. https://doi.org/10.1016/j.cpr.2018.05.007

Gold JM, Strauss GP, Waltz JA, Robinson BM, Brown JK, Frank MJ (2013) Negative symptoms of schizophrenia are associated with abnormal effort-cost computations. Biol Psychiat 74(2):130–136. https://doi.org/10.1016/j.biopsych.2012.12.022

Green MF, Horan WP, Barch DM, Gold JM (2015) Effort-based decision making: a novel approach for assessing motivation in schizophrenia. Schizophr Bull 41(5):1035–1044. https://doi.org/10.1093/schbul/sbv071

Hartmann MN, Hager OM, Tobler PN, Kaiser S (2013) Parabolic discounting of monetary rewards by physical effort. Behav Proc 100:192–196. https://doi.org/10.1016/j.beproc.2013.09.014

Heron CL, Holroyd CB, Salamone J, Husain M (2019) Brain mechanisms underlying apathy. J Neurol Neurosurg Psychiatry 90(3):302–312. https://doi.org/10.1136/jnnp-2018-318265

Hogan PS, Galaro JK, Chib VS (2019) Roles of ventromedial prefrontal cortex and anterior cingulate in subjective valuation of prospective effort. Cereb Cortex 29(10):4277–4290. https://doi.org/10.1093/cercor/bhy310

Huang J, Yang X, Lan Y, Zhu C, Liu X, Wang Y, Cheung EFC, Xie G, Chan RCK (2016) Neural substrates of the impaired effort expenditure decision making in schizophrenia. Neuropsychology 30(6):685–696. https://doi.org/10.1037/neu0000284

Husain M, Roiser JP (2018) Neuroscience of apathy and anhedonia: a transdiagnostic approach. Nat Rev Neurosci 19(8). https://doi.org/10.1038/s41583-018-0029-9

Kotov R, Krueger RF, Watson D, Achenbach TM, Althoff RR, Bagby RM, Brown TA, Carpenter WT, Caspi A, Clark LA, Eaton NR, Forbes MK, Forbush KT, Goldberg D, Hasin D, Hyman SE, Ivanova MY, Lynam DR, Markon K, Zimmerman M (2017) The hierarchical taxonomy

of psychopathology (HiTOP): a dimensional alternative to traditional nosologies. J Abnorm Psychol 126(4):454–477. https://doi.org/10.1037/abn0000258
Kwapil TR, Barrantes-Vidal N (2015) Schizotypy: looking back and moving forward. Schizophr Bull 41(suppl_2):S366–S373. https://doi.org/10.1093/schbul/sbu186
Le Heron C, Apps MAJ, Husain M (2018) The anatomy of apathy: a neurocognitive framework for amotivated behaviour. Neuropsychologia 118:54–67. https://doi.org/10.1016/j.neuropsychologia.2017.07.003
Levy R, Dubois B (2006) Apathy and the functional anatomy of the prefrontal cortex-basal ganglia circuits. Cereb Cortex 16(7):916–928. https://doi.org/10.1093/cercor/bhj043
Levy R (2012) Apathy: a pathology of goal-directed behaviour. A new concept of the clinic and pathophysiology of apathy. Rev Neurol 168(8):585–597. https://doi.org/10.1016/j.neurol.2012.05.003
Luther L, Fischer MW, Firmin RL, Salyers MP (2019) Clarifying the overlap between motivation and negative symptom measures in schizophrenia research: a meta-analysis. Schizophr Res 206:27–36. https://doi.org/10.1016/j.schres.2018.10.010
Mai B, Sommer S, Hauber W (2012) Motivational states influence effort-based decision making in rats: the role of dopamine in the nucleus accumbens. Cogn Affect Behav Neurosci 12(1):74–84. https://doi.org/10.3758/s13415-011-0068-4
Marder SR, Galderisi S (2017) The current conceptualization of negative symptoms in schizophrenia. World Psychiatry 16(1):14–24. https://doi.org/10.1002/wps.20385
Marin R (1990) Differential diagnosis and classification of apathy. Am J Psychiatry 147:22–30. https://doi.org/10.1176/ajp.147.1.22
McCarthy JM, Treadway MT, Blanchard JJ (2015) Motivation and effort in individuals with social anhedonia. Schizophr Res 165(1):70–75. https://doi.org/10.1016/j.schres.2015.03.030
McCutcheon RA, Reis Marques T, Howes OD (2020) Schizophrenia—An overview. JAMA Psychiat 77(2):201–210. https://doi.org/10.1001/jamapsychiatry.2019.3360
Moran EK, Culbreth AJ, Kandala S, Barch DM (2019) From neuroimaging to daily functioning: a multimethod analysis of reward anticipation in people with schizophrenia. J Abnorm Psychol 128(7):723–734. https://doi.org/10.1037/abn0000461
Park IH, Lee BC, Kim J-J, Kim JI, Koo M-S (2017) Effort-based reinforcement processing and functional connectivity underlying amotivation in medicated patients with depression and schizophrenia. J Neurosci 37(16):4370–4380. https://doi.org/10.1523/JNEUROSCI.2524-16.2017
Pessiglione M, Vinckier F, Bouret S, Daunizeau J, Le Bouc R (2018) Why not try harder? Computational approach to motivation deficits in neuro-psychiatric diseases. Brain 141(3):629–650. https://doi.org/10.1093/brain/awx278
Prettyman GE, Kable JW, Didier P, Shankar S, Satterthwaite TD, Davatzikos C, Bilker WB, Elliott MA, Ruparel K, Wolf DH (2021) Relationship of ventral striatum activation during effort discounting to clinical amotivation severity in schizophrenia. Npj Schizophr 7(1). https://doi.org/10.1038/s41537-021-00178-9
Reddy LF, Horan WP, Barch DM, Buchanan RW, Dunayevich E, Gold JM, Lyons N, Marder SR, Treadway MT, Wynn JK, Young JW, Green MF (2015) Effort-based decision-making paradigms for clinical trials in schizophrenia: part 1—Psychometric characteristics of 5 paradigms. Schizophr Bull 41(5):1045–1054. https://doi.org/10.1093/schbul/sbv089
Reddy LF, Horan WP, Green MF (2016) Motivational deficits and negative symptoms in schizophrenia: concepts and assessments. In: Simpson EH, Balsam PD (Eds.), Behavioral neuroscience of motivation, pp 357–373. Springer International Publishing. https://doi.org/10.1007/7854_2015_379
Rigoli F, Pezzulo G (2022) A reference-based theory of motivation and effort allocation. Psychon Bull Rev 29(6):2070–2082. https://doi.org/10.3758/s13423-022-02135-8
Ristic B, Batinic B (2022) Correlation between social functioning and quality of life in patients diagnosed with schizophrenia. Eur Psychiatry 65(S1):S195–S195. https://doi.org/10.1192/j.eurpsy.2022.513

Robert P, Lanctôt KL, Agüera-Ortiz L, Aalten P, Bremond F, Defrancesco M, Hanon C, David R, Dubois B, Dujardin K, Husain M, König A, Levy R, Mantua V, Meulien D, Miller D, Moebius HJ, Rasmussen J, Robert G, Manera V (2018) Is it time to revise the diagnostic criteria for apathy in brain disorders? The 2018 international consensus group. Eur Psychiatry 54:71–76. https://doi.org/10.1016/j.eurpsy.2018.07.008

Salamone JD (1996) The behavioral neurochemistry of motivation: Methodological and conceptual issues in studies of the dynamic activity of nucleus accumbens dopamine. J Neurosci Methods 64(2):137–149. https://doi.org/10.1016/0165-0270(95)00125-5

Salamone JD, Steinpreis RE, McCullough LD, Smith P, Grebel D, Mahan K (1991) Haloperidol and nucleus accumbens dopamine depletion suppress lever pressing for food but increase free food consumption in a novel food choice procedure. Psychopharmacology 104(4):515–521. https://doi.org/10.1007/BF02245659

Salamone JD, Cousins MS, Bucher S (1994a) Anhedonia or anergia? Effects of haloperidol and nucleus accumbens dopamine depletion on instrumental response selection in a T-maze cost/benefit procedure. Behav Brain Res 65(2):221–229. https://doi.org/10.1016/0166-4328(94)90108-2

Salamone JD, Cousins MS, McCullough LD, Carriero DL, Berkowitz RJ (1994b) Nucleus accumbens dopamine release increases during instrumental lever pressing for food but not free food consumption. Pharmacol Biochem Behav 49(1):25–31. https://doi.org/10.1016/0091-3057(94)90452-9

Salamone JD, Correa M, Farrar A, Mingote SM (2007) Effort-related functions of nucleus accumbens dopamine and associated forebrain circuits. Psychopharmacology 191(3):461–482. https://doi.org/10.1007/s00213-006-0668-9

Salamone JD, Yohn SE, López-Cruz L, San Miguel N, Correa M (2016) Activational and effort-related aspects of motivation: neural mechanisms and implications for psychopathology. Brain 139(5):1325–1347. https://doi.org/10.1093/brain/aww050

Saleh Y, Jarratt-Barnham I, Fernandez-Egea E, Husain M (2021) Mechanisms underlying motivational dysfunction in schizophrenia. Front Behav Neurosci 15. https://doi.org/10.3389/fnbeh.2021.709753

Saleh Y, Jarratt-Barnham I, Petitet P, Fernandez-Egea E, Manohar SG, Husain M (2023) Negative symptoms and cognitive impairment are associated with distinct motivational deficits in treatment resistant schizophrenia. Mol Psychiatry, pp 1–11. https://doi.org/10.1038/s41380-023-02232-7

Sauvé G, Brodeur MB, Shah JL, Lepage M (2019) The prevalence of negative symptoms across the stages of the psychosis continuum. Harv Rev Psychiatry 27(1):15. https://doi.org/10.1097/HRP.0000000000000184

Scheffer JA, Cameron CD, Inzlicht M (2022) Caring is costly: people avoid the cognitive work of compassion. J Exp Psychol Gen 151(1):172–196. https://doi.org/10.1037/xge0001073

Sesack SR, Grace AA (2010) Cortico-basal ganglia reward network: microcircuitry. Neuropsychopharmacology 35(1). https://doi.org/10.1038/npp.2009.93

Shukla DK, Chiappelli JJ, Sampath H, Kochunov P, Hare SM, Wisner K, Rowland LM, Hong LE (2019) Aberrant frontostriatal connectivity in negative symptoms of schizophrenia. Schizophr Bull 45(5):1051–1059. https://doi.org/10.1093/schbul/sby165

Starkstein SE (2000) Apathy and withdrawal. Int Psychogeriatr 12(S1):135–137. https://doi.org/10.1017/S1041610200006918

Starkstein SE, Leentjens AFG (2008) The nosological position of apathy in clinical practice. J Neurol Neurosurg Psychiatry. https://doi.org/10.1136/jnnp.2007.136895

Strauss GP, Whearty KM, Morra LF, Sullivan SK, Ossenfort KL, Frost KH (2016) Avolition in schizophrenia is associated with reduced willingness to expend effort for reward on a progressive ratio task. Schizophr Res 170(1):198–204. https://doi.org/10.1016/j.schres.2015.12.006

Strauss GP, Zamani Esfahlani F, Sayama H, Kirkpatrick B, Opler MG, Saoud JB, Davidson M, Luthringer R (2020) Network analysis indicates that avolition is the most central domain for the successful treatment of negative symptoms: evidence from the roluperidone randomized clinical trial. Schizophr Bull 46(4):964–970. https://doi.org/10.1093/schbul/sbz141

Strauss GP, Bartolomeo LA, Luther L (2023) Reduced willingness to expend effort for rewards is associated with risk for conversion and negative symptom severity in youth at clinical high-risk for psychosis. Psychol Med 53(3):714–721. https://doi.org/10.1017/S003329172100204X

Strauss GP, Bartolomeo LA, Harvey PD (2021) Avolition as the core negative symptom in schizophrenia: relevance to pharmacological treatment development. Npj Schizophr 7(1). https://doi.org/10.1038/s41537-021-00145-4

Suzuki S, Lawlor VM, Cooper JA, Arulpragasam AR, Treadway MT (2021) Distinct regions of the striatum underlying effort, movement initiation and effort discounting. Nat Hum Behav 5(3). https://doi.org/10.1038/s41562-020-00972-y

Terenzi D, Mainetto E, Barbato M, Rumiati RI, Aiello M (2019) Temporal and effort cost decision-making in healthy individuals with subclinical psychotic symptoms. Sci Rep 9(1). https://doi.org/10.1038/s41598-018-38284-x

Treadway MT, Buckholtz JW, Schwartzman AN, Lambert WE, Zald DH (2009) Worth the 'EEfRT'? The effort expenditure for rewards task as an objective measure of motivation and anhedonia. PLoS ONE 4(8):e6598. https://doi.org/10.1371/journal.pone.0006598

Treadway MT, Peterman JS, Zald DH, Park S (2015) Impaired effort allocation in patients with schizophrenia. Schizophr Res 161(2):382–385. https://doi.org/10.1016/j.schres.2014.11.024

Vassena E, Holroyd CB, Alexander WH (2017) Computational models of anterior cingulate cortex: at the crossroads between prediction and effort. Front Neurosci 11. https://doi.org/10.3389/fnins.2017.00316

Walton ME, Chau BK, Kennerley SW (2015) Prioritising the relevant information for learning and decision making within orbital and ventromedial prefrontal cortex. Curr Opin Behav Sci 1:78–85. https://doi.org/10.1016/j.cobeha.2014.10.005

Wang J, Huang J, Yang X, Lui SSY, Cheung EFC, Chan RCK (2015) Anhedonia in schizophrenia: deficits in both motivation and hedonic capacity. Schizophr Res 168(1):465–474. https://doi.org/10.1016/j.schres.2015.06.019

Wang J, Huang J, Lui SSY, Cheung EFC, Yue X, Chan RCK (2018) Motivation deficits in individuals with social anhedonia. Psychiatry Res 261:527–534. https://doi.org/10.1016/j.psychres.2018.01.033

Wang Y, Wang Y, Huang J, Sun X, Wang X, Zhang S, Zhu G, Lui SSY, Cheung EFC, Sun H, Chan RCK (2022) Shared and distinct reward neural mechanisms among patients with schizophrenia, major depressive disorder, and bipolar disorder: an effort-based functional imaging study. Eur Arch Psychiatry Clin Neurosci 272(5):859–871. https://doi.org/10.1007/s00406-021-01376-3

Whitton AE, Merchant JT, Lewandowski KE (2020) Dissociable mechanisms underpinning effort-cost decision-making across the psychosis spectrum. Schizophr Res 224:133–140. https://doi.org/10.1016/j.schres.2020.09.014

Wolf DH, Satterthwaite TD, Kantrowitz JJ, Katchmar N, Vandekar L, Elliott MA, Ruparel K (2014) Amotivation in schizophrenia: integrated assessment with behavioral, clinical, and imaging measures. Schizophr Bull 40(6):1328–1337. https://doi.org/10.1093/schbul/sbu026

Yang X, Huang J, Harrision P, Roser ME, Tian K, Wang D, Liu G (2021) Motivational differences in unipolar and bipolar depression, manic bipolar, acute and stable phase schizophrenia. J Affect Disord 283:254–261. https://doi.org/10.1016/j.jad.2021.01.075

Zou Y-M, Ni K, Wang Y-Y, Yu E-Q, Lui SSY, Zhou F-C, Yang H-X, Cohen AS, Strauss GP, Cheung EFC, Chan RCK (2020) Effort–cost computation in a transdiagnostic psychiatric sample: differences among patients with schizophrenia, bipolar disorder, and major depressive disorder. PsyCh Journal 9(2):210–222. https://doi.org/10.1002/pchj.316

Chapter 17
Not Black or White—Ethical Decision Making in Complex Medical Situations

Lukas Radbruch, Marta Przyborek, and Séverine M. Surges

Abstract In Germany, as in many other countries, any medical decision is based on two tiers: Firstly, the intervention requires a medical indication, and secondly, if indicated, the intervention requires patient consent. Patient consent should be grounded in participatory decision making, i.e., a process with input from both physician and patient, using a collegial model of the physician–patient-relationship. However, in many, if not most cases, there are no simple black or white solutions, and decision-making is complex and cumbersome. Medical decisions more often than not do not follow a textbook approach. Settings are complex, with a range of confounding factors, and without a clear-cut, easy solution. Medical standards and guidelines provide guidance on many illnesses, but decisions have to be tailored to the specific situation of the individual patient. Physicians should understand decision-making as a process and not as a singular event. This requires the ability and willingness to enter a relationship with the patient, building trust, repeating discussions, enduring ambiguities and tolerating ambivalence. This relationship should be at eye-level, with mutual respect between patient and physician as experts in their specific fields. The physician is the medical expert who knows about the available treatment options and their benefit and burden, and the patient is the expert for himself or herself, and more specifically for his or her personal preferences and values.

Keywords Medical indication · Futility · Shared decision-making · Health literacy · Wish for hastened death

17.1 Medical Decisions

If you watch TV shows such as Emergency Room or Grey's Anatomy, you most probably will have got the impression that decisions by medical practitioners are always clear-cut, straightforward processes, with a particular set of symptoms leading immediately to a specific diagnosis. It presupposes that the patient does not omit any

---

L. Radbruch (✉) · M. Przyborek · S. M. Surges
Department of Palliative Medicine, University Hospital Bonn, Bonn, Germany
e-mail: Lukas.Radbruch@ukbonn.de

U. Ettinger et al. (eds.), *Decision Making*, Studies in Neuroscience, Psychology and Behavioral Economics, https://doi.org/10.1007/978-3-032-00880-0_17

critical information, and that medical staff are competent and diligent. It is not how decision-making works in real clinical life, however.

The following case reports are real cases from the palliative care or ethics consultation experience of the authors and are representative of typical situations in medical decision-making.

Example A 94-year-old woman is treated in a neurology ward. She suffered a stroke a few weeks ago. She is unable to talk, but can understand and communicate nonverbally. The next of kin is her nephew, who holds lasting power of attorney for her. As the patient has problems swallowing, the insertion of a percutaneous endoscopic gastrostomy (PEG) is indicated. However, the patient cannot indicate clearly whether she consents to the procedure or not. The nephew is leaning to decide against PEG placement, given that her aunt always wanted to remain as self-sufficient as possible and not being able to feed herself might not be a life worth living for her. On the other hand, there are clear indicators that she is still enjoying life to some degree. When she is asked in the ethics consultation who should decide about the PEG placement, she points at the ward physician. The physician follows the medical indication and supports a decision in favour of the PEG placement.

The example case illustrates typical problems with medical decision-making: impaired communication, unclear information status, problems to elicit patient preferences and values and problems to identify and balance the risks and benefits of medical interventions for the specific situation of the patient.

The demands on medical decision-making are high. In Germany, as in many other countries, any medical decision is based on two tiers: Firstly, the intervention requires a medical indication, and secondly, if indicated, the intervention requires patient consent. Patient consent should be grounded in participatory decision making, as process with input from both physician and patient, using a cooperative model of the physician–patient-relationship. Using this model, patients have the right to make decisions about their own healthcare in line with their own values and preferences, even if those decision may seem irrational to healthcare practitioners. However, in real life, all of these foundations of medical decision-making are fraught with problems.

17.2 The Role of the Physician

17.2.1 Medical Indication

Medical interventions should only be offered as options to patients if they are indicated. However, there is no clear definition of a medical indication. The German civil or penal code does not include any clear-cut definition of medical indications. The German Medical Association has defined "medical indication" as a doctor's assessment that a specific medical measure is required in order to achieve a specific treatment goal (Bundesärztekammer 2015). This is described as a core element of the medical profession. Three sources deliver input to this assessment: the medical professional's assessment, moral value and the patient's individual preferences. Medical professionals need to balance benefits and risks of an intervention and use professional standards and guidelines for orientation. Moral values provide the foundation for goals for medical interventions, for example, the relief of suffering. Individual preferences resist any standardization, but instead reflect the personal characteristics of a patient, his or her values and biographical situation. This individual dimension does not establish a patient's right to request an intervention. Instead, it provides a first filter as to whether an intervention should be offered to the patient at all. If a physician's assessment does not find that an intervention is indicated, it should not even be offered to the patient. Interventions with a medical indication for an individual patient in a specific setting are offered to the patient, who then can select from among these options or reject them. Patients can reject any intervention, but they cannot demand an intervention that has no medical indication in their specific situation.

Balancing benefits and risks can be difficult. Physicians who perform an intervention regularly may be overly enthusiastic about the benefits it bestows, and may be inclined to acknowledge a medical indication for this intervention too often. This has been named the law of the instrument, according to Abraham Maslow: "it is tempting, if the only tool you have is a hammer, to treat everything as if it were a nail" (Quoteinvestigator 2014). Physicians who are not familiar with this intervention may be biased in the other direction.

As a consequence, assessment of a medical indication may benefit from team discussions or multidisciplinary consultations. Comprehensive cancer centers have implemented multidisciplinary tumour boards for these discussions, to minimize bias in cancer treatment decisions. Team discussions can relieve the burden of decision-making in other settings as well, specifically for withholding or discontinuation of therapies. A survey of specialist palliative care teams on use of antibiotics found that the decision to initiate antibiotic therapy was usually made by the physician alone, whereas the decision to withhold or discontinue most often required involvement of other team members (Stiel et al. 2012).

17.2.2 Medical Futility

Conflicts regarding medical indications may really come to a head if life sustaining therapies are at stake, and especially if the patient requests such an intervention. The medical indication for life sustaining therapies is considered absent if the intervention is considered futile. However, medical futility is a concept that is ill-defined. A much-cited definition states that "when physicians conclude (either through personal experience, experiences shared with colleagues, or consideration of published empiric data) that in the last 100 cases a medical treatment has been useless, they should regard that treatment as futile" (Schneiderman et al. 1990). Other authors have proposed even stricter definitions; for example, Waisel and Truog have recommended to restrict the use to physiological futility, that is, a situation in which an intervention cannot achieve its physiological goal, as any broader definition would include ethical values that might not be shared by others (Waisel and Truog 1995). The restriction of the concept of futility to the rare occasions where an intervention cannot achieve the physiological goal was adopted in a consensus paper of several medical professional associations, providing an example of cardiopulmonary resuscitation being only futile if cardiac circulation or ventilation are not possible because of cardiac tamponade or cardiac rupture (Bosslet et al. 2015). For all other scenarios, the term "potentially inappropriate treatments" was recommended, i.e., if interventions still have some—however small—chances of success, but carry a burden of serious ethical concerns.

In clinical practice, broader definitions of medical futility are used (Aghabarary and Dehghan Nayeri 2016), often stating that interventions are futile if they do not offer reasonable hope to achieve the treatment goal, provide a benefit or result in a quality of life that would be satisfactory to the patient.

Example A 21-year old woman with melanoma of the vulva is transferred from the oncology department of another hospital to the palliative care unit of the university hospital. Cancer was diagnosed three years ago, and she has been treated with a broad range of chemotherapy, immunotherapy and surgical procedures. In spite of this, cancer has spread locally and metastasized in the abdomen, causing gastrointestinal obstruction, ascites, pleural effusions and renal insufficiency. The patient and her mother are requesting maximal medical therapy, including cardiopulmonary resuscitation if needed. The pretreating hospital, following an ethics consultation and a rather ambiguous statement from a lawyer, had agreed to provide all that. With admission, the team discussed the medical indications for resuscitation, continuation of antibiotic therapy and parenteral nutrition and fluids, as the prognosis was considered to be only a few days or less with the advanced and progressive stage of the disease. The team consented that resuscitation would be futile, and that an increase of the parenteral nutrition and fluids (as patient and mother had requested) would be non-beneficial and most probably would harm the patient by increasing ascites and pleural effusions, which then could exacerbate breathlessness and pain. However, unchanged continuation of parenteral nutrition and antibiotics was considered at least not to harm the patient, and was thus kept in place. The patient was too weak and exhausted to protest, but delegated decision-making to her mother, who fought the decision to not resuscitate continuously, causing much distress to the palliative care team. The patient died four days later without any resuscitation attempts. Two retrospective ethics consultations were needed to clarify whether medical indication and futility of parenteral nutrition and antibiotics had been considered adequately, whether the patient should have been shielded from her mother, who often also declined symptom relieving medication such as opioids, and whether the team would have needed more protection from the aggressive behaviour of the mother. In spite of the challenging course of treatment, the team agreed again that resuscitation would have been futile and thus had not been indicated.

This case example describes the challenges with medical indication and futility, not only for resuscitation measures, but also for other interventions that might be harmful for the imminently dying patient. Team discussions and ethical consultations demonstrated a range of diverging ethical values in the team, but finally provided consensus on which interventions should have been initiated, continued or withheld.

17.2.3 Shared Decision Making

In an ideal cooperative relationship, patient and physician would collaborate to identify realistic treatment goals and select the best available treatment option. Such shared decision-making (Alsulamy et al. 2020; Belanger et al. 2011; Brock 1991; Covvey et al. 2019; Ernst et al. 2014; Moumjid et al. 2007) requires considerable trust, with patient and physician discussing at eye-level, and respecting each other as experts in their specific fields. The physician is the medical expert who knows about the available treatment options and their benefit and burden. The patient is the expert for himself or herself, and, more specifically, for his or her personal preferences and values. The physician offers treatment options that are medically indicated, and the patient selects the option best suited to his or her preferences. The physician can even contribute to the value-driven evaluation of treatment options, either using his or her own perspective ("If I were in your place, I probably would prefer to…") or his or her understanding of the patient's perspective ("As far as I have come to know you, I would think that you probably would prefer to…").

A paternalistic patient-physician relationship is outdated. A contractual relationship, with the physician merely delivering what the patient has ordered, also seems incompatible with the four principles of biomedical ethics, which balances autonomy with the principles of beneficence, non-malevolence and justice.[1] However, there are patients who wish to delegate as much decision-making as possible to the physician, and for whom a more paternalistic approach might be appropriate. In our experience, most physicians do not seem to have a problem with such a shift. On the other end of the spectrum are patients with a very high degree of autonomy, who want to be in charge of decision-making. Even though these patients might benefit from a contractual relationship with their physician, this usually meets much more resistance from the physician's side.

In the cooperative relationship the physician informs about all available treatment options, with the prerequisite that only options with a medical indication are offered to the patient. Considering the grey areas around medical indications, physicians quite often will include such treatment options in their discussions with the patient even though they think these interventions will not benefit or unduly burden the patient. They would expect that the patient would realize this inappropriate benefit-risk ratio and reject these options. However, patients may find large risks and small

[1] Beauchamps T, Childress J (2012) The principles of biomedical ethics. 7th ed. Oxford University Press, New York.

benefits acceptable, and might select these options. Instead of a clear statement that there is no indication for a specific intervention, it then seems easier to delegate the decision to the patient, by painting the picture darker: “I will just discuss this with the patient again, this time describing the disadvantages and risks more forcefully and in great detail, until the patient does not want the intervention anymore” (as a participant in a palliative care course explained). This is profoundly unfair, as it burdens the patient with a decision that the physician should have made, and leaves all the distress of the decision-making with the patient.

17.3 The Patient’s Perspective

17.3.1 Health Literacy

Shared decision-making requires adequate cognitive capacity of the patient, with the ability to understand the information, to be aware of the consequences of the decision and to express the decision.

A basic understanding of how medicine works is also desirable. From the layperson’s perspective, a medical problem has a cause that needs to be identified and treated, leading to rapid improvement of symptoms and function. However, this is the exemption rather than the rule in real life. A broad range of differential diagnoses has to be considered during diagnosis, and the most frequent diseases may be diagnosed by exclusion. To give an example: Arterial hypertension is a widespread disease, affecting around one third of the German population (Hengel et al. 2022), but a specific aetiology can be found only for 5–10% of the patients (Tziomalos 2020). The vast majority is left with the diagnosis of essential hypertension, i.e. no known cause.

“Watch and wait” is an important concept in medicine. A cough most often is related to the common cold, and (with or without treatment) will disappear within a few days. Only if coughing persists for more than four weeks, detailed diagnosis is required to exclude serious illnesses such as lung cancer. This means that often patients and physicians are left with a tentative diagnosis, awaiting confirmation or differential diagnosis from further tests or sometimes from the course of the disease.

On the other end of the spectrum, with progressive and advanced illness, experts from intensive care, geriatrics and palliative care can all testify that monocausal aetiologies are the exception in patients with severe illness. More often multimorbidity, side effects from extensive medication and multiple organ dysfunctions all merge in a kind of “black box”, where cause-effect relationships are difficult to interpret. This means that the effects of any medical intervention cannot be completely predicted. As a consequence, changes of the therapeutic regimen have to be done regularly and carefully, and one step at a time, whilst closely observing the resulting positive and negative changes.

"Watch and wait" can be an important factor in decision-making as well. Ethical consultations in intensive care sometimes lead to no consensus on whether to continue or even escalate medical care or switch to withholding and discontinuing life-sustaining therapies, as the prognosis is unclear and team members do not agree on the predicted disease trajectory. In such cases, continuation of intensive care for a predetermined length of time, for example two weeks, might be the best option, followed by re-evaluation with another ethics consultation. This "watch and wait"-period often results in a consensus as the course of the illness becomes clearer.

This basic understanding of the ways and means of medicine should be part of the health literacy of patients. Health literacy also includes a basic understanding of how the human body functions, as well as knowledge of health promoting factors and risk factors of illnesses, and tell-tale signs of diseases. Health literacy was found to be low in 59% of the German population, with low levels of education, social status, literacy or poverty as well as old age as risk factors for lower health literacy (Schaeffer et al. 2021). Lower health literacy correlated significantly with the presence of multiple chronic diseases. Improving health literacy with public campaigns or as part of the general education seems paramount to improving public health in general and facilitating shared decision-making more specifically.

A specific example of poor health literacy is the belief that palliative care is only for the imminently dying (Alcalde and Zimmermann 2022; Cherny 2009). As a physician in the palliative care consultation service in the hospital, introducing oneself to a patient in the oncology department often is met with fear, as the palliative care expert is seen as some kind of death angel, foreshadowing imminent death. This myth is hard to overcome, despite all efforts to implement the early integration of palliative care, aiming for information on symptom relief and other palliative care options early on in the disease trajectory.

17.3.2 Individual Values

Most patients will have very specific individual values that influence their hopes and fears related to health care. This will affect their position related to medical decisions. As an example from palliative care, some patients facing imminent death wish to prevent suffering at all costs, even if the medication renders them drowsy or even unconsciousness, while others want to keep a clear mind for as long as possible, sometimes even rejecting opioid medications and bearing the pain in order to stay awake. For a health care professional, trying to understand the individual values, preferences and priorities is a prerequisite of shared decision-making.

This understanding is not always easy to get. Patient autonomy includes the patient's right to make seemingly irrational decisions, for example rejecting life-sustaining therapies. Even life-sustaining therapies such as mechanical ventilation for a patient with an advanced neurological disease such as amyotrophic lateral sclerosis (ALS) has to be discontinued, if the patient decides so, no matter how much the medical team finds them effective. This "letting die" is not only allowed in Germany, but is a patient's right that he or she can legally enforce.

Example A 54-year-old man had been diagnosed with an HIV infection, but had rejected antiretroviral therapy (ART). He did not want to elaborate, but stated clearly and repeatedly that he did not want to start this treatment. Over several years, the infection progressed to AIDS with a range of infections, but he still refused ART. After three years, following admission to a palliative care unit for pain management, he suddenly started a discussion whether this had been the right decision, and whether he should now start ART. However, the disease had progressed further, leading to his death before ART could be initiated.

Apparently irrational decisions such as the decision not to use ART often lead to irritations and anger among the therapeutic team. However, this should be met with reflection, inter- or supervision in the team, and efforts to better understand the values of the patient. Even if irrational decisions are accepted and supported by the medical team, this can lead to a dilemma later on when the patient begins questioning his original decision, but has reached a point in his disease trajectory where this treatment option is no longer viable.

Example A 62-year old Turkish woman had been admitted to the internal medicine ward with a tentative diagnosis of inflammation of the bile ducts. However, diagnostic tests raised the suspicion of pancreatic cancer. The patient knew only a few German words, and communication was mostly via her daughter, who together with two other siblings also had medical power of attorney for her. The daughter rejected a professional translator and insisted that she would be the only one to translate any communication with the patient. She also insisted that for psychological reasons, death, dying or cancer diagnosis should not be mentioned at all in the presence of the patient. An advance directive was presented, but it was in German language and thus did not seem to be relevant considering the patient's lack of knowledge of the German language. The health care team was in a catch-22 between the requests from the daughter and the need to discuss invasive diagnostic procedures and chemotherapy options with the patient. During an ethics consultation, this conflict was discussed with the daughter in the presence of the patient, who suddenly stated that she wanted "no chemotherapy". Asked about this, the daughter explained that they had discussed the medical issues in the family using the metaphor of a "malignant inflammation", which might require chemotherapy, instead of the cancer terminology. Using this terminology, the mother had stated clearly that she did not want chemotherapy. The medical team accepted this terminology, as it would allow them to discuss diagnostic and therapeutic options with the patient.

In this case the family had decided that any mentioning of a potential cancer diagnosis would burden the patient too much, would sap her strength and maybe even hasten death. Not talking about this tentative diagnosis would allow her to retain hope.

Irrational decisions may be linked to irrational hopes, for example hoping for a miracle in the face of imminent death. This requires careful elicitation of such hopes, not with the aim of squashing them, but if possible recalibrating hope towards a more achievable goal.

Something that seems like an unrealistic hope and may be the result of a different evaluation of chances and risks of therapeutic options, which may make a lot of sense from the patient's perspective.

Example A 42-year of women had been admitted to the palliative care unit with far advanced and metastatic cancer of the ovaries. First- und second-line chemotherapy had been futile, and the oncologist had stated that third-line chemotherapy would have a chance of 7% to prolong life expectancy, but probably only for a short period of time. As treating physician, and considering the severe side effects and the risk of complication with chemotherapy as well as the low chance of effectiveness, I assumed that she would reject this option in favour of more time and better quality of life for the remaining life span. However, the patient insisted on starting the chemotherapy, explaining to me that 7% was better than nothing, and she felt that she owed it to her children (aged 10–15 years) that she would grab any therapeutic option, no matter how slim the chances. She explained that her slogan was "never say die, better to go down with the flags flying than to give up".

Denial of the seriousness of the illness, or even denial of diagnosis and prognosis shortening the remaining lifespan suddenly so much does not necessarily mean a lack of understanding or comprehension, or unreasonableness. As Kübler-Ross (1970) has described, denial (as well as aggression, bargaining, depression or acceptance) are perfectly normal reactions to sudden loss (of life expectancy and hope) and grief. Urgent requests of further interventions may be an audible expression of these reactions.

17.4 Decision-Making as a Process

17.4.1 Complexity

The issues explained above show that decision-making is complex and cumbersome in many, if not most, cases. Medical decisions more often than not do not follow a textbook approach. Settings are complex, with a range of confounding factors, and without clear-cut, easy solutions. Medical standards and guidelines provide guidance on many illnesses, but decisions have to be tailored to the specific situation of the individual patient. With serious and progressive illness many factors have to be taken into account.

Example A 69-year old patient with lung cancer is admitted to the palliative care unit of a university hospital. At admission, liver metastases are diagnosed. The patient is cachectic with a body weight of 52 kg at 180 cm body height. Second-line chemotherapy had been discontinued the week before because it lacked effectiveness. He is asking repeatedly for new chemotherapy or other anticancer interventions:

"You have to do something!" The consulted oncologist estimated that a third-line chemotherapy would have only a 5% chance of prolonging life. Please answer the following questions: (a) Is a third-line chemotherapy indicated?; (b) How appropriate would rejection of further chemotherapy be?; and c) Are there any alternative options?

This simple case example, compiled as a combination of several real cases, can lead to long discussions in palliative care courses, for example on the medical indication of further chemotherapy (Radbruch 2019) (Table 17.1).

This list of pro and contra arguments can be expanded. Offering another chemotherapy intervention may be motivated more by a feeling of helplessness in the care team than realistic hope of effectiveness. The treating physician may find it hard to reject the patient's or the family's request for help. Offering another chemotherapy regimen seems to provide help and signifies competence, even if the benefits for the

Table 17.1 Pro and con arguments for chemotherapy

Pro chemotherapy	Contra chemotherapy
Some newer chemotherapy substances are well tolerated with few and mild side effects	Chemotherapy often causes massive adverse events that add to the patient's burden
Some newer regimens can be administered in the outpatient setting, some can be administered orally	Initiation of a new chemotherapy requires that the patient has to spend a lot of time on the treatment regimen, which is not available anymore for family or other individual priorities at the end of life
Seven percent is better than nothing	Chances of prolonging life are minimal
Sometimes even a short prolongation of life may be worthwhile for the patient, for example if there is a specific goal such as a silver wedding anniversary	The percentage has been estimated as chance of prolongation of life, not cure. This requires more information on the extend of this prolongation. Sometimes clinical studies report prolongation of a few weeks only
	Is the information from clinical studies relevant for this patient? Liver metastases or cachexia often are listed among the exclusion criteria of clinical trials, and for these patient groups chances of effective treatment are much smaller to non-existent
Should chemotherapeutic regimen be offered in spite of all contra arguments, as the patient should not lose hope?	Even with chemotherapy further progression and deterioration of well-being is to be expected, and the patient has to cope with the resulting loss of hope sooner or later
Is there an option not to initiate chemotherapy if the patient insists?	Any medical intervention (including chemotherapy) requires a medical indication. If the physician does not ascertain such an indication, the interventions is not even offered to the patient, and the patient does not have a right to request interventions that are not indicated

patient are minuscule. The request "you have to do something" is certainly a call for action, but from the palliative care perspective, this should be the initiation of a conversation about realistic goals of care. Such a conversion needs to provide ample opportunity for the patient and his or her significant others to express their fears and concerns, including the fear of imminent dying. A realistic goal of care in an advanced stage of illness more often than not will focus on optimal quality of life for the remaining life span, and not on extension of survival at all cost.

However, the intrinsic value of ambivalence for patients in this process should not be underestimated. From the health care professional's perspective, rapid decisions—either for or against specific interventions—are paramount, as this allows clear arrangements, not only for this decision but also for subsequent ones. A patient's decision on withdrawing from chemotherapy may lead to rapid discharge. This normative pressure from the care team towards rapid and clear decisions neglects completely that not deciding may be much easier, and maybe much less burdening, for the patient. Even if chemotherapy has so many burdensome side effects that the patients does not want to experience them again with the next course of therapy, he or she might want to keep the option for further chemotherapy alive for the near future, if his or her well-being improves by then. Such ambivalence keeps all options open for him or her and prevents that he or she would have to surrender a lot of hope, or finds himself or herself suddenly much nearer to the end of life.

There are many other pivotal situations in the disease trajectory requiring decisions on either starting or withholding potentially life-sustaining interventions, each prone to this ambiguity dilemma. The health care professionals push for clear decisions, but the patient prefers to stay ambivalent, so that he or she does not have to face the finality of life with the withholding of the intervention.

Inadequate preparation on the side of the health care professionals with patchy information, misunderstandings, inadequate knowledge of the patient's values and preferences, as well as a lack of consensus on the therapeutic goals within the health care team further complicate and impede shared decision-making. Health care professionals often lack communication skills and find it difficult to communicate about difficult and complex topics, including breaking bad news.

The real-life setting of healthcare services provides additional barriers to shared decision-making, with financial pressures, limited resources and a general lack of time. The German healthcare system uses diagnosis-related flat rate reimbursements for inpatient treatment. This incentivizes rapid discharge, with a focus on the diagnosis that has led to the inpatient admission, and without searching for a more holistic consideration of the patient's physical and psychological performance status and well-being.

17.4.2 Consultation

Communication between physician and patient is not a bidirectional exchange of facts, followed by a common understanding of these facts. There is a saying "said is not yet heard", which means that the healthcare professional may have explained the diagnostic facts, the available therapeutic options and their chances of success and tolerability, but more often than not, the patient will not be able to hear or will prefer not to hear it. "Said is not yet heard" is part of a longer quote on physician–patient communication:

> Meant is not yet said,
> Said is not yet heard,
> Heard is not yet understood,
> Understood is not yet agreed to,
> Agreed to is not yet applied,
> Applied is not yet maintained.

Physicians may think they have expressed themselves clearly, but fail to do so, and patients may not be able or willing to take any of the subsequent steps (hearing, understanding, agreeing, applying and maintaining) towards a common understanding of facts and consequences.

This implies that shared decision-making in complex situations requires more than one conversation. It has to be based on a series of discussions, with repeated provision of information and leaving adequate time between talks for digestion of information and reflection on available options.

It is important to acknowledge the amount of emotional distress that complex decisions may place on patients. This can be alleviated by taking the "sense of coherence concept" into account (Antonovsky and Sagy 1986; Eriksson and Lindstrom 2006). Three domains contribute to the sense of coherence: comprehensibility, meaningfulness and manageability. Health care professionals can improve comprehensibility by repeatedly offering information on diseases, problems and interventions, using layperson language and good communication skills. They also can improve the feeling of manageability by focusing more on remaining options and resources than on problems, and highlighting the patient's power to make decisions, including decisions to withhold or withdraw life-sustaining interventions. Fostering the sense of coherence by improving comprehensibility and manageability will improve the physical and mental well-being of the patient (Swenne and Skytt 2014).

Discussions on treatment goals and decisions can also make good use of concepts such as meaning in life or resilience. Health care professionals should not restrict their conversation to medical facts and options, but can ask patients what they find important in life, what legacy they would like to leave behind, and what they perceive to provide meaning in life. Areas relevant for the patient's meaning in life can then again be explored for manageability to foster the sense of coherence.

Similarly, patients can be asked about previous times of crises in their lives, and how they managed to cope with them. Most patients will be able to tap into these experiences as a source of resilience. Resilience should not be understood in a purely mechanistic way as springing back to the previous form, which is not possible with advanced life-limiting disease, but rather as a coping process that allows patients to adapt to the new situation (Peusquens et al. 2023).

17.4.3 Competency

Shared decision-making is complex in advanced life-limiting diseases, but gets even more complex when patients are incapacitated due to cognitive impairment. Patients with mild cognitive impairment are still competent to participate in shared decision-making, and there is no clear-cut threshold when cognitive impairment becomes so severe that it renders informed consent impossible. The evaluation of cognitive function may need expert input, for example, from psychiatry or ethics consultations. Evaluation of decision-making capacity may be facilitated by a multiprofessional team approach, bringing together the perspectives of physicians, nurses and other health care professionals.

Patients who are unable to give consent need a representative in the decision-making process. This surrogate decision-maker needs to understand the patient's values and preferences to decide in the patient's best interest. A common source of bias is that the proxy decision-maker weighs any health care decision according to his or her own preferences and not according to those of the patient. A communication method to avoid this bias is the use of circular questions: "If we could talk to the patient right now and ask him or her this question, how do you think he or she would decide?", thus shifting the focus back on the patient's values.

An advance directive is valuable if the patient has issued one. However, only a minority of patients has advanced directives (Wurm et al. 2023), and many of those are too general or ambiguous to provide clear guidance for the situation at hand. Even with all efforts to improve advance care planning in Germany there is not much progress (Gotze et al. 2023). There are even some problems raised by advance directives, e.g., the validity of so-called "Ulysses contracts" (Brenna et al. 2023; Marks and Ahronheim 2020; Widdershovenand Berghmans 2001).

Example Text excerpt from a notarised declaration: "I have drawn up this directive after careful consideration. It is an expression of my right to self-determination. For this reason, I do not wish that a change of my will is insinuated in the specific situation of lacking decision-making capacity until I have expressly revoked it (in writing or demonstrably verbally) […]. I do not want my directive to be revoked merely on the basis of a stronger will to live, spontaneous situational behaviour or words and gestures, even though this would be possible under the statutory provisions."

Similar to Ulysses, who wanted to listen to the sirens and asked his comrades to tie him to the mast and not release him, no matter how much he insists that they should do so, the author of this advance directive does not want it to be disregarded, even if he should convey the impression that he has changed his mind later on in the disease trajectory. If, for example, a patient with advanced dementia has lost much of his or her intellectual capability due to cognitive decline, but still seems to enjoy life and finds emotional comfort for example with eating or in contact with pet animals, such a Ulysses directive would require to neglect these apparent positive experiences in favour of the values and preferences of the patient's former will. This leads to ethical dilemmas for family members as well as for health care professionals.

17.4.4 Wish for Hastened Death

Patients wishing for hastened death or requesting suicide assistance pose an enormous challenge for healthcare professionals, especially in palliative care, who have to find out whether it is a deliberate call for action or rather a cry for help to alleviate suffering.[2] The wish for hastened death can manifest in a broad range of forms, from the acceptance of imminent death in the face of a progressive and life-limiting illness, to the hypothetical wish for an escape plan if things get really bad ("exit strategy"), to patients who clearly ask for assisted suicide or killing on demand.

Frequent reasons for the wish for hastened death are the feeling of loss of control, dignity or meaning in life, but also altruism ("I don't want to be a burden on my loved ones"). However, in palliative care, the request for assisted suicide is more often understood as a cry for help ("I don't want to live like this anymore!"), and in most patients the wish to die is not absolute and constant, but often occurs simultaneously or alternately with the will to live (Pestinger et al. 2015).

The wish for hastened death occurs in patients with serious physical illnesses, but suicidal behaviour and requests for assisted suicide also occur in psychiatric illnesses (e.g., depression, psychoses, addiction, post-traumatic stress disorders, personality disorders), where they are usually to be understood as a symptom of the illness and not as an expression of free will.

The desire for assisted suicide can also be an expression of old-age loneliness in the elderly who do not find any prospects of meaning in the continuation of life.

However, the majority of suicides and attempted suicides in Germany are not related to physical or psychiatric illnesses, but take place in professional, family or private life crises (Nationales Suizidpräventionsprogramm 2021). In these exceptional psychological situations, suicide appears to be the only way out for those affected.

Patients with a wish for hastened death thus should be evaluated based on severity (acceptance of imminent death, desire to end or forgo life-sustaining therapies, desire for assisted suicide or suicidal tendencies), motivation (illness, life crisis, life exhaustion) and the time dimension (hypothetical for the future, concrete plans, acute suicidal tendencies). This requires the willingness to enter a trusting relationship with the patient, communication skills and access to expert consultations for example with psychiatric or ethics experts. For our department, we have formulated six steps for communication about wishes for hastened death (Klinisches Ethikkomitee Universitätsklinikum Bonn 2023) (Table 17.2).

---

[2] Balaguer A, Monforte-Royo C, Porta-Sales J et al. (2016) An International Consensus Definition of the Wish to Hasten Death and Its Related Factors. PLoS One 11(1): e0146184.

Table 17.2 Six steps for communication about wishes for hastened death (Klinisches Ethikkomitee Universitätsklinikum Bonn 2023)

Addressing the issue	Expression of suicidal thoughts or requests for assisted suicide should always be taken seriously Questions should be asked about the motives and the suffering behind the wish for hastened death If a patient alludes to a wish for hastened death or suicidal ideation, it can be helpful to clearly address the implied content ("Did I understand you correctly that you are thinking about suicide?")
Additional information	• Kind of wish for hastened death (acceptance of death, ending life-sustaining therapies, suicidal ideation) • Time dimension (general consideration, hypothetical plan if the situation becomes unbearable, concrete plan) • Most severe problem or symptom ("What is bothering you the most?")
Respecting the decision	Acknowledgement of the patient's suffering and distress Respecting the patient's decision, even if it seems to be opposed to one's own values
Offering alternatives	Requests for assisted dying can be rejected, without abandoning the patient, if alternatives are being offered Anticipated suffering may be alleviated with information on palliative care, including options such as palliative sedation for refractory suffering Patients have a right to forgo or discontinue any sustaining therapy, including mechanical ventilation Voluntarily stopping eating and drinking is an option that does not require any assistance from other persons, and can be controlled completely by the patient
Building a trusting relationship	A one-off discussion will not provide any change in the wish for hastened death. The first talk should thus only be the starting point of building a trusting relationship with the patient This should be expressed clearly to the patient ("we will not leave you alone with this issue") Open issues at the end of the discussion have to be endured, for example when patients repeat their request for assisted dying
Debriefing and reflection	• Talking to patients about wishes for hastened death can be challenging for healthcare professionals • Debriefing is necessary to reflect on the challenges and to alleviate distress • Ideally debriefing should be done before subsequent discussions with the patient • This can be done as informal discussions with team members, moral case deliberations, supervision or ethical consultations

17.5 Conclusion and Vision for the Future

Shared decision-making is a well-acknowledged standard in healthcare. However, there are a number of challenges and problems with shared decision-making in clinical practice, both from health care professional and patient sides.

From a physician's perspective, it is important to understand how shared decision-making can be applied in the face of the complexities of medical decisions. In many, if not most cases there are no simple black or white solutions, but complex situations requiring complex decision-making processes. This requires a reflection on the own external (from the health care system settings) and internal (prompt decisions leading to rapid resolution of uncertainties) demands. Physicians should understand decision-making as a process and not as a singular event. This requires the ability and willingness to enter a relationship with the patient, building trust, repeating discussions, enduring ambiguities and tolerating ambivalence.

From the patient's perspective, decision-making may benefit from acknowledgement of the patient's expertise as an expert for himself or herself. The patient's competence can further be strengthened, for example, by increasing awareness of available resources instead of focusing on problems or by exploring options for patients to maintain a sense of control and manageability or by eliciting and ranking health-related priorities and preferences of the patient.

Building a trusting relationship between healthcare professionals and the patient may require compromises, for example, with interventions that the patient or family members have requested but the physician deems ineffective. These compromises will help to build a bridge of trust. Once that bridge is strong enough, it will support difficult decision to be made in the shared process between physician and patient.

References

Aghabarary M, Dehghan Nayeri N (2016) Medical futility and its challenges: a review study. J Med Ethics Hist Med 9:11

Alcalde J, Zimmermann C (2022) Stigma about palliative care: origins and solutions. E-Cancer Med Sci 16:1377. https://doi.org/10.3332/ecancer.2022.1377

Alsulamy N, Lee A, Thokala P, Alessa T (2020) What influences the implementation of shared decision making: an umbrella review. Patient Educ Couns. https://doi.org/10.1016/j.pec.2020.08.009

Antonovsky H, Sagy S (1986) The development of a sense of coherence and its impact on responses to stress situations. J Soc Psychol 126(2):213–225

Belanger E, Rodriguez C, Groleau D (2011) Shared decision-making in palliative care: a systematic mixed studies review using narrative synthesis. Palliat Med 25(3):242–261. https://doi.org/10.1177/269216310389348

Bosslet GT, Pope TM, Rubenfeld GD, Lo B, Truog RD, Rushton CH, Society of Critical C (2015) An official ATS/AACN/ACCP/ESICM/SCCM policy statement: responding to requests for potentially inappropriate treatments in intensive care units. Am J Respir Crit Care Med 191(11):1318–1330. https://doi.org/10.1164/rccm.201505-0924ST

Brenna CTA, Chen SS, Cho M, McCoy LG, Das S (2023) Steering clear of akrasia: an integrative review of self-binding ulysses contracts in clinical practice. Bioethics 37(7):690–714. https://doi.org/10.1111/bioe.13197

Brock DW (1991) The ideal of shared decision making between physicians and patients. Kennedy Inst Ethics J 1(1):28–47

Bundesärztekammer (2015) Medizinische Indikationsstellung und Ökonomisierung. Deutsches Ärzteblatt 112:A836. Retrieved from https://www.bundesaerztekammer.de/fileadmin/user_upload/_old-files/downloads/pdf-Ordner/Stellungnahmen/Stn_Medizinische_Indikationsstellung_und_OEkonomisierung.pdf

Cherny NI (2009) Stigma associated with "palliative care": getting around it or getting over it. Cancer 115(9):1808–1812. https://doi.org/10.1002/cncr.24212

Covvey JR, Kamal KM, Gorse EE, Mehta Z, Dhumal T, Heidari E, Zacker C (2019) Barriers and facilitators to shared decision-making in oncology: a systematic review of the literature. Support Care Cancer 27(5):1613–1637. https://doi.org/10.1007/s00520-019-04675-7
Eriksson M, Lindstrom B (2006) Antonovsky's sense of coherence scale and the relation with health: a systematic review. J Epidemiol Community Health 60(5):376–381. https://doi.org/10.1136/jech.2005.041616
Ernst J, Brahler E, Weissflog G (2014) Beteiligung von Patienten an medizinischen Entscheidungen–ein Uberblick zu Patientenpraferenzen und Einflussfaktoren. Gesundheitswesen 76(4):187–192. https://doi.org/10.1055/s-0033-1361150
Gotze K, Feddersen B, In der Schmitten J, Marckmann G (2023) Advance care planning in Germany. Z Evid Fortbild Qual Gesundhwes 180:127–132. https://doi.org/10.1016/j.zefq.2023.05.005
Hengel FE, Sommer C, Wenzel U (2022) Arterielle Hypertonie–Eine Ubersicht fur den arztlichen Alltag. Dtsch Med Wochenschr 147(7):414–428. https://doi.org/10.1055/a-1577-8663
Klinisches Ethikkomitee Universitätsklinikum Bonn (2023) Umgang mit Patientenwünschen oder -forderungen nach Suizidhilfe am Universitätsklinikum Bonn. Retrieved from https://www.ukbonn.de/site/assets/files/43404/sop_umgang_mit_suizidwunschen.pdf
Kübler-Ross E (1970) On death and dying. Simon & Schuster/Collier Books, New York
Marks AD, Ahronheim JC (2020) Advance directive as Ulysses contract: the application of stopping of eating and drinking by advance directive. Am J Hosp Palliat Care 37(11):974–979. https://doi.org/10.1177/1049909120912951
Moumjid N, Gafni A, Bremond A, Carrere MO (2007) Shared decision making in the medical encounter: are we all talking about the same thing? Med Decis Making 27(5):539–546. https://doi.org/10.1177/0272989X07306779
Pestinger M, Stiel S, Elsner F, Widdershoven G, Voltz R, Nauck F, Radbruch L (2015) The desire to hasten death: using grounded theory for a better understanding "When perception of time tends to be a slippery slope." Palliat Med 29(8):711–719. https://doi.org/10.1177/0269216315577748
Peusquens F, Maus K, Geiser F, Jaspers B, Radbruch L (2023) Wer hat Angst vor "Ockham's Razor"?: Eine Diskursanalyse zur Resilienz in der Palliativversorgung (2000–2021). Schmerz 37(2):107–115. https://doi.org/10.1007/s00482-023-00703-y
Quoteinvestigator (2014) If your only tool is a hammer then every problem looks like a nail. Retrieved from https://quoteinvestigator.com/2014/05/08/hammer-nail/
Radbruch L (2019) Palliativmedizin. Nova Acta Leopoldina NF 424:167–185
Schaeffer D, Berens EM, Vogt D, Gille S, Griese L, Klinger J, Hurrelmann K (2021) Health literacy in Germany–findings of a representative follow-up survey. Dtsch Arztebl Int 118(43):723–728. https://doi.org/10.3238/arztebl.m2021.0310
Schneiderman LJ, Jecker NS, Jonsen AR (1990) Medical futility: its meaning and ethical implications. Ann Intern Med 112(12):949–954. https://doi.org/10.7326/0003-4819-112-12-949
Stiel S, Krumm N, Pestinger M, Lindena G, Nauck F, Ostgathe C, Elsner F (2012) Antibiotics in palliative medicine–results from a prospective epidemiological investigation from the HOPE survey. Support Care Cancer 20(2):325–333. https://doi.org/10.1007/s00520-011-1084-1
Nationales Suizidpräventionsprogramm (2021) Informationen über Suizidalität und Suizid. Retrieved from https://www.suizidpraevention.de/informationen-ueber-suizid/
Swenne CL, Skytt B (2014) The ward round–patient experiences and barriers to participation. Scand J Caring Sci 28(2):297–304. https://doi.org/10.1111/scs.12059
Tziomalos K (2020) Secondary hypertension: novel insights. Curr Hypertens Rev 16(1):11. https://doi.org/10.2174/1573402115666190416161116
Waisel DB, Truog RD (1995) The cardiopulmonary resuscitation-not-indicated order: futility revisited. Ann Intern Med 122(4):304–308. https://doi.org/10.7326/0003-4819-122-4-199502150-00011
Widdershoven G, Berghmans R (2001) Advance directives in psychiatric care: a narrative approach. J Med Ethics 27(2):92–97
Wurm S, Spuling SM, Reinhard AK, Ehrlich U (2023) Prevalence of living wills among older adults in Germany. J Health Monit 8(3):55–60. https://doi.org/10.25646/11665

MIX
Papier aus verantwortungsvollen Quellen
Paper from responsible sources
FSC® C105338

If you have any concerns about our products,
you can contact us on
ProductSafety@springernature.com

In case Publisher is established outside the EU,
the EU authorized representative is:
Springer Nature Customer Service Center GmbH
Europaplatz 3, 69115 Heidelberg, Germany

Printed by Libri Plureos GmbH
in Hamburg, Germany